Chemistry of Higher Heterocyclic Compounds

Chemistry of Higher Heterocyclic Compounds

Dr. Mahima Srivastava
M.Sc. [Gold Medalist], Ph.D.
Associate Professor, Department of Chemistry,
D.B.S. (PG) College Dehradun,
H.N.B. Garhwal Central University,
Uttarakhand, India

CWP
Central West Publishing

This edition has been published by Central West Publishing, Australia

For more information about the books published by Central West Publishing, please visit https://centralwestpublishing.com

Disclaimer

Every effort has been made by the publisher, editors and authors while preparing this book, however, no warranties are made regarding the accuracy and completeness of the content. The publisher, editors and authors disclaim without any limitation all warranties as well as any implied warranties about sales, along with fitness of the content for a particular purpose. Citation of any website and other information sources does not mean any endorsement from the publisher, editors and authors. For ascertaining the suitability of the contents contained herein for a particular lab or commercial use, consultation with the subject expert is needed. In addition, while using the information and methods contained herein, the practitioners and researchers need to be mindful for their own safety, along with the safety of others, including the professional parties and premises for whom they have professional responsibility. To the fullest extent of law, the publisher, editors and authors are not liable in all circumstances (special, incidental, and consequential) for any injury and/or damage to persons and property, along with any potential loss of profit and other commercial damages due to the use of any methods, products, guidelines, procedures contained in the material herein.

NATIONAL LIBRARY OF AUSTRALIA

A catalogue record for this book is available from the National Library of Australia

ISBN (print): 978-1-922617-52-1

I dedicate this book to my parents, brother Anuj Anand, son Daksh Srivastava, my teachers and everybody who has helped me in being what I am today

जवाहरलाल नेहरू उन्नत वैज्ञानिक अनुसंधान केंद्र
Jawaharlal Nehru Centre For Advanced Scientific Research
JNCASR

Professor C.N.R. Rao, F.R.S.
Linus Pauling Research Professor and Honorary President

Jakkur P.O., Bengaluru - 560064
Karnataka, INDIA

+91 80 23653075; 22082761 (Off)
+91 80 23601410 (Res)

cnrrao@jncasr.ac.in

www.jncasr.ac.in/cnrrao

An Autonomous Institution under Department of Science and Technology, Govt. of India.

An Institution Deemed-to-be-University

FOREWORD

Higher heterocyclic compounds show extensive structural diversity and applications. Along with this they often possess remarkable biological activity. Medicinal chemists have espoused this property in designing most of the drugs in use today. This book offers readers a fundamental understanding of the details of higher heterocyclic compounds and their applications in diverse areas in addition to natural products such as amino acids, DNA, vitamins, and antibiotics.

Thus, the importance of heterocyclic chemistry continues to increase and the book by Dr. Mahima Srivastava is a welcome addition to the available literature on the subject. Its scope places it in a useful niche among all the learners of this category. The author has retained the tested and tried classical approach but has successfully placed her own individual whirl on her arrangement. She has put together a selected range amongst the most important of the vast array of content available. The matter is structured in a perspicuous, logical, and lucid pattern over the whole book. The book includes content dealing with comprehensive aspects of higher heterocycles such as Six – membered compounds (Pyridine, Pyridazine, Pyrimidine, Pyrazine, Quinolinium ion, Naphthyridines, Pyrido diazines, Pteridines, Pyrans, Thiopyrans, Oxazines and Thiazine), seven – membered heterocyclic compounds (Azepine, Oxepine, Thiepine) and heterocyclic compounds containing elements other than nitrogen, sulphur and oxygen. Applied aspects concerning these compounds have also been taken care of in detail.

The present work should be of value to the students and practitioners of heterocyclic chemistry at all levels. It is also endorsed to researchers working in the fields of heterocyclic chemistry and students in the field of organic chemical science. It will be of assistance in presenting a fortifying and invigorating view of the subject to those who use heterocyclic chemistry.

I heartily congratulate Dr. Mahima Srivastava for the coruscating job she has done.

C.N.R. Rao

Table of Contents

Preface

The University Grants Commission has revised and proposed the syllabus for B.Sc. and M.Sc. courses in Chemistry keeping up with the developments that have taken place in the subject over the last few decades. This scheme of examination and syllabus is a credit-based system programme.

I feel immense pleasure in presenting this book to the teachers and students of undergraduate and postgraduate level. This book is the outcome of my long teaching experience of more than a decade at graduate and postgraduate level.

A long span of interaction with students has guided me to give special attention to dark areas which students find difficult to understand. The book has been written in a simple and easy-to-understand language. The topics pertaining to the syllabus have been covered to full extent. Some salient features of the book are:

- Many illustrations and examples have been provided to make the reading interesting and engaging.
- Maximum methods of synthesis and reactions for every compound have been discussed.
- The Compounds in every category have been discussed in detail for clear understanding.
- General methods of synthesis and reactions has also been elaborated along with specific methods of synthesis and reactions so that basic pattern in which the compound behaves may be followed.
- The uses of every compound have been mentioned and general applications have been discussed in a separate chapter.

I am thankful to the management and editorial team of the Publishing House for the help and support in the publication of this book.

I believe that the book contains all that is needed to understand the subject in a systematic manner. However, I also believe that no work is perfect and there is always scope for improvement. I would welcome suggestions from the teaching fraternity and students for further improvement of the book.

Symbols and Abbreviations

Å = Angstrom is a metric unit of length equal to 10^{-10} m. The unit is named after the Swedish physicist Anders Jonas Ångström

D = Dipole Moment

α and β carbons = The first carbon atom after the carbon atom bonded to a unctional group is known as alpha carbon and further carbon in a chain are known as beta, gamma, etc.

λ_{max} = wavelength of maximum absorbance

ε = molar absorptivity or molar absorption coefficient

Me = Methyl functional group

Et = Ethyl functional group

Ph = Phenyl functional group

IUPAC = International Union of Pure and Applied Chemistry

I = Inductive effect

M = Mesomeric effect

TEA = Triethyl amine

Et al = and others

CHAPTER 1

INTRODUCTION

Higher heterocyclic compounds find wide usage in varied sectors. Pyridine is an important solvent and intermediate used to make other compounds. Pharmaceutically important pyridines include the tuberculostat isoniazid (isonicotinic acid hydrazide), the anti-AIDS-virus drug nevirapine, the vasodilator nicorandil, used for treating angina, the urinary-tract analgesic phenazopyridine, and the anti-inflammatory sulfa drug. 1-(1-phenylcyclohexyl) piperidine (PCP, phencyclidine) was originally used as an anaesthetic, but its powerful hallucinogenic properties have led to abuse. Diquat, paraquat, clopyralid, and diflufenican are well-known pyridine derivatives used as herbicides.

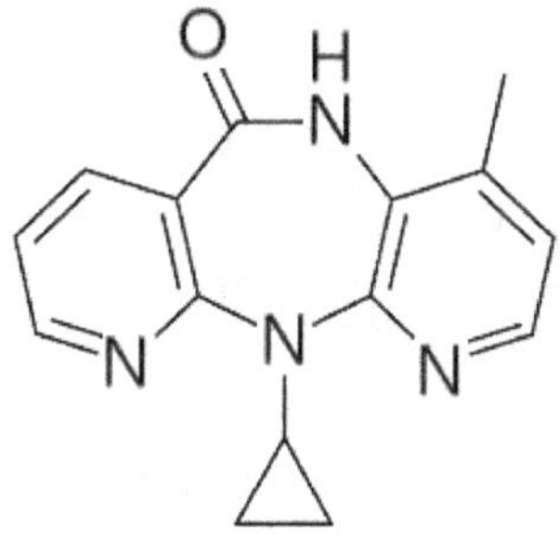

NEVIRAPINE

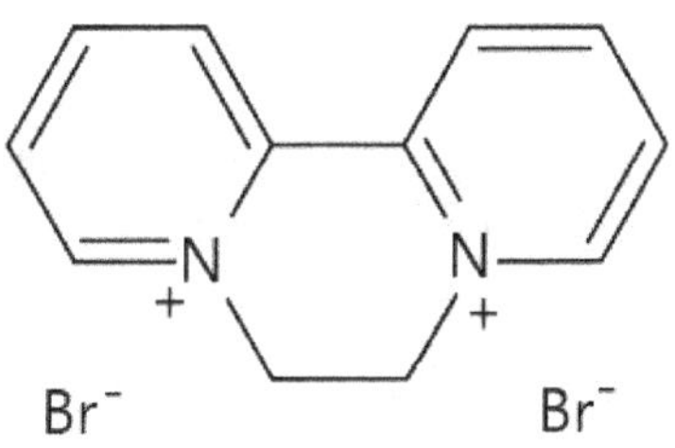

DIQUAT

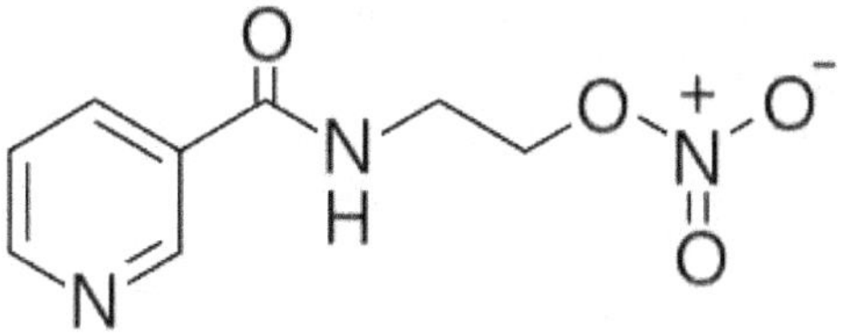

NICORANDIL

Important synthetic derivatives of the benzo- and dibenzo pyridines include cyanine dyes, used as sensitizers in silver halide photographic emulsions, and quinophthalone dyes, applied in plastics, polymer textiles, paper, cosmetics, transfer-printing processes, electronic photography, and laser dyes. Other derivatives are useful bacteria-staining agents for microscopy, antiseptics such as the coal-tar dye acriflavine, antimalarial agents such as mepacrine (quinacrine) and chloroquine, antibacterial agents such as ciprofloxacin, and the reagent oxine (8-hydroxyquinoline or 8-quinolinol), used in analytical chemistry.

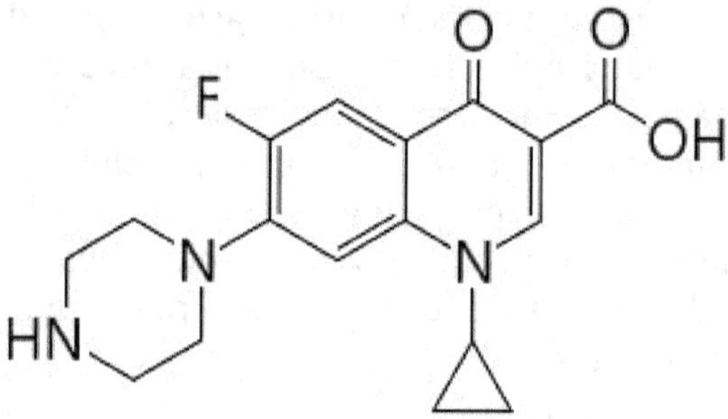

CIPROFLOXACIN

The pyridazine derivative maleic hydrazide is an herbicide, and some pyrazines occur naturally—the antibiotic aspergillic acid, for example. The pyrazine ring is a component of many polycyclic compounds of biological or industrial significance. Important members of the pyrazine family include pteridines, alloxazines, and phenazines. Biologically and pharmacologically, however, the most important diazines are the pyrimidines. The biological significance of pteridine compounds as pigments of butterfly wings is well known. One example is the yellow pigment 2-amino-4,6-pteridinedione (xanthopterin). Folic acid, also a pteridine, is a B-complex vitamin and an important growth factor.

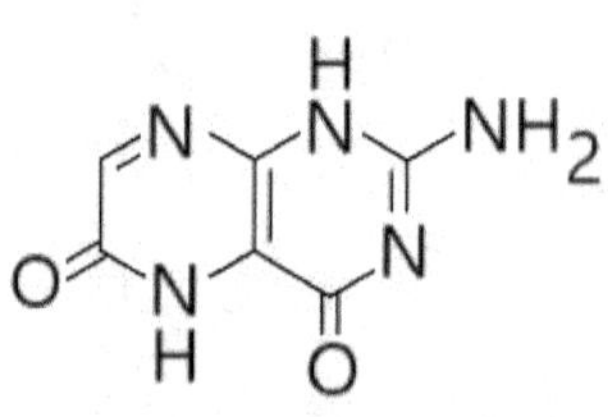

XANTHOPTERIN

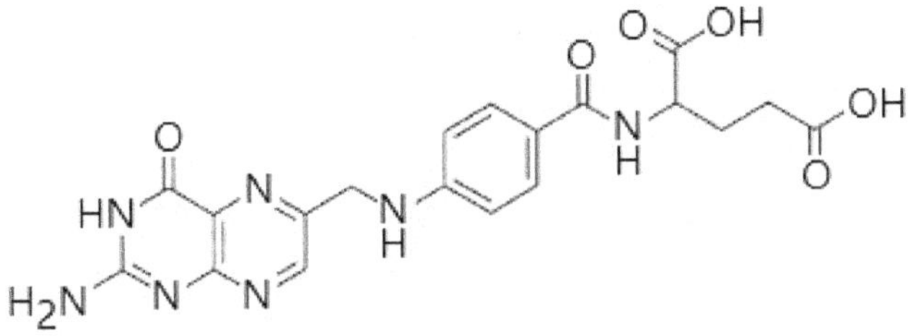

FOLIC ACID

As the size of the ring increases, the range of compounds that can be obtained by varying the number, type, and location of the heteroatoms increases enormously. Nevertheless, the chemistry of heterocyclic compounds with rings seven-membered or larger is much less developed than that of five- and six-membered ring heterocycles, although these compounds are usually stable and some of them have found practical application. Of the seven-membered ring compounds, one-heteroatom heterocycles—azepines, oxepines, and thiepines—and their derivatives are the most comprehensively studied.

The increase in ring size constrains these compounds to be nonplanar in order to lessen the ring strain. Nonplanarity, however, affects aromaticity, so these heterocycles react as cyclic polyenes (compounds with noninteracting, alternating single and double bonds). Azepine and Oxepine rings are important constituents of numerous naturally occurring alkaloids and metabolic products of marine organisms. The azepine derivative caprolactam is produced commercially in bulk for use as an intermediate in the manufacture of nylon-6 and in production of films, coatings, and synthetic leather Many heterocycles containing selenium (Se) atoms are known. Selenium shows much similarity in behaviour to sulphur; hence, selenophene, with the structure shown, resembles thiophene quite closely. Carbon-selenium and selenium-selenium bonds are considerably weaker than the corresponding carbon-sulphur and sulphur-sulphur bonds; consequently, the reactions of selenium compounds tend to involve the heteroatom. In industry, selenium heterocycles are used almost exclusively in the preparation of "organic metals"—organic materials that have high electrical conductivity, much like metals. Tellurium (Te) heterocycles are rarer and even less stable than selenium heterocycles. One of the first such compounds, prepared in 1971, is dibenzo tellurophene.

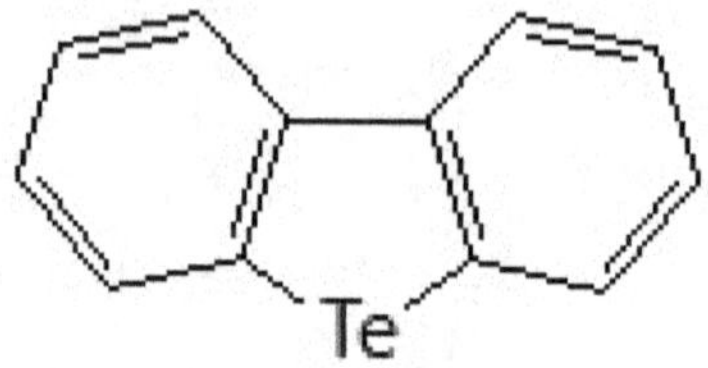

DIBENZOTELLUROPHENE

Although many organophosphorus compounds are used in medicine or as insecticides and herbicides, they include few phosphorus-containing heterocycles. Two phosphorus heterocycles of practical importance are the anticancer drug cyclophosphamide and the insect chemo sterilant apholate; the latter compound also contains six three-membered aziridine rings. The arsenic heterocycle 10,10′-oxybisphenoxarsine is used as a fungicidal and bactericidal additive to plastics.

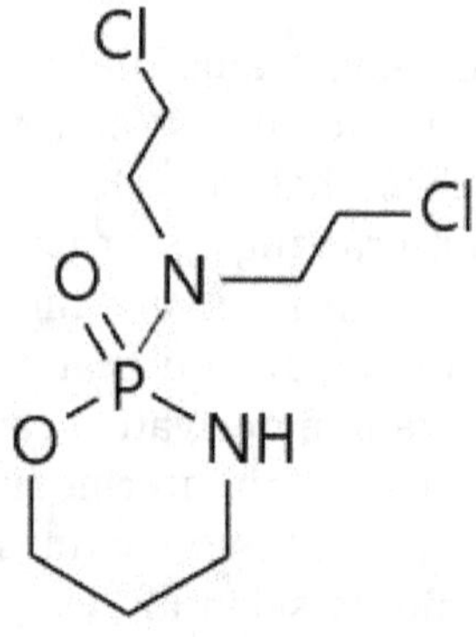

CYCLOPHOSPHAMIDE

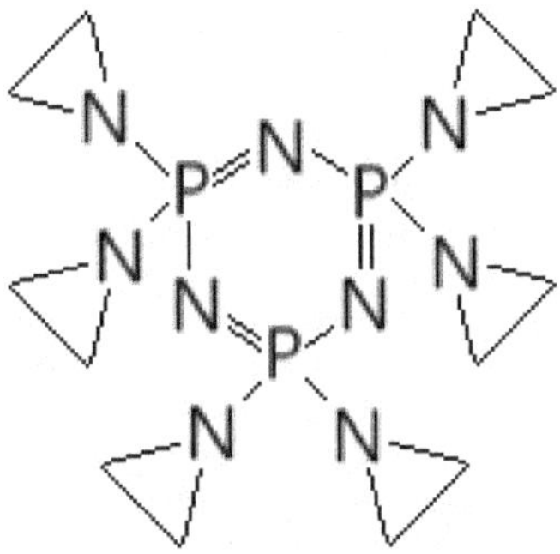

APHOLATE

Silicon (Si), germanium (Ge), tin (Sn), and lead (Pb) tend to form stronger bonds with nitrogen and, especially, with oxygen than with carbon. Alternating replacement of all the carbon in carbocyclic compounds with silicon and oxygen results in silicates, which are inorganic analogs of organic heterocycles.

A large variety of heterocycles with five-, six-, or seven-membered rings containing boron (B) have been prepared and studied. Several saturated boron heterocycles were found to be more stable than their open-chain analogs, suggesting that the boron-containing cyclic structure itself favours stability. One of the best-known saturated heterocyclic boranes, widely used in organic chemistry as a reducing agent, is 9-borabicyclo [3.3.1]nonane (9-BBN).

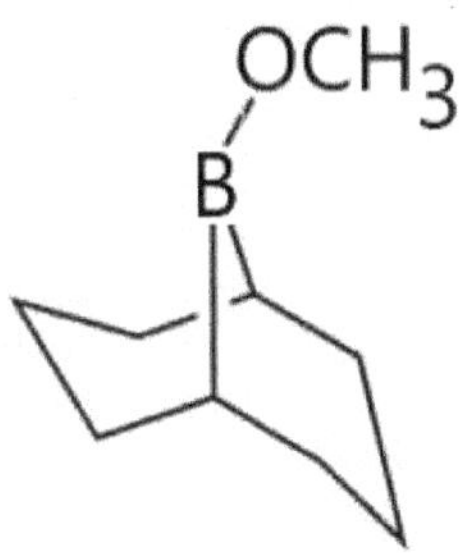

9 - BBN

CHAPTER 2

SIX-MEMBERED HETEROCYCLIC COMPOUNDS

2.1 INTRODUCTION

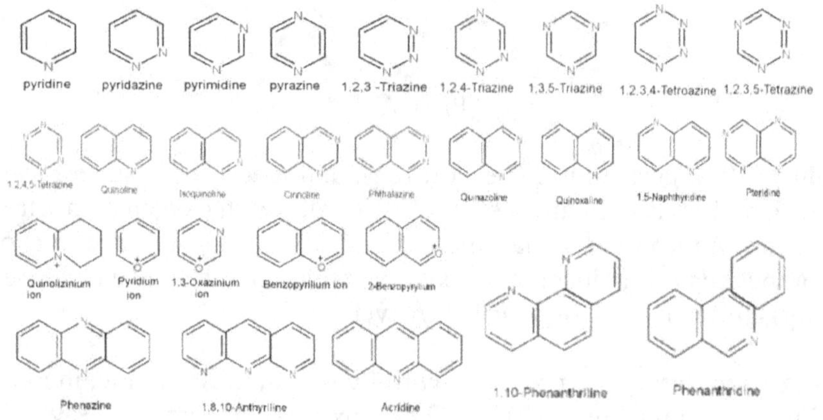

2.2 PYRIDINE

2.2.1 SYNTHESIS OF PYRIDINE

The replacement of hydroxyl group takes place on **oxidation** of hydroquinone by ferricyanide results in the formation of 6-hydroxyquinoline.

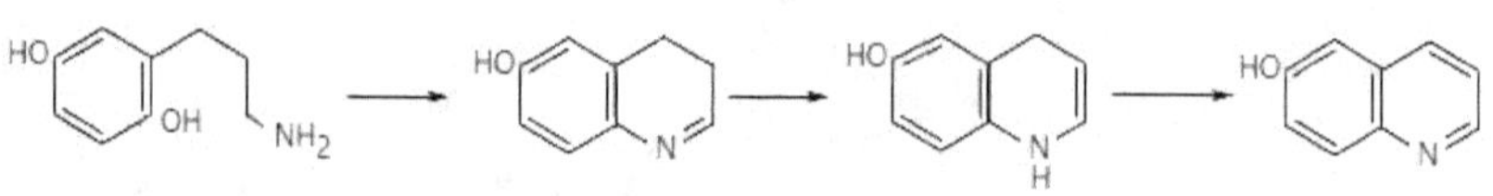

Likewise, sorbic acid, when added to methyl amine, forms an intermediate, methylaminopiperid-2-one (A) that results in the formation of dihydropyrid-2-one on heating.

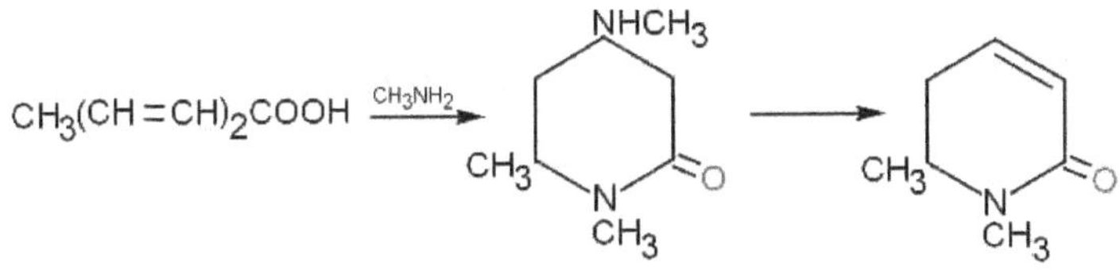

Through a variety of chemical reactions, pyridine derivatives maybe formed from cyanoacetate when removal of oxygen in the presence of triethyl phosphite takes place; then o-nitro benzylidene cyano-acetate shows **cyclization** reaction to form (B). However, on **reduction** with palladium, 2-aminoquinoline N-oxide (C) or 2-aminoquinoline (D) is obtained and in the presence of sodium borohydride, 3-cyanoquinol-2-one or its hydroxy derivative is formed.

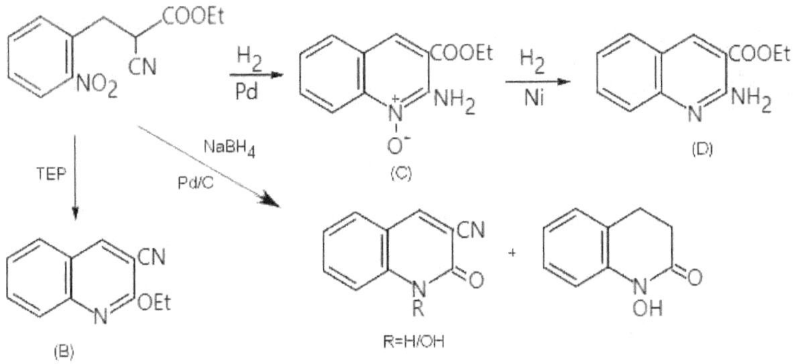

Six membered nitrogen heterocyclic compounds maybe easily obtained by **cyclization** in acid or by Grignard's reagent.

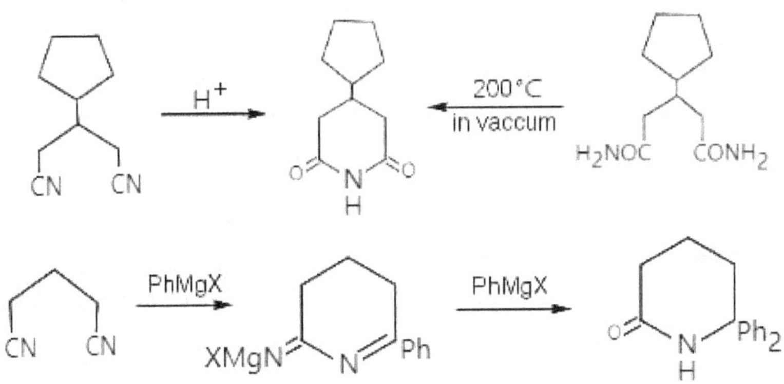

Pyridone-4-carboxamide is obtained by cyclization of citric acid tri-amide in presence of base at high temperature.

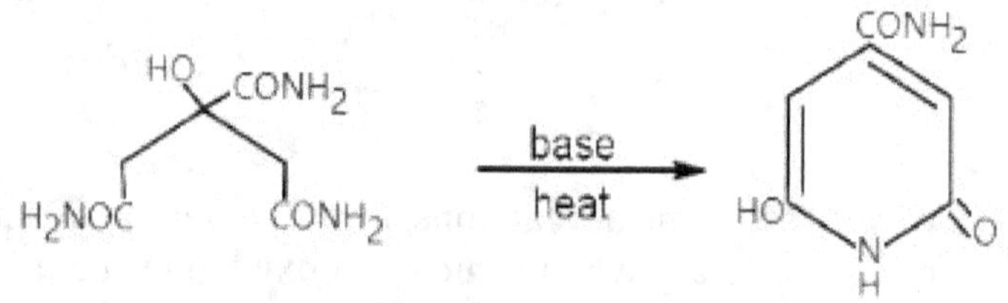

When a nitro group and active methylene group condense **intramolecularly**, then quinoline N-oxide is obtained. In the same fashion, phenanthridine N-oxide is also formed.

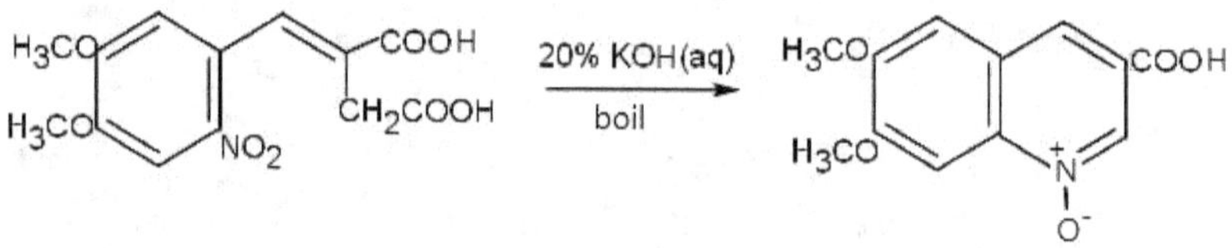

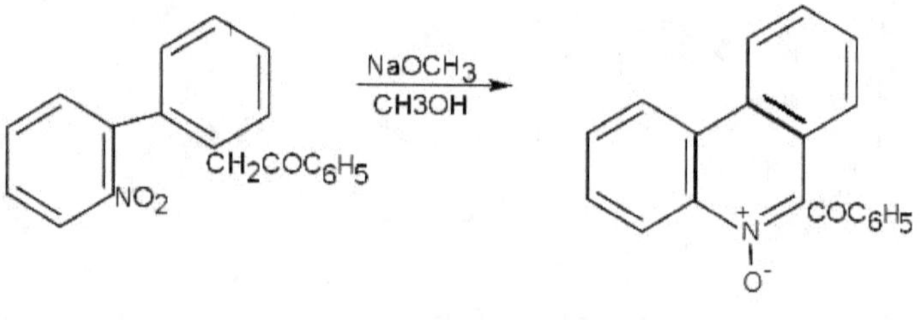

Isocyanates, as intermediate, also **cyclize** to yield pyrid-2-ones.

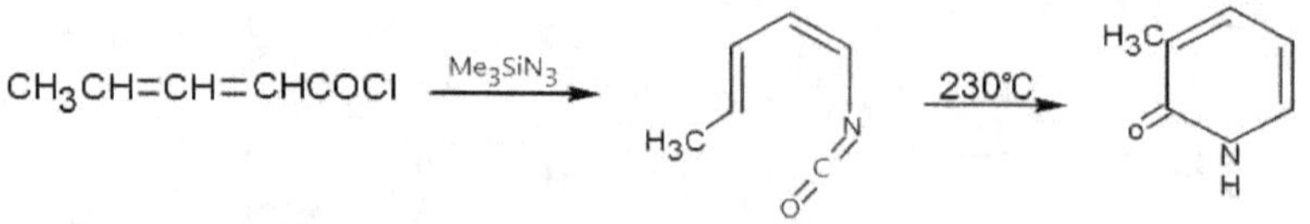

In the presence of trimethyl oxonium fluoroborate, isothiocyanate is **cyclized** to form 1-methylthio-3,4-dihydroisoquinoline, (E); however, thioamide is cyclized in the presence of mercuric chloride.

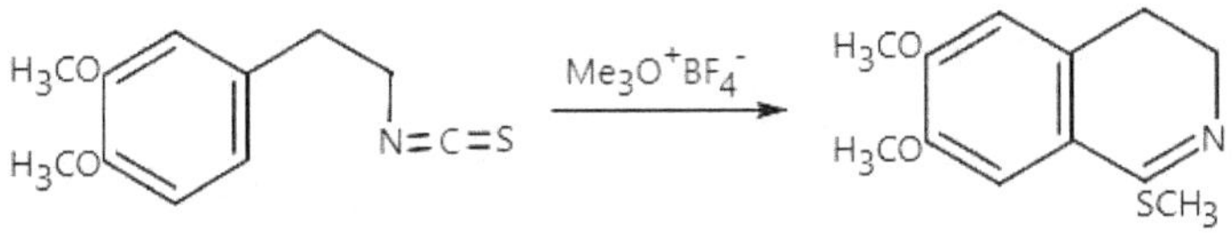

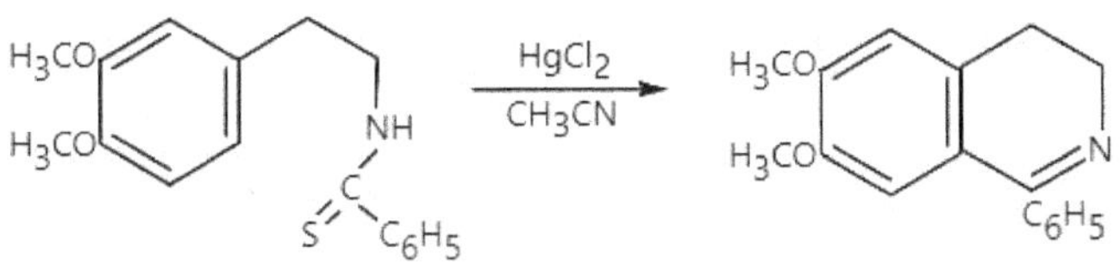

When cyclization of o- amino styrene derivative takes place in phosphoryl chloride, then quinoline is produced.

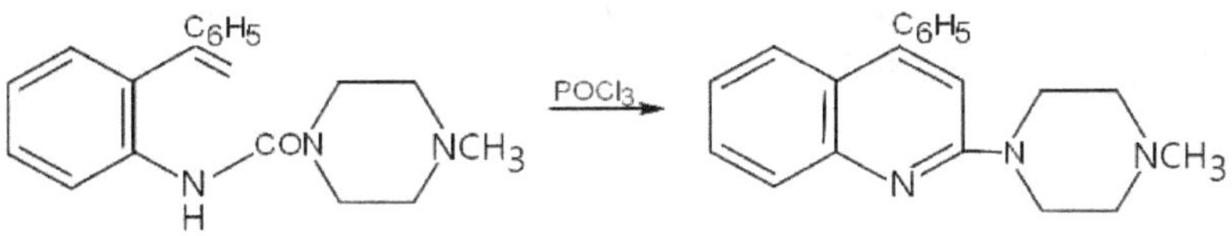

In **Conrad-Limpach synthesis**, malonic acid anilides or dianilides convert to form 4-hydroxy quinol-2-ones on cyclization.

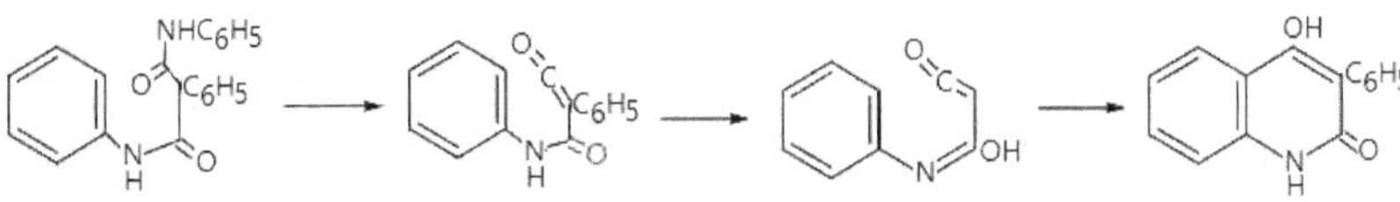

Acridines are obtained in acidic environment by fusion of rings.

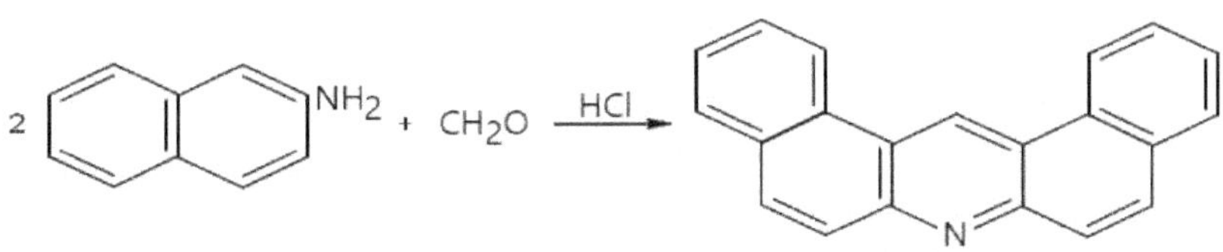

Dibenz [a,h] acridine

Likewise, in yet another method, when dichloromethane (R = H), 1,1-dichloroethane (R = CH_3) or benzal chloride (R = C_6H_5) react with 1-naphthylamine in appropriate solvent, dibenz[c, h] acridine is formed on heating.

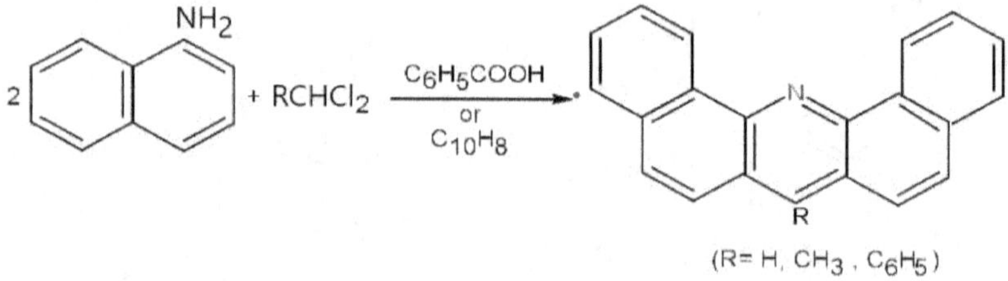

Furan derivatives have been used to synthesize pyridines. In a sequence of reactions, furan firstly undergo **electrolytic methylation** resulting in the formation of 2,5-dimethoxy-2,5-dihydro furan (F) which after modifications changes into 3-hydroxypyridine (G). Also, **reduction** of (F) gives tetrahydrofuran (H) which finally yields 3-hydroxypyrid-2-one (I).

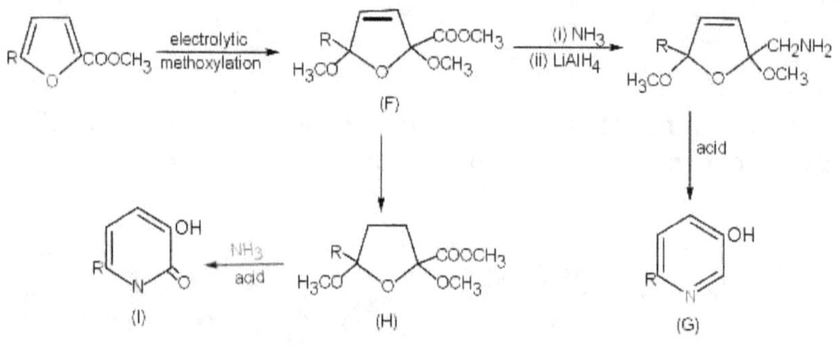

Pyrimidine derivatives can also be converted to pyridines. On **heating** tetrahydro pyrimidine with silica and alumina, we get dihydropyridine.

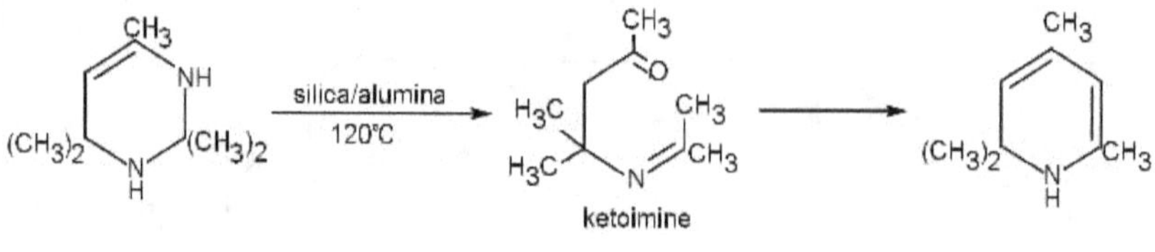

However, in aqueous ammonia/methylamine at 190°C, pyrimidine gives 2-methyl-5-ethylpyridine.

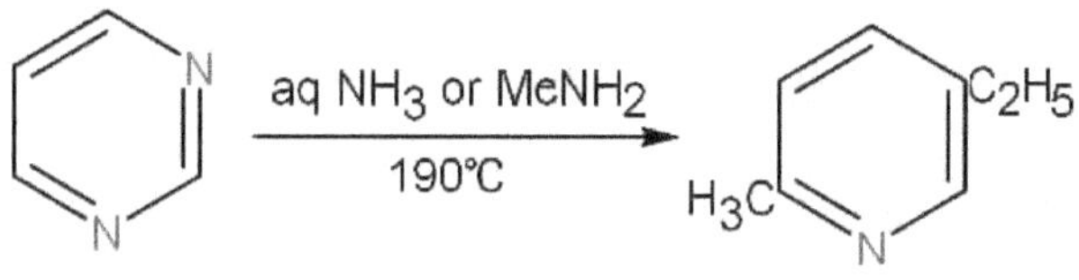

When azepinone undergoes Clemmensen **reduction**, piperidine is obtained.

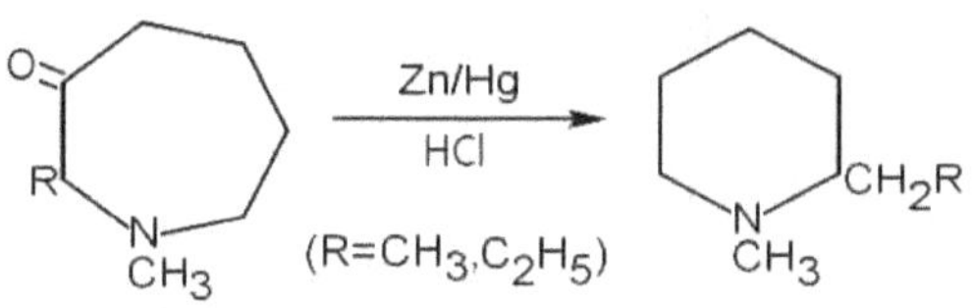

Likewise, benzo fused seven membered compounds also gives good yields. Dioxo-benzazepine results in the formation of 4-hydroxy quinol-2-one.

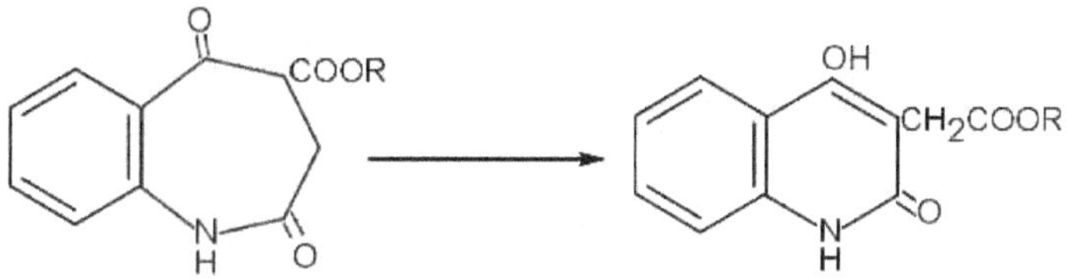

1,3-Oxazpine, under the effect of heat, result in the formation of 3-hydroxy-2-phenylpyridine.

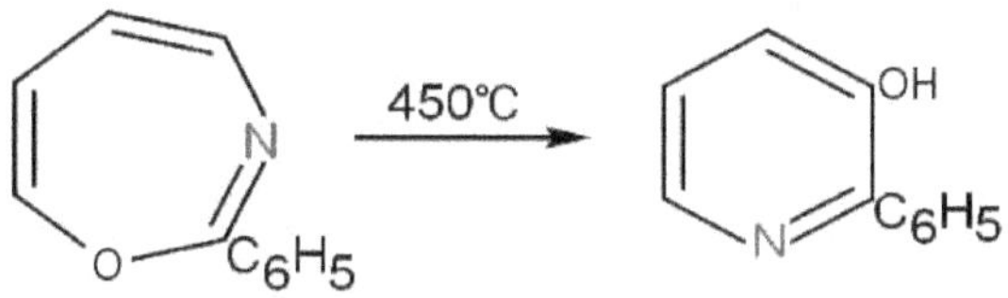

The benzoxazepine gets converted to diamino-quinoline in the presence of methylamine.

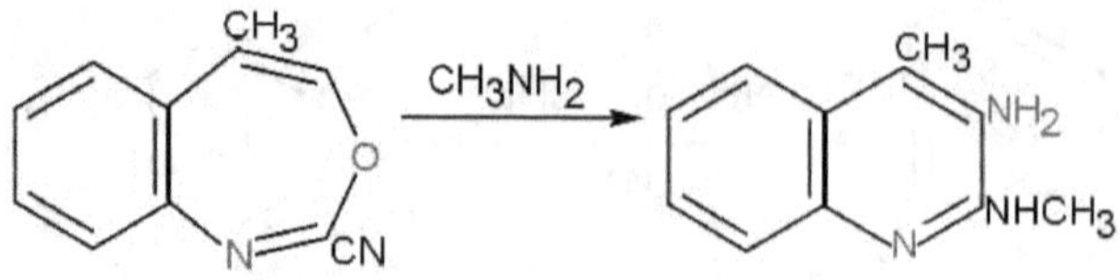

2.2.2 PROPERTIES OF PYRIDINE

When pyridine is exposed to halides of boron, aluminium, tin and antimony then adduct is formed.

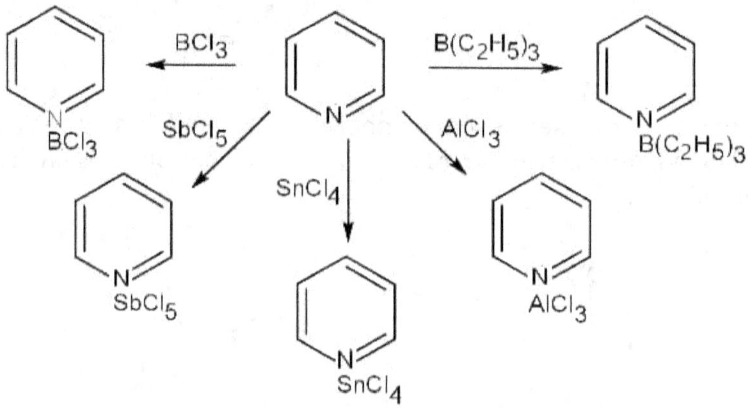

Electrophilic attack at pyridine takes place in the presence of alkyl halide. Pyridine gives rigorous reactions with methyl iodide, benzyl halide and diphenyl methyl halide via **SN_1 mechanism**. Triphenyl-methyl chloride reacts with pyridine at high pressure and results in the formation of triphenylmethyl pyridinium chloride.

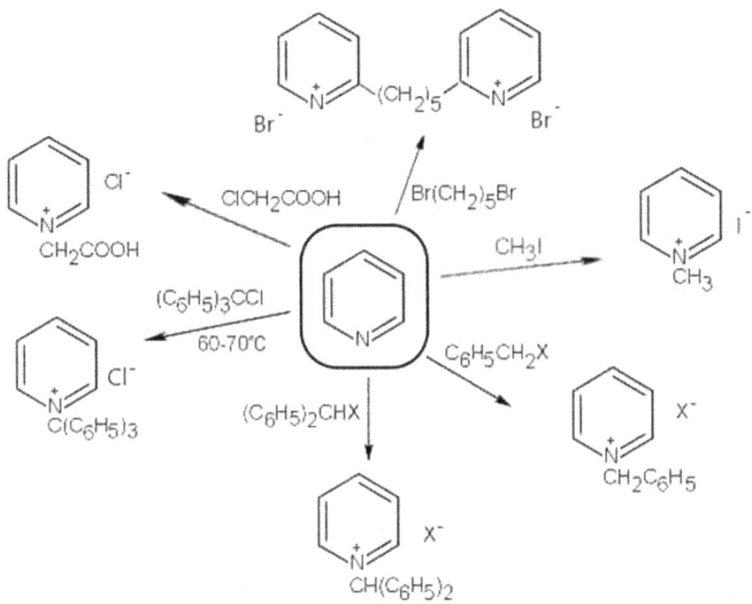

When acridine reacts with fluorosulphonate, then at high temperature, **quaternization reaction** takes place.

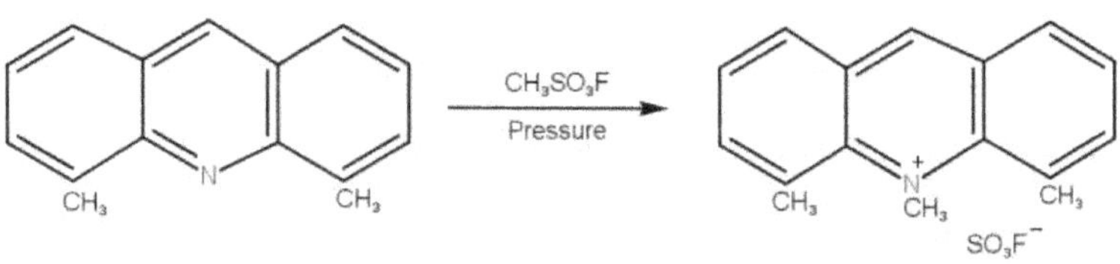

On **methylation** of 2,4,6-triphenylpyridine-2-one in the presence of methanol, formation of ortho methyl product takes place.

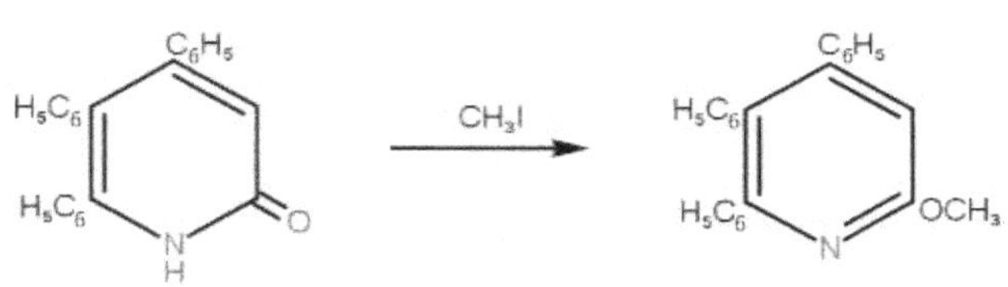

N - methylation takes place in both pyridin-2-one as well as pyridin-4-one. However, in pyridin-2-one, N-methylation is achieved by conversion of the 2-oxo group to trimethylsilyl ether prior to methylation. Also, methyl iodide brings about **N-methylation** in 3-nitropyridin-4-one (A).

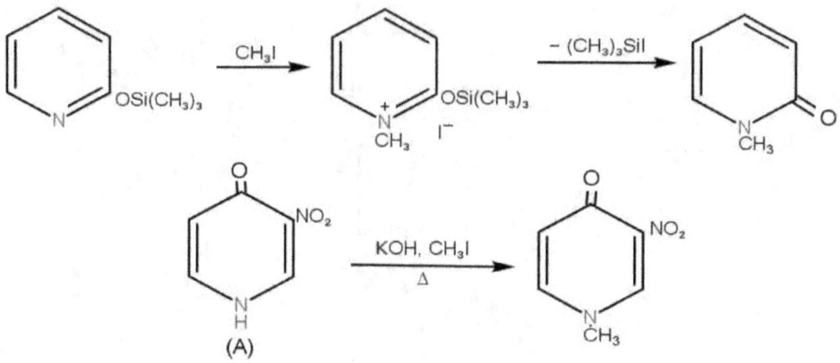

In an interesting set of reactions, when pyridine is treated with chlorine in presence of aluminium chloride, then 4-pyridylpyridinium salt **(B)** is obtained. However, in the presence of iodine chloride, pyridinium hydrochloride gets converted to 2-pyridylpyridinium salt **(C)** at 250°C.

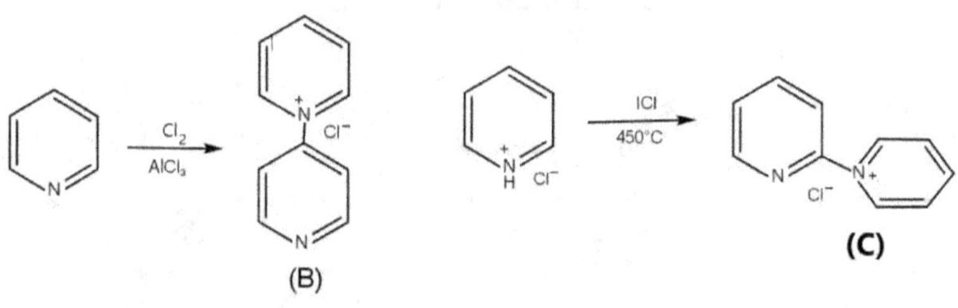

2,6-Dimethoxypyridine-1-oxide results in 3-nitro derivative; however, prolonged exposure to **nitrating mixture** yield substitution at position-5 also.

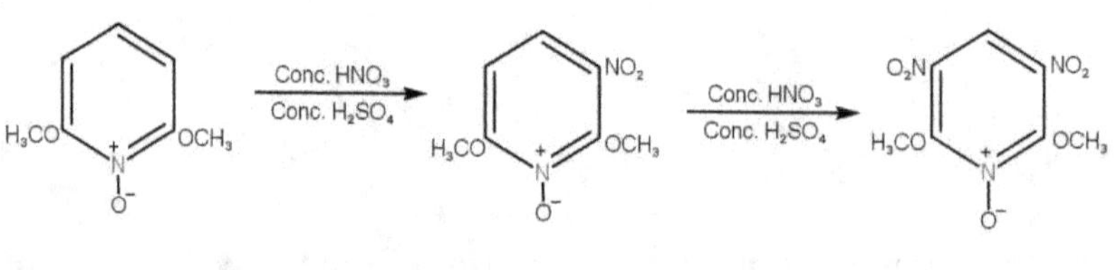

When position-3 in pyridine is occupied by hydroxy or methoxy groups, then **nitration** takes place at position-2 and good yield of product is obtained. However, if there is substitution at position-2, then nitration occurs in position-4.

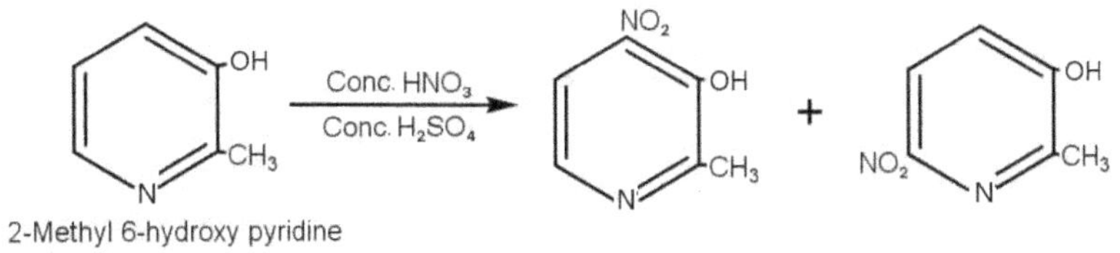

Nitration of 4-aminopyridine, via intermediate **N-nitro amine**, finally results in the formation of 4-amino-3-nitropyridine.

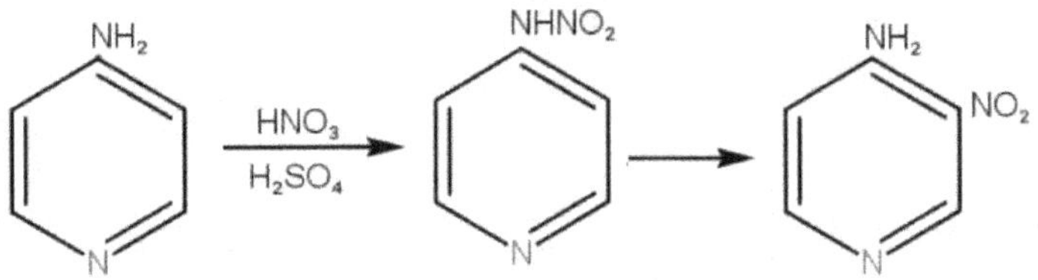

Halogenation takes place in position-5 in 2-aminopyridine when exposed in the presence of sulphuric acid.

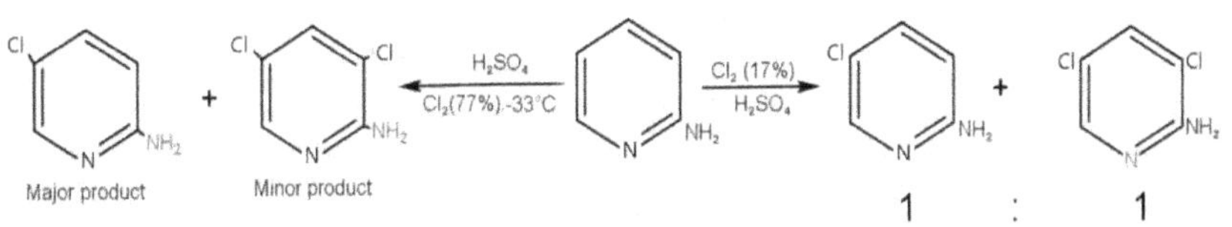

If 3-hydroxypyridine is exposed to potassium salt at high temperatures in carbon dioxide atmosphere, then carboxylic acid functional group occupies position-2.

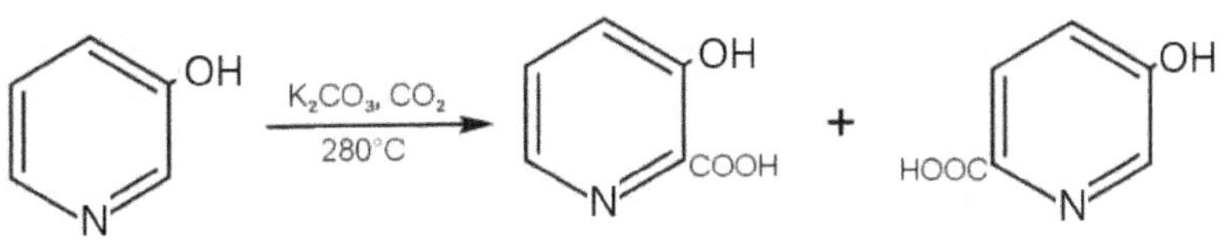

Chlorination at exclusive position can be achieved easily. When 4-carboxamidopyrimidine -1- oxide is exposed to a mixture of phosphorus oxychloride & phosphorus pentachloride, then 2-chloroisonicotinonitrile is obtained.

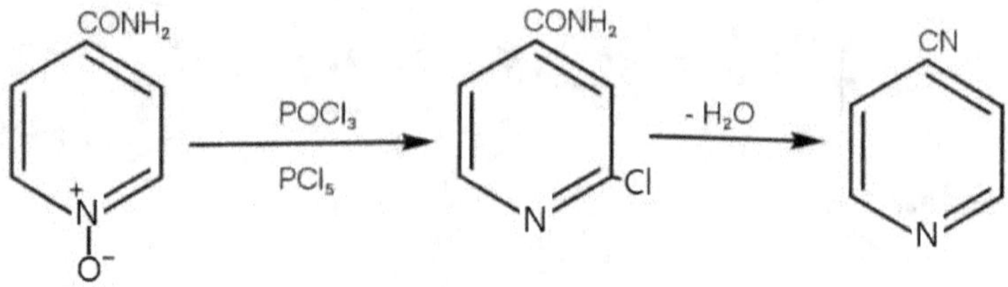

Mechanism of formation of 3-chloro product on exposure of isonicotinonitrile-1-oxide to a mixture of phosphorus oxychloride and phosphorus pentachloride is shown in two possible ways.

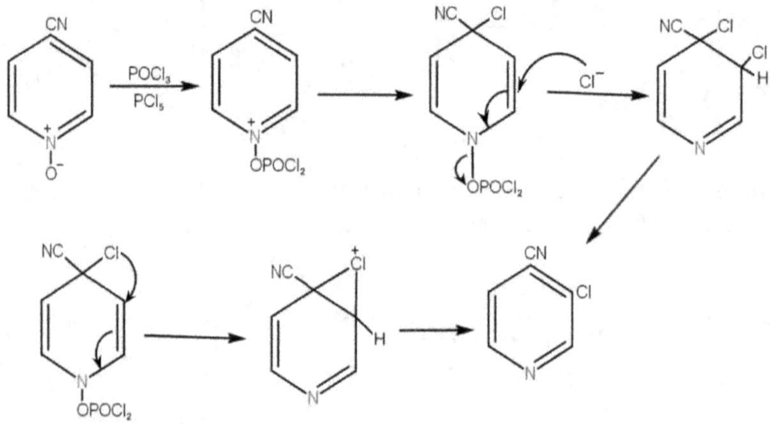

In presence of sulphuryl chloride, **poly-chlorination** takes place at high temperatures.

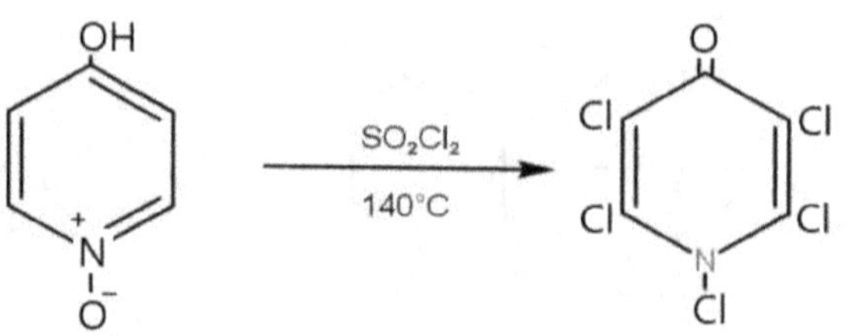

Ozone brings about **nucleophilic** attack in quinoline-1-oxide to produce hydroxamic acid (A) along with ether product.

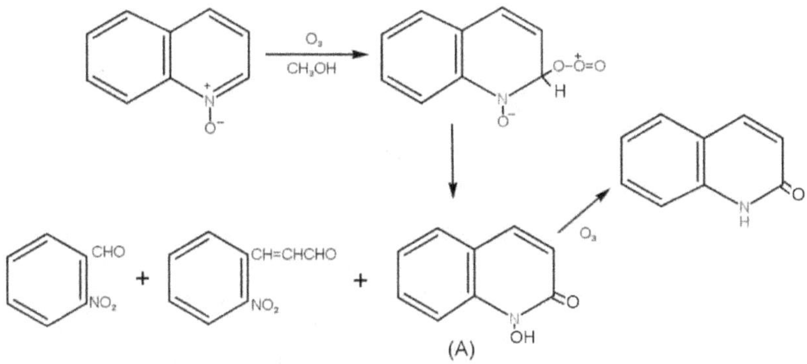

When acridine undergoes **amination** in presence of dimethylaniline, then 9,9'-biacridanyl is obtained as main product along with 9-aminoacridine.

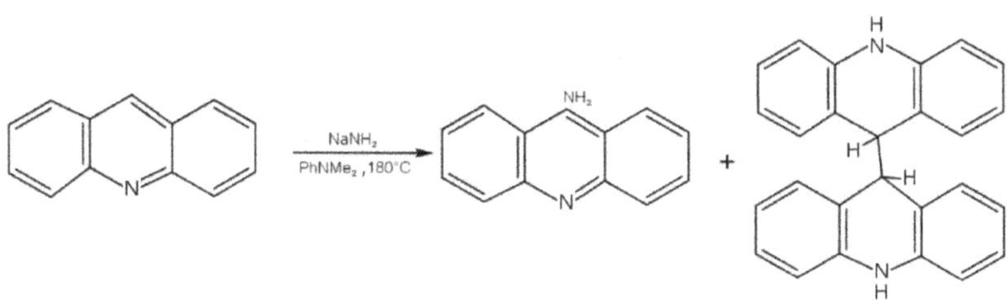

Iso-quinolines and quinolines undergo **arylation** whereby substitution is facilitated at C-4 instead of C-2.

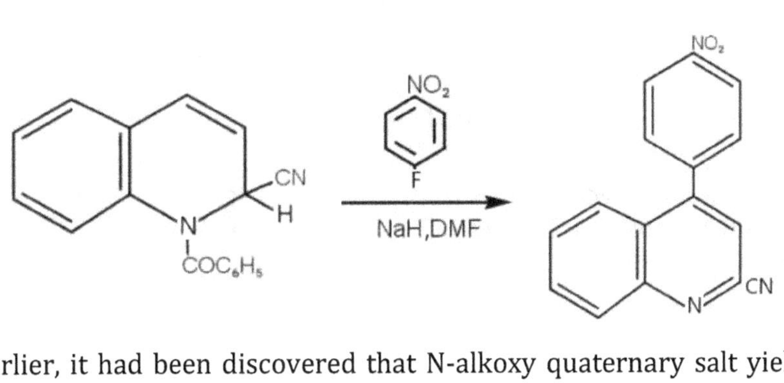

Earlier, it had been discovered that N-alkoxy quaternary salt yield α- and Y-cyano derivatives along with an alcohol when reacted with cyanide ion. Now the same principle is used in the **synthesis** of 4-cyano-2-methyl pyridine from reactant 1-ethoxy-2-methyl pyridine.

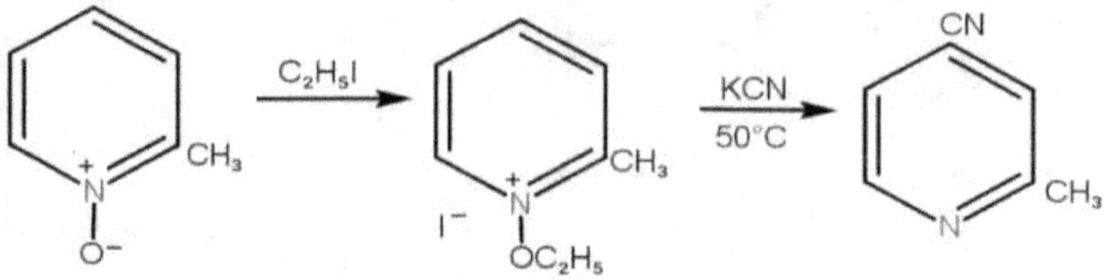

Organolithium compounds bring about **alkylation** in pyridines more easily than Grignard's reagent. At 0°C, pyridine forms a complex with phenyl-lithium which gets converted to 2-phenyl pyridine **(B)** on heating. Hydrolysis of the same complex gives dihydro product **(C)** from which **(B)** can be obtained by **oxidation**.

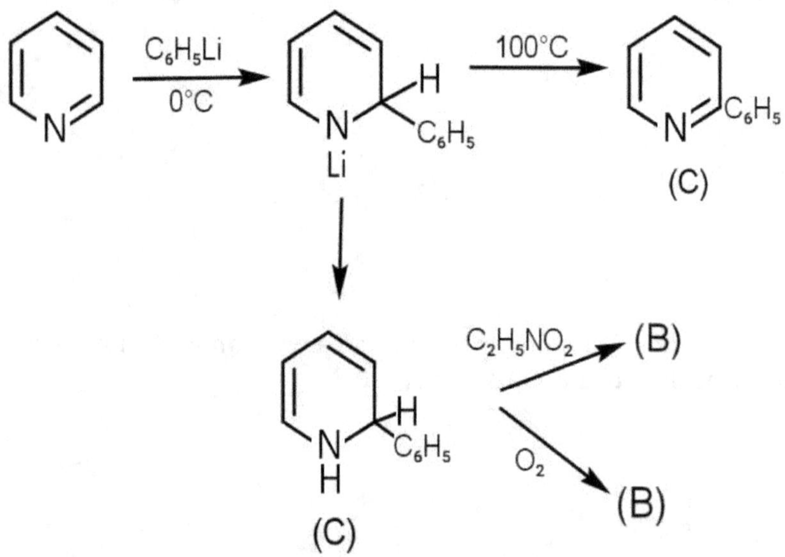

When N-alkyl pyridinium salt reacts with Grignard's reagent, then **alkylation** takes place at the α-positions however **arylation** occurs only at C-6.

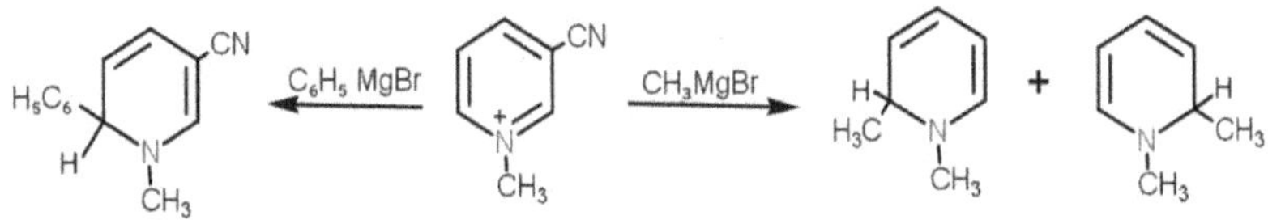

Hydroxy methylation reaction comes under the category of special reactions to be brought about in the presence of ammonium persulphate and methanol whereby substitution takes place at C-2 and C-4.

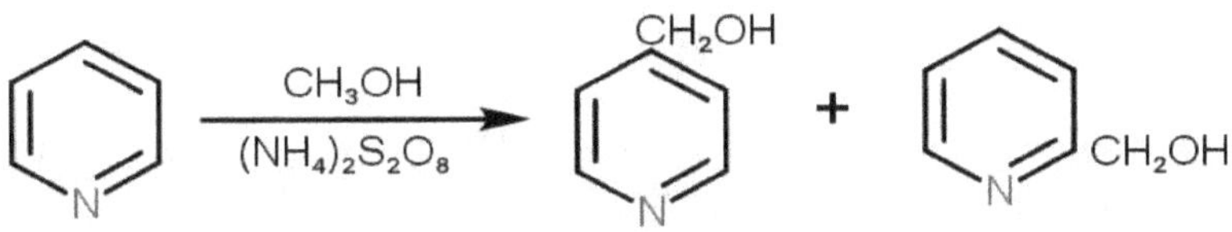

Photolysis of pyridine takes place in the presence of chlorine and carbon tetrachloride. However, bromination occurs in vapour phase with bromine chloride to result in the formation of 2- bromopyridine as major product and 1:1 ratio of 2-bromopyridine and 3-bromopyridine in bromine atmosphere.

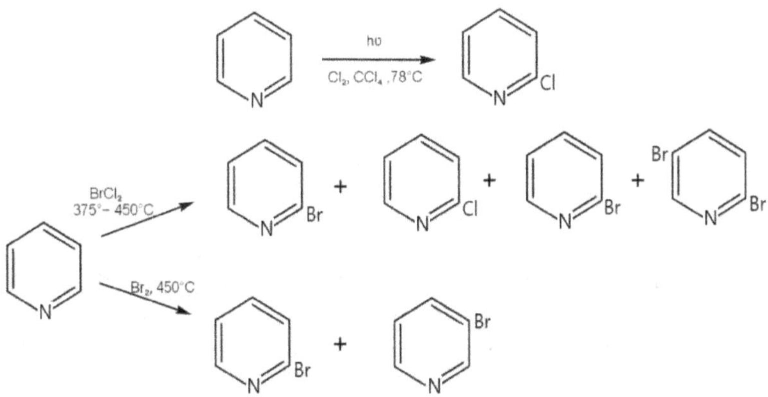

Furnishing sulphuric acid at temperature 90° C when act on quinoline result in the formation of quinoline-8-sulphuric acid (D) while 5- sulfonic acid (E) product is obtained at 170° C in the presence of necessary or its salt however it still higher temperature (300 °C) quinoline bi sulfonic acid (F) obtained.

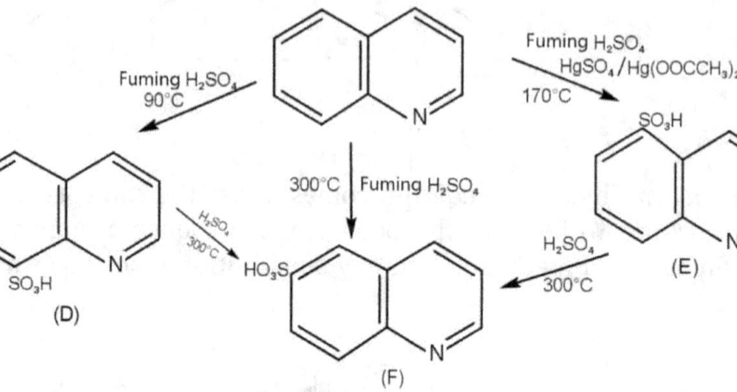

Isoquinoline–5–sulfonic acid is obtained when isoquinoline is put in 40% oleum or excess of SO_3 at temperature up to 180°C. However, at temperatures higher to this, 4–hydroxyquinoline gives 8–sulfonic acid product.

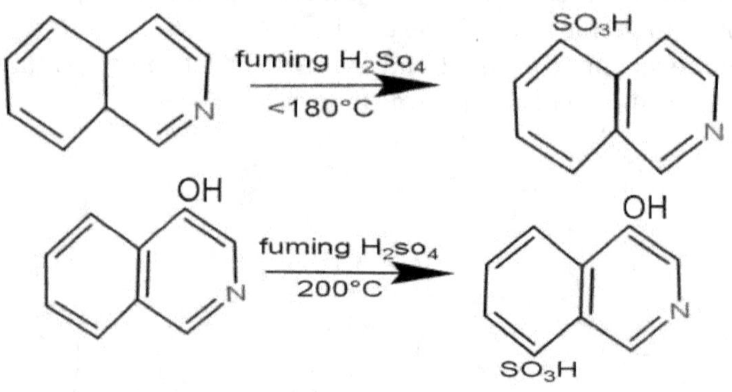

Substitution by nucleophiles does not take place in fused benzene ring but in heterocyclic ring; however, if the benzene ring is activated by nitro – group halogen then benzene ring substitution takes place.

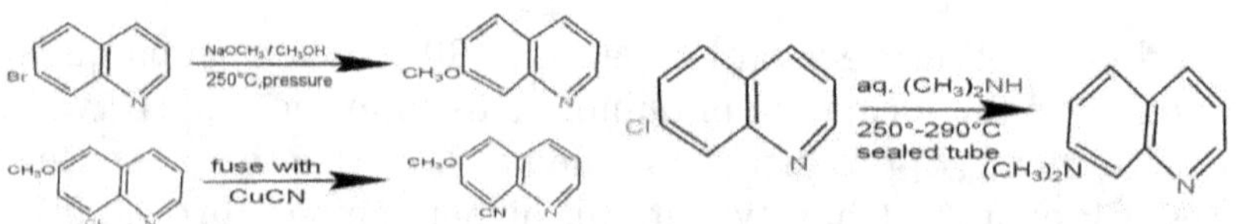

Iso quinoline is attacked by nucleophiles easily at heterocyclic ring; attack at benzene ring fused to heterocyclic ring is brought about either by **displacement** of existing substituent or in place of hydrogen.

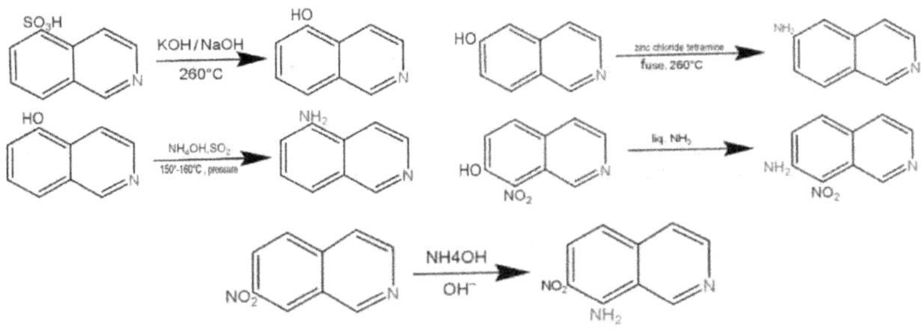

When quinoline reacts with alkaline potassium permanganate, then quinolinic acid is obtained which forms nicotinic acid on heating.

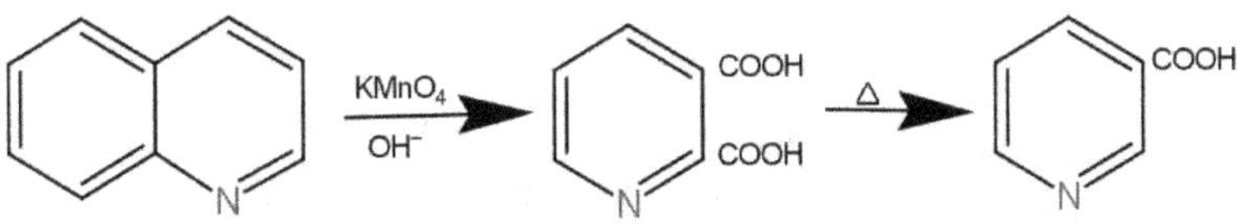

Reduction of benzene ring takes place at high temperature. 2,3,4-trimethyl quinoline gets reduced in the presence of nickel at 155-160 °C to produce 5,6,7,8-tetrahydro-2,3,4-trimethyl quinoline. However, quinoline gets reduced in presence of platinum is trifluoroacetic acid to yield 5,6,7,8 – tetrahydroquinoline.

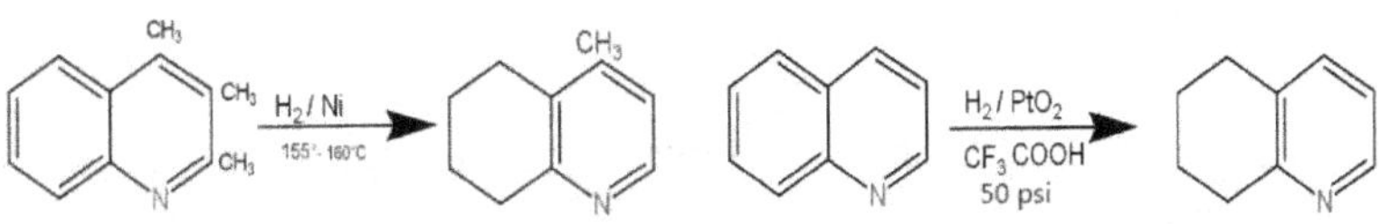

Substituted pyridine-2-sulfonic acid gets converted to pyridone in the presence of dilute mineral acid. Also, pyridine-4-sulfonic acid shows displacement at position-4 in the presence of aqueous ammonia.

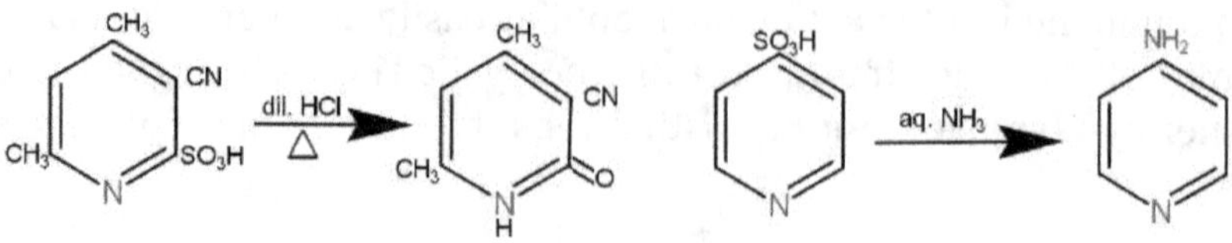

Quinoline–8–sulfonic acid show **displacement** in presence of phosphorus pentachloride.

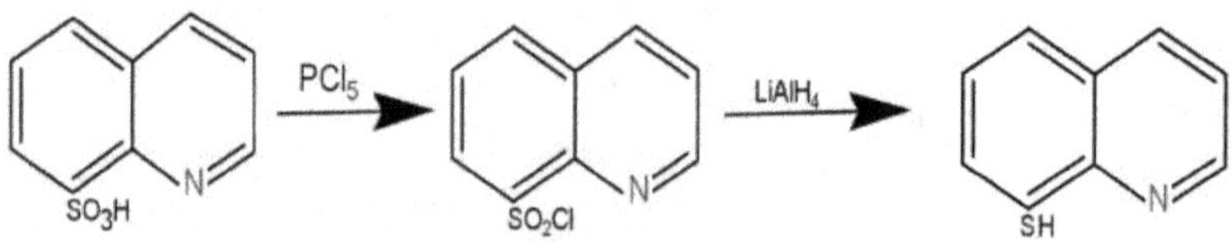

2.2.3 APPLICATIONS OF PYRIDINE

There are numerous applications of pyridine and its benzo - derivatives.

1. An important group of alkaloids in natural products is formed by pyridines, for example – nicotine, ricinine, nor nicotine, etc.

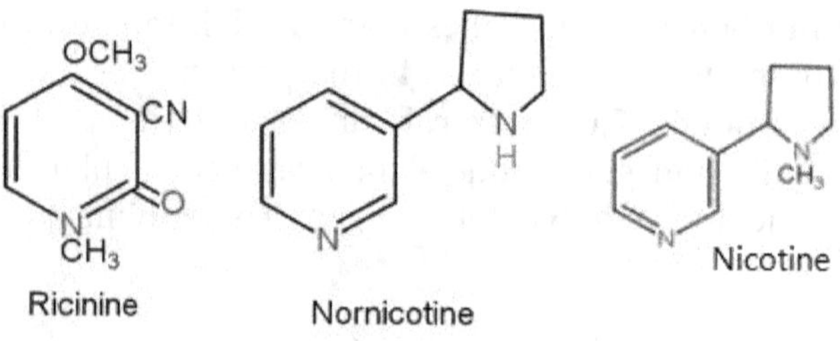

2. B_6 vitamins are also the same group of structurally inter-related natural products.

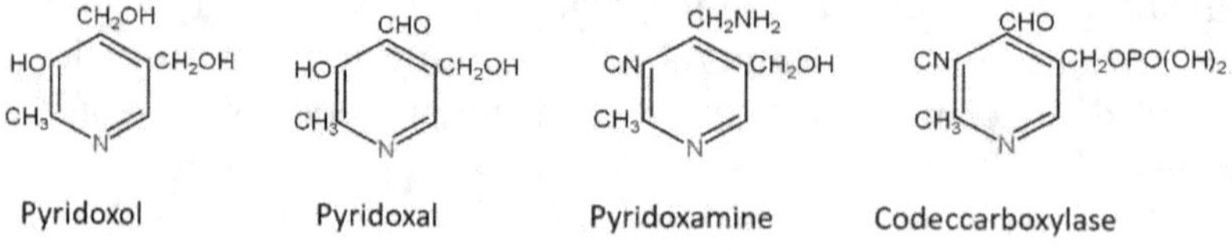

3. **Nicotinamide** adenine dinucleotide (NAD^{+}) takes part in biological redox reactions.

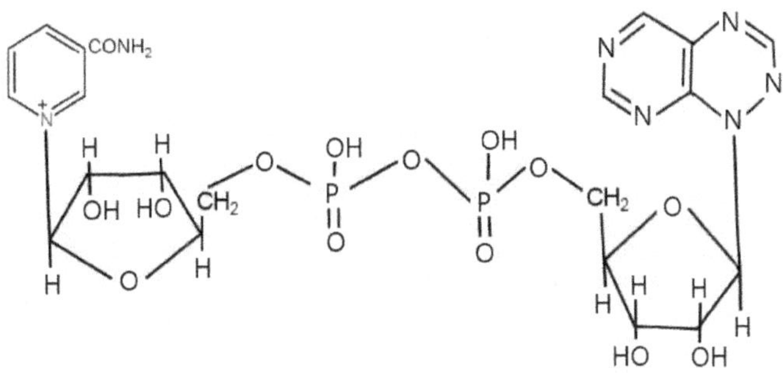

4. Medicinal importance of a group of iso-quinoline alkaloids like **corydaldine** has been established. Likewise, benzyl – iso-quinoline alkaloid, **Papaverine** is a coronary vasodilator.

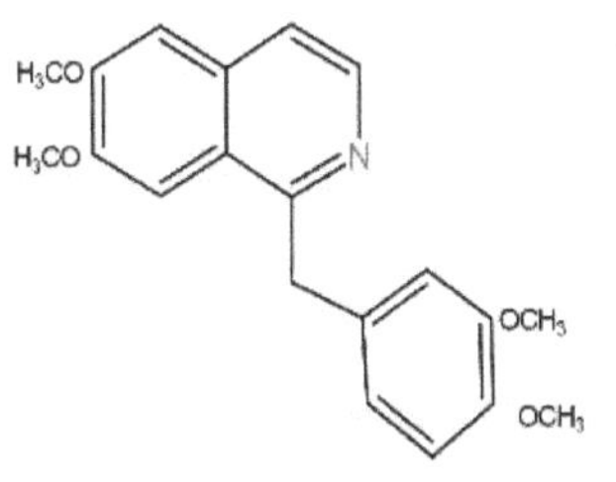

Papaverine

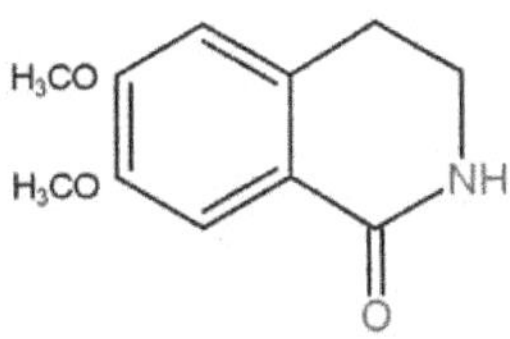

Corydaldine

5. A variety of halo pyridines act as good herbicides for weeds and various other bush and tree species.

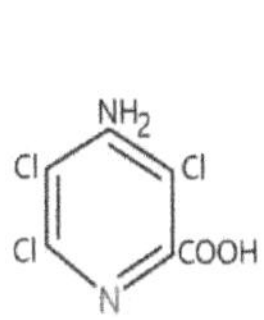

4- amino - 3,5,6 - trichloropicolinic acid

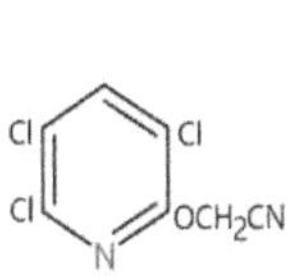

(3,5,6- trichloro -2- pyridinyl - oxy) acetic acid

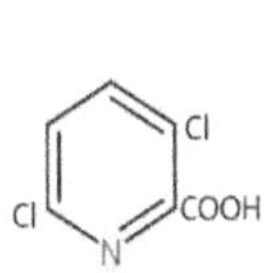

3,6- dichloropicolinic acid

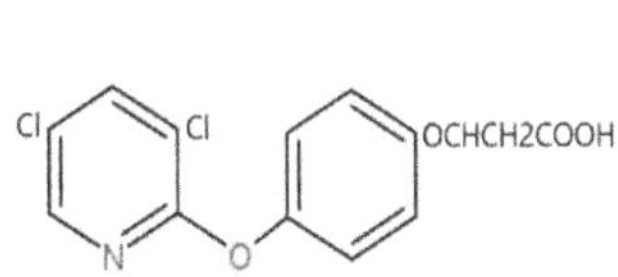

Halophenoxypyridine compound

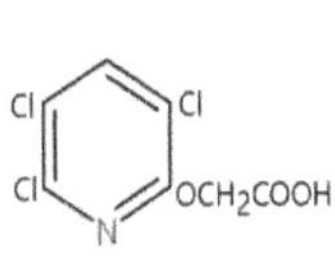

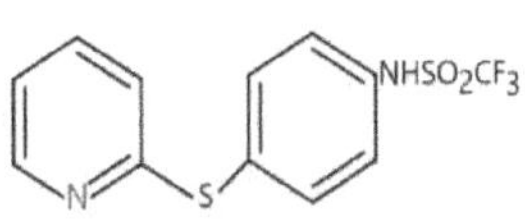

Sulfanilide

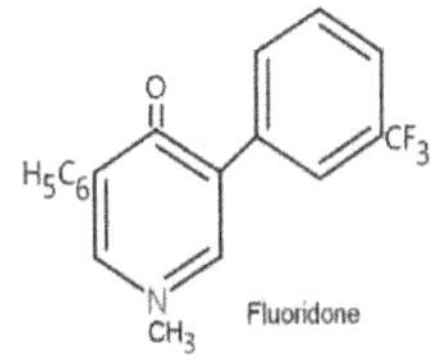

Fluoridone

6. **8-Hydroxyquinoline** in a significant benzo-pyridine fungicide.

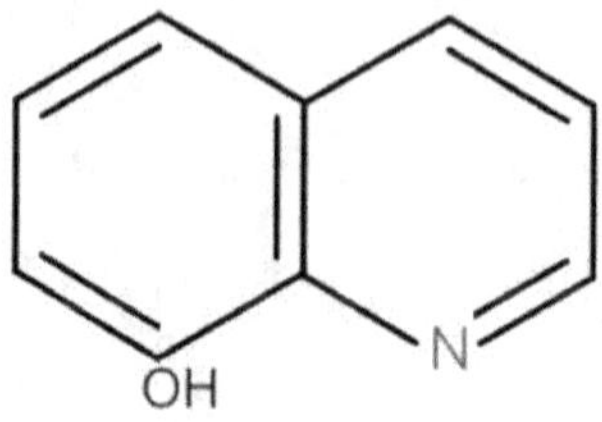

7. There are a variety of veterinary products of pyridine and quinoline derivatives. For example, **2-vinylpyridine** shows Michael acceptor properties and anti-nematode agent (hence used is veterinary medicine). Niacin, which is **pyridine-3-carboxylic acid,** is an important component in the diet of both animal and humans (preventing diseases likes pellagra) hence it acts as a food enriching agent.

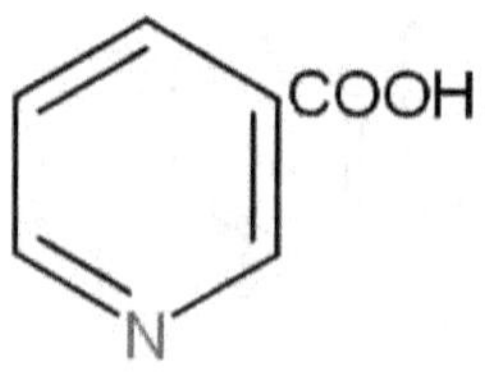

Pyridine-3-carboxylic acid

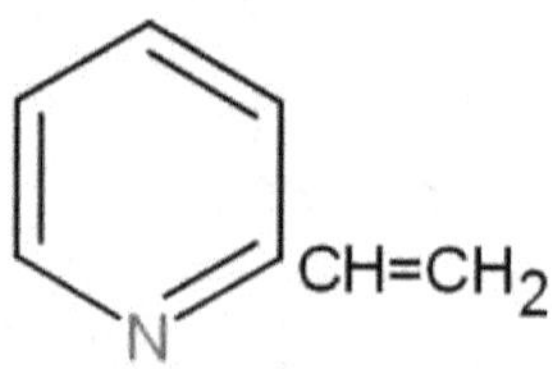

2-vinyl pyridine

8. 8-Amino-6-methoxyquinoline acts as antimalarial medicine in the name of **Pamaquine** and was first developed in Germany in the 1930.

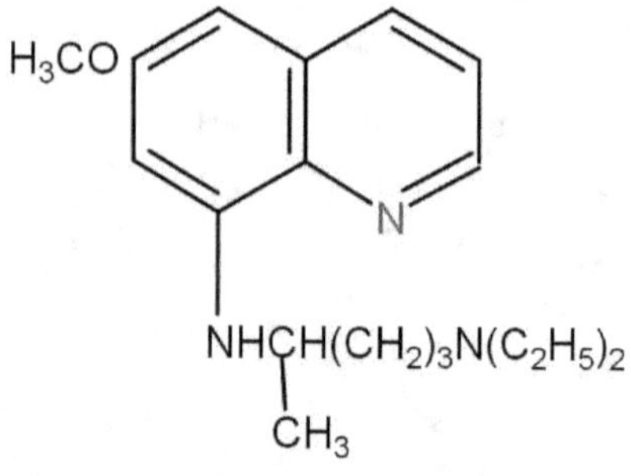

9. There are various aminopyridine and benzo-pyridine derivatives that show medicinal activity.

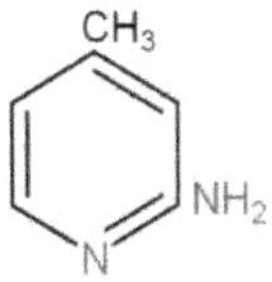

2-amino-4-picoline

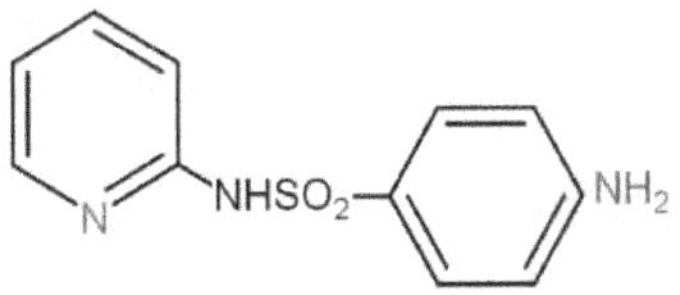

Sulphapyridine

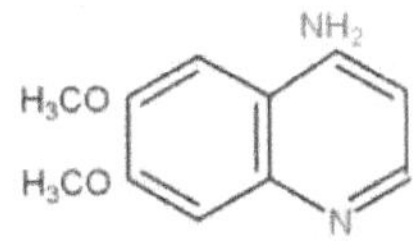

4-amino - 6,7- dimethoxy quinoline

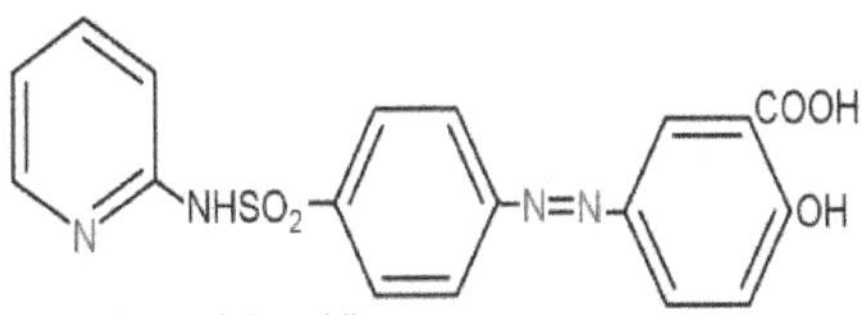

Salazo-sulphapyridine

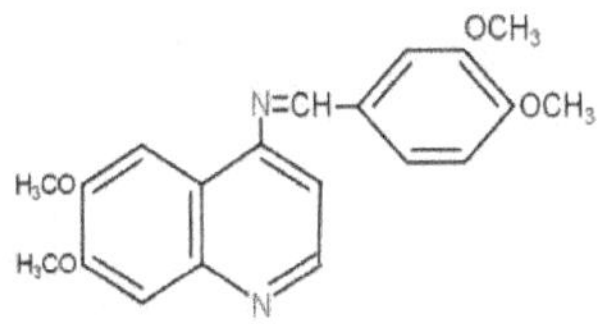

Leniquinsin

Amquinsin (4-amino-6,7-dimethoxyquinoline) and **Leniquinsin**, its condensation products are hypotension agents. 2-Amino-4-picoline acts as a good analgesic. **Sulphapyridine** is an antibacterial agent and **Salazo-sulphapyridine** is used in the treatment of intestinal disorders.

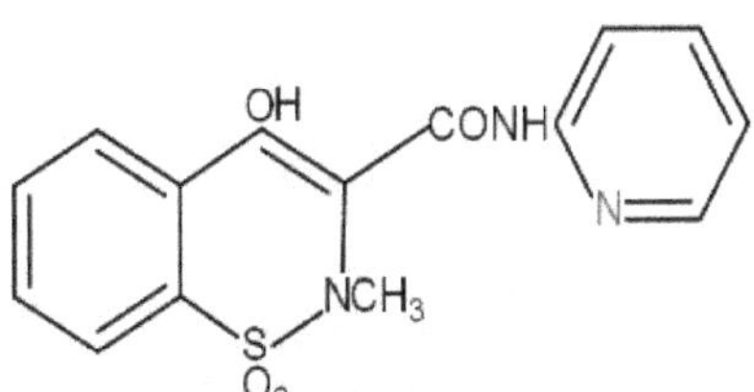

Piroxicam

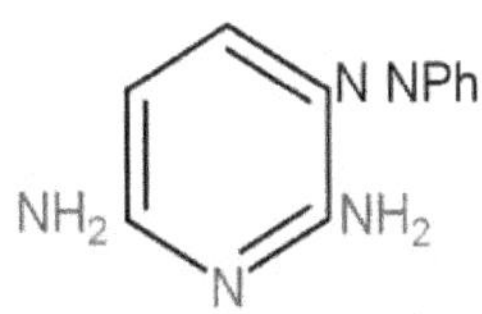

Phenazopyridine

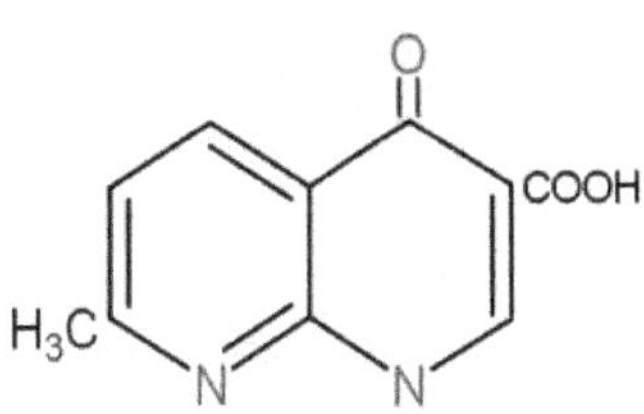

Nalidixic acid

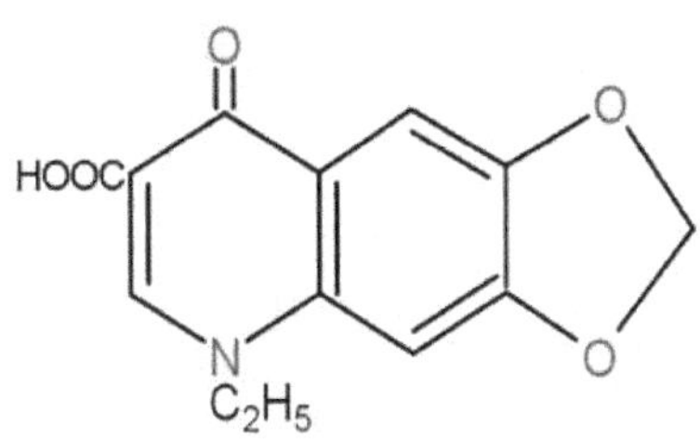

Oxolinic acid

Phenazopyridine is used as an antibacterial agent used to treat urinary infections along with being a urinary tract analgesic. Likewise, **Nalidixic acid** also shows antibacterial activity. **Oxolinic acid** also shows similar properties. **Piroxicam** shows good anti–inflammatory properties and treatment of rheumatoid arthritis.

10. Anticancer activity is depicted by some derivatives of pyridine.

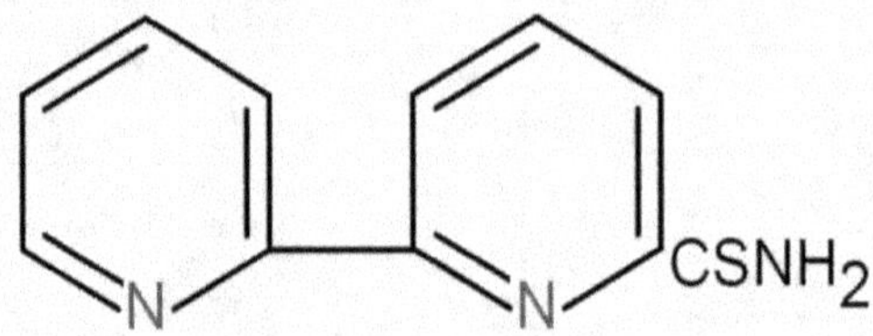

11. The antibiotic properties have been exhibited by **Cephaloridine**, which contains a quaternized ring and by **Cepharin Sodium**, which is based on pyridine-4-thioglycollic acid.

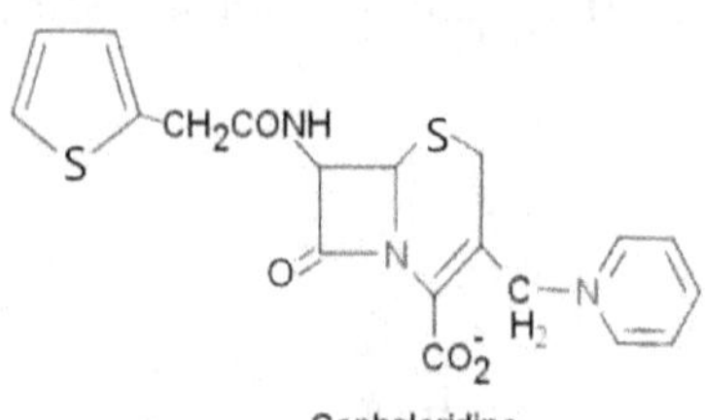

Cephaloridine

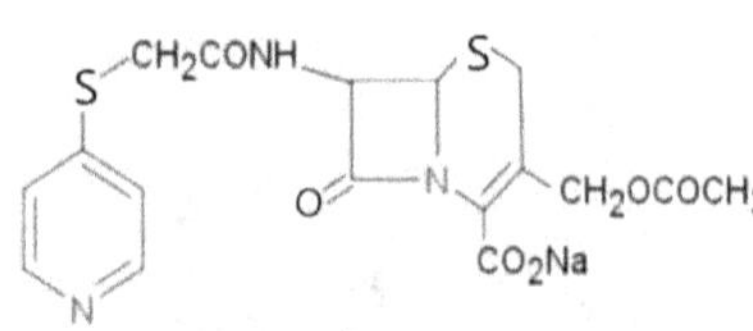

Cepharin Sodium

12. Compound **Anginin** is used to reduce lipids and cholesterol levels in blood systems. **Nifedipine** is used to treat angina.

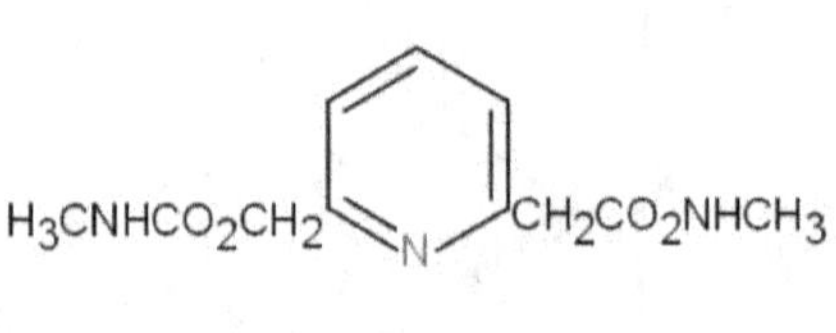

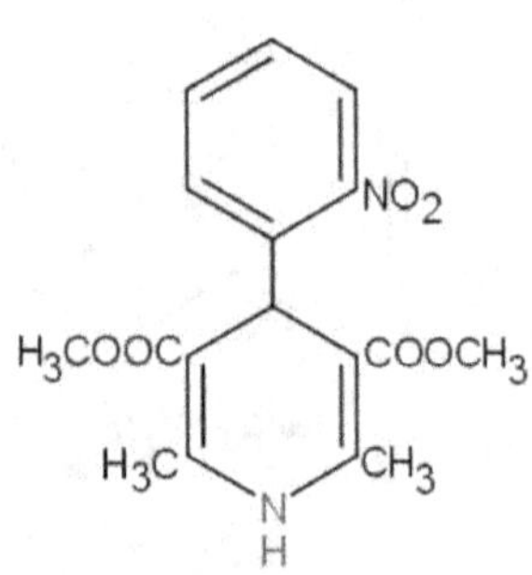

13. A special application of **4-alkyl-2-mercaptopyridine**, found in floatation process for recovery of metal concentrates, is in the form of metal-bearing ores.

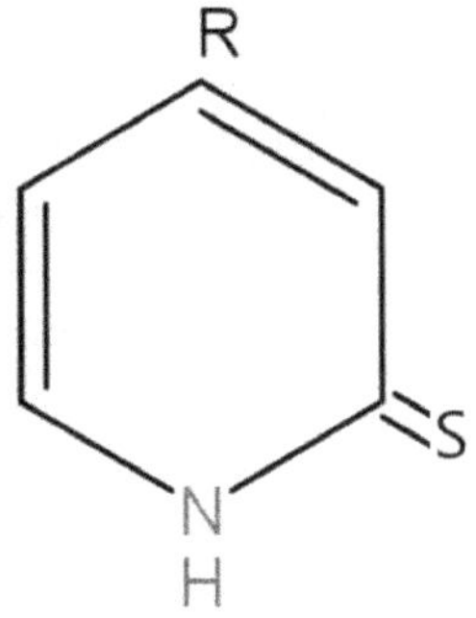

2.3 PYRIDAZINES (1,2-diazine) AND ITS DERIVATIVES

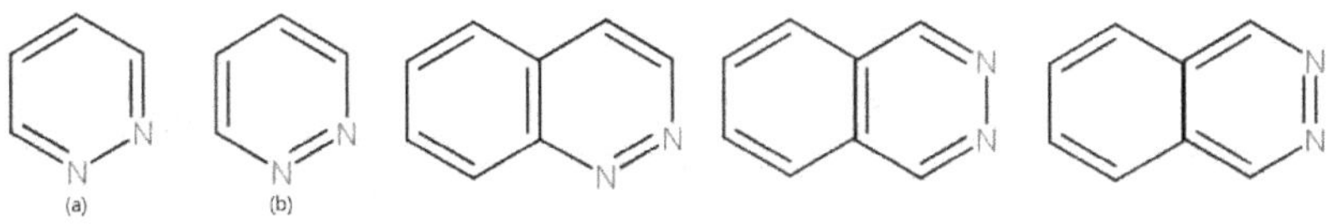

Pyridazine is assumed to be a planner six membered ring and is represented as a resonance hybrid of two structures (a) and (b) with a greater contribution form canonical structure (a).

2.3.1 SYNTHESIS OF PYRIDAZINE

Cyclization of tricyano-hydrazone in the presence of heat gives pyridazines.

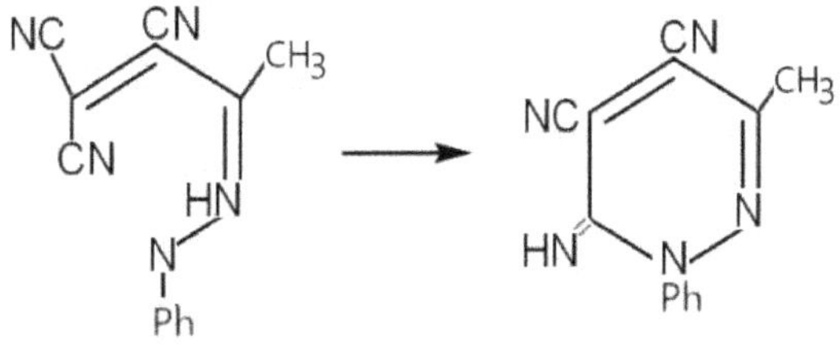

Widman – Stoermer synthesis is the conversion of diazotized o – aminoarylethylene results in substituted bicyclic systems.

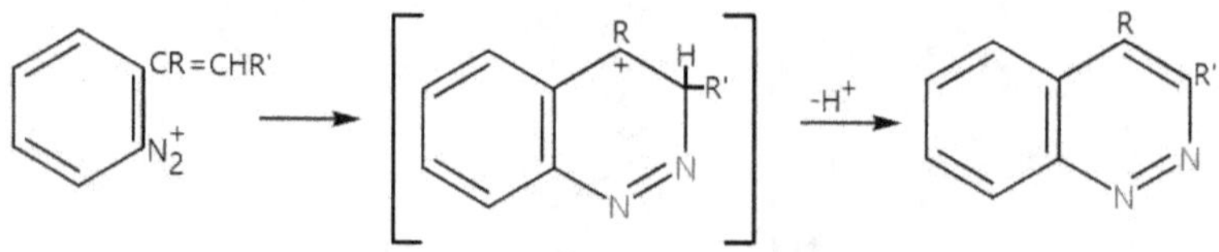

There are two more reactions of the same type: **Richter synthesis** and Neber – Bossel synthesis.

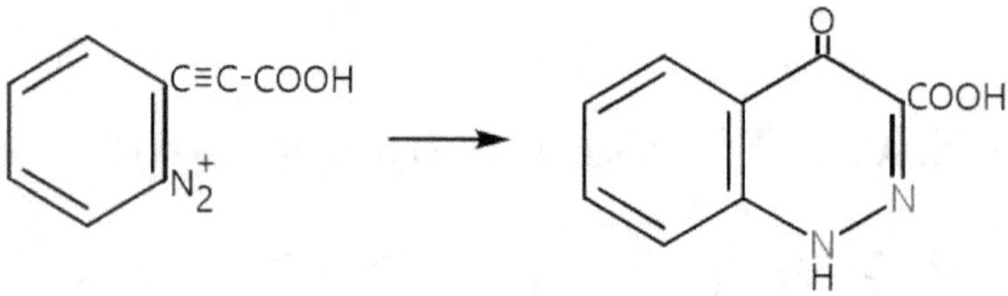

In **Neber – Bossel synthesis**, conversion of diazonium group to hydrazine group takes place, in the first step, followed by cyclization.

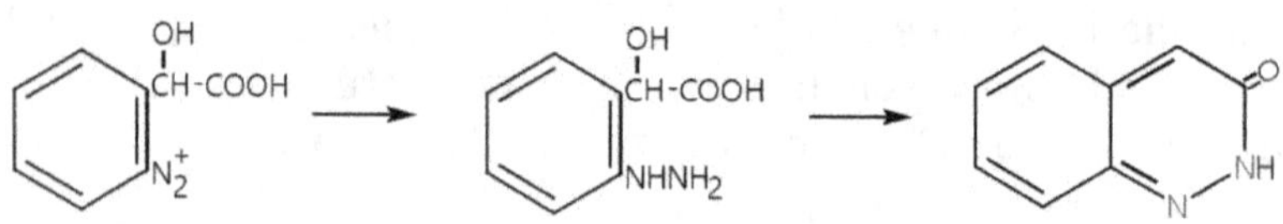

1,4-Diamino phthalazine, which is obtained by the **conversion** of phthalo dinitrile by methanolic hydrazine, gets converted to 1,4-dihydrazino phthalazine more than hydrazine in dioxane–acetic acid.

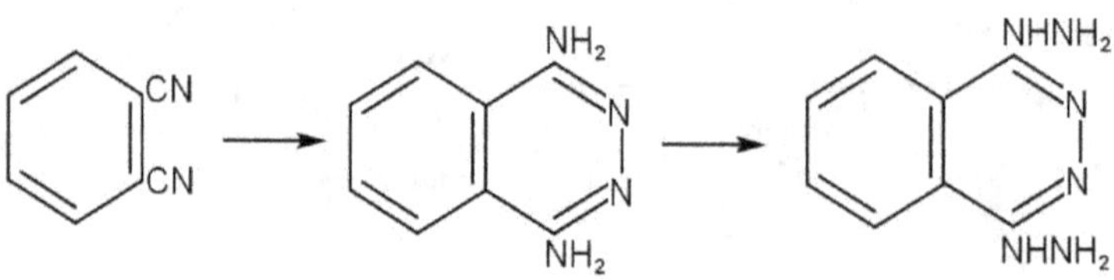

When 2–amino phthalimide on heating with dilute alkali/acid **rearranges** to 4–hydroxy phthalazin–1(2H)–one.

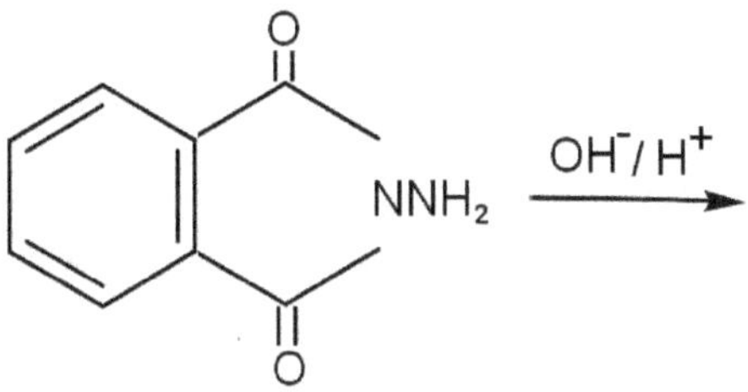

2.3.2 PROPERTIES OF PYRIDAZINE

The substituent at position–6 decides the **thermolysis** product of 3–azidopyridazine-2–oxide.

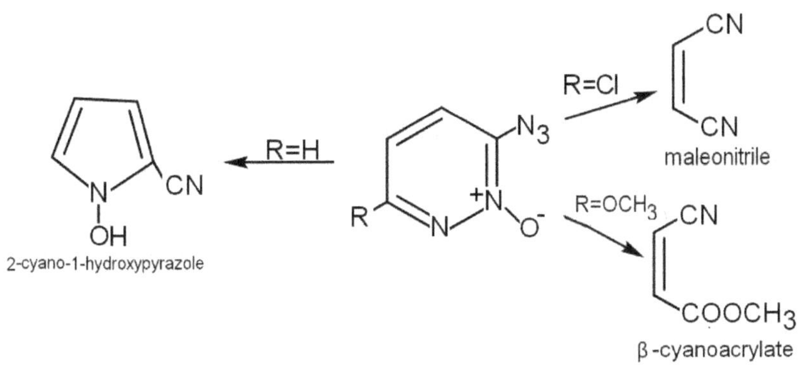

Rearrangement of 4–diazo–5,6–dioxo–1–phenyl-1,4,5,6-tetrahydropyridazine at room temperature in anhydrous conditions, via intermediates, forms pyrazolo acetone. This compound has resonance structures. This compound converts to 5–hydroxy-1–phenyl pyrazole-4-carboxylic acid in moisture and on decarboxylation yields 5–hydroxy-1-phenylpyrazole.

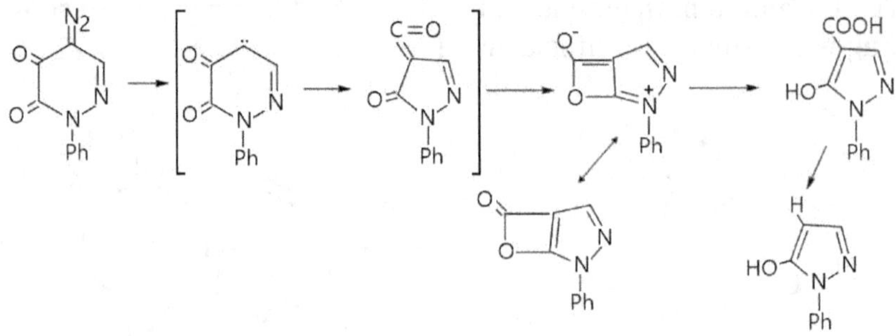

Photochemical reactions of pyridazines can best be depicted by the following ones. Pyridazin-3(2H)one gives two products on **irradiation** in neutral methanol.

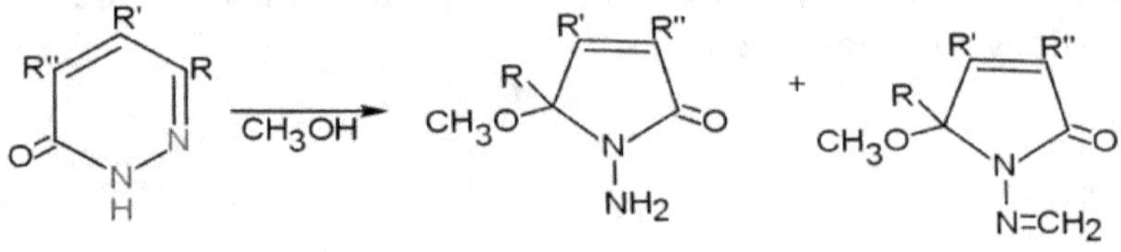

In the same way, 5–amino-4–chloro-2–phenylpyridazin-3(2H)–one on **photolysis** via intermediate results in bis-pyridazino-pyrazine.

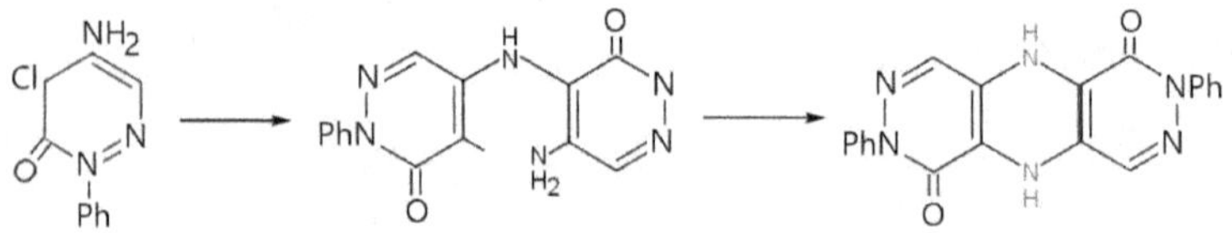

There are also a variety of reactions throwing light on the **photochemical** aspect.

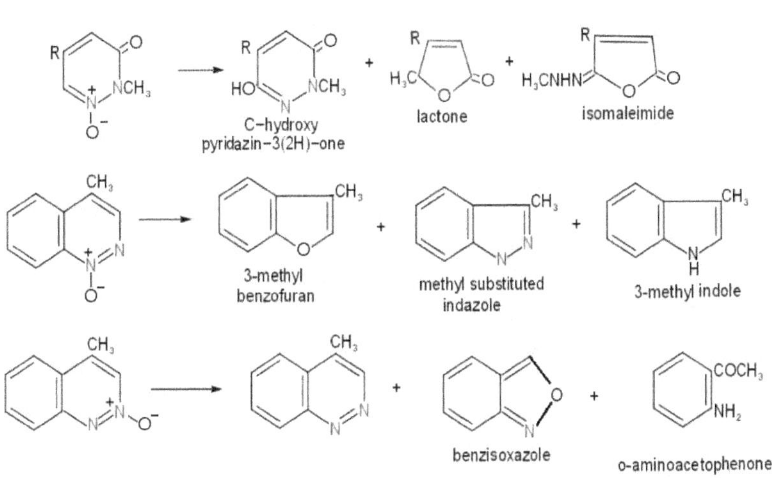

N–alkylated products can be obtained by treatment with alkyl halides; α–halo acids, etc. A mixture of ketone and betaine can be obtained from 3–alkyl-5-amino-pyridazin-6-(1H) one in presence of dimethyl sulphate.

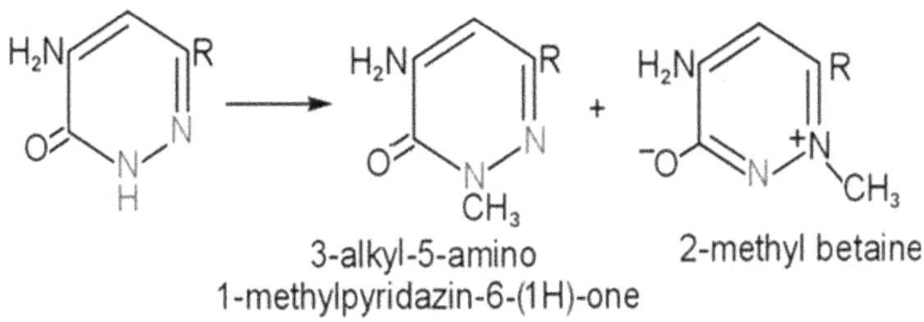

On reaction of diazomethane with 5–ethoxycarbonyl-3-methyl–6–phenylpyridazin-4(4H)–one, N–methyl,4–methoxy and obtain derivatives are obtained.

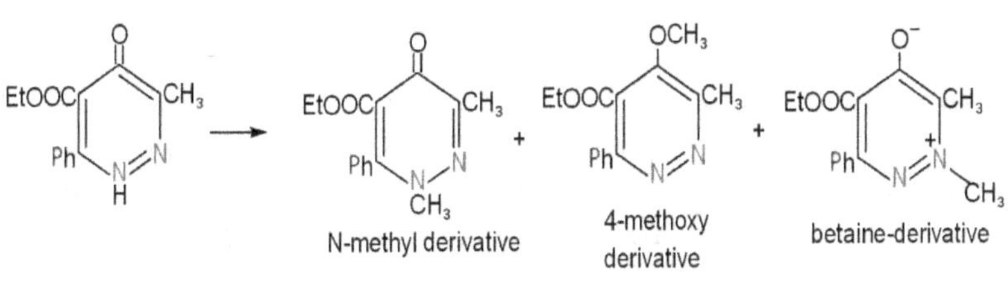

A series of **methylation** products are obtained under different conditions.

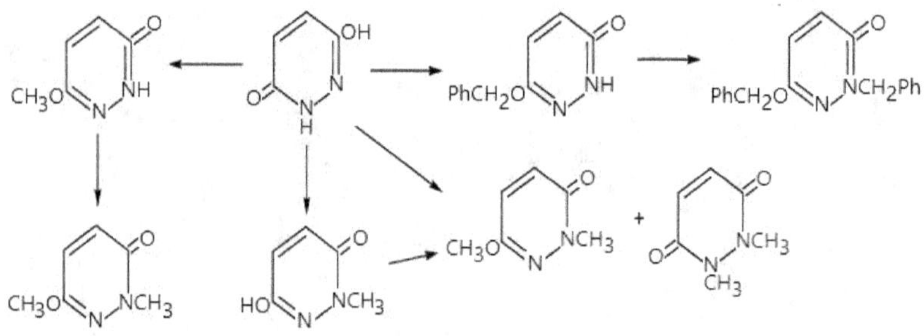

Electrophilic attack at carbon can best be shown in pyridazine-1-oxide which have substitution either at position-3 or at position-3 and position-6. In such cases, position-5 gets substituted by nitro group.

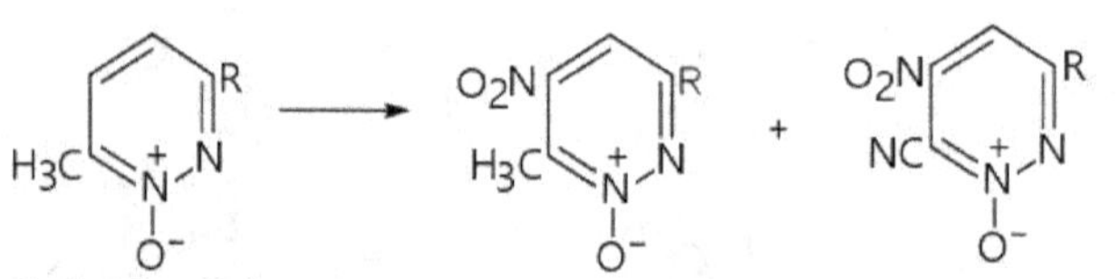

3-alkyl-6-methyl pyridazine-1-oxide; 5-nitroderivative; 6-cyano-5-nitro derivative

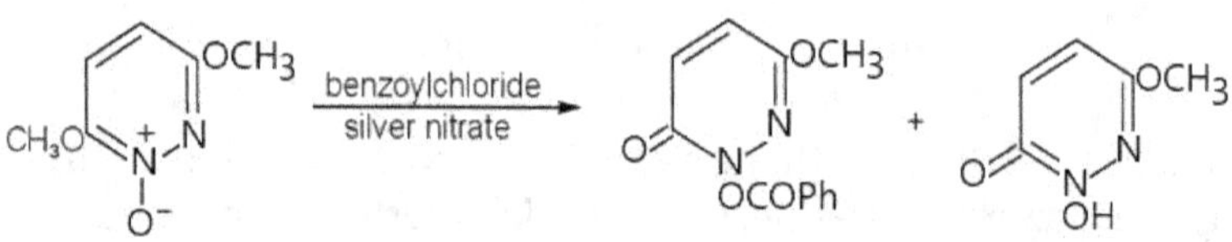

1-benzoyloxy-3-methoxypyridazin-6-(1H)-one; 1-hydroxy-3-methoxypyridazin-6-(1H)-one

Nucleophilic attack at carbon can be illustrated when 3-methoxy pyridazin-1-oxide gives reaction with methyl β–amino crotonate in the presence of benzoyl chloride and yields α-(6-methoxy-3-pyridazinyl)β-amino crotonate and later on hydrolysis to acetate.

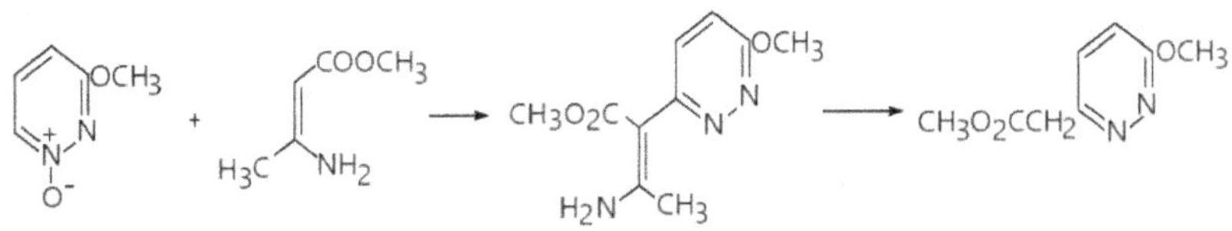

Valence isomerization occurs when other groups are attached at position–4 of 3,6-dichloropyridazine and hence azides/tetrazolo[1,5-b]pyridazines are obtained.

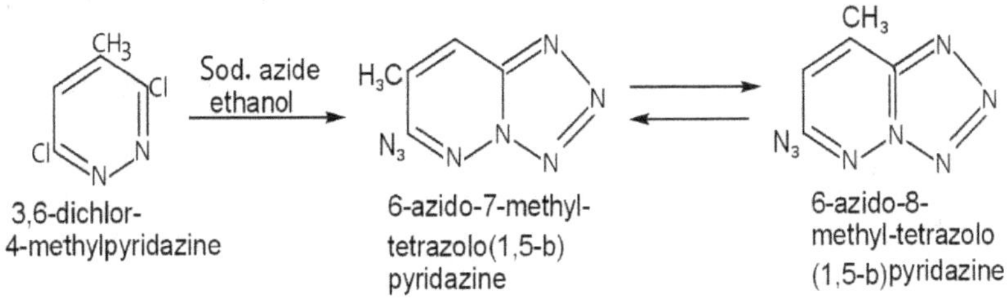

2.4 PYRIMIDINE AND ITS BENZO DERIVATIVES

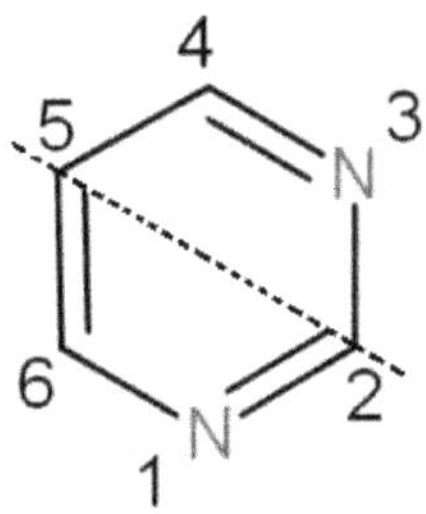

2.4.1 SYNTHESIS OF PYRIMIDINE

Uracil is obtained via **Rinke's synthesis** in which malediamide is treated with sodium hypochlorite. The reaction proceeds by isocyanate intermediate that cyclises and results in the formation of uracil.

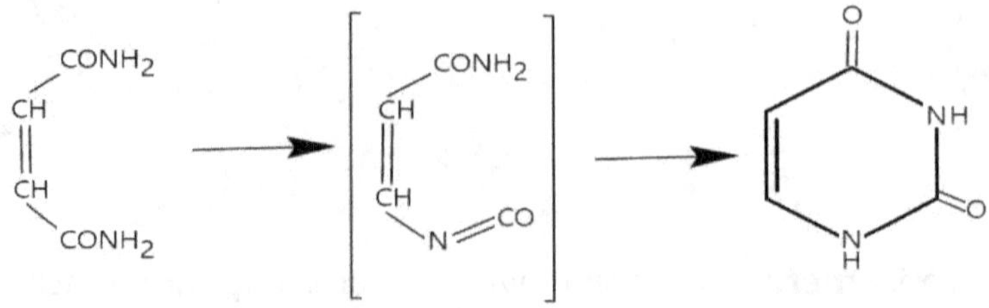

When benzaldehyde reacts with 2-ethyl-2-phenylmalondiamide, 5-ethyl-2,5-diphenyl-2,5-dihydropyrimidine-4,6-(1H,3H)–dione is obtained. Also, when o–benzyl amino benzyl amine is treated with boiling formic acid, then 1-benzyl-1,4-dihydroquinazoline is prepared.

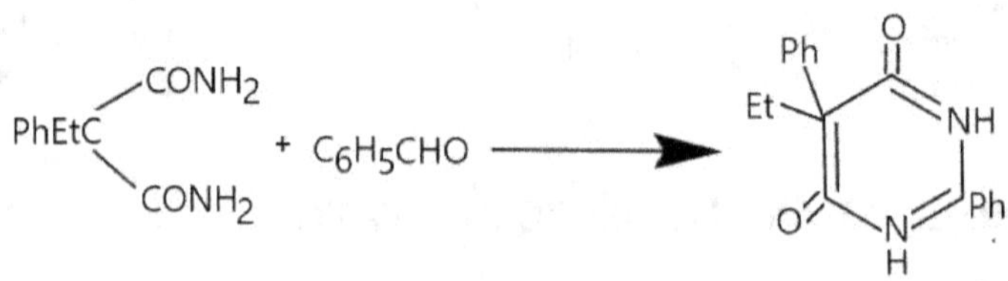

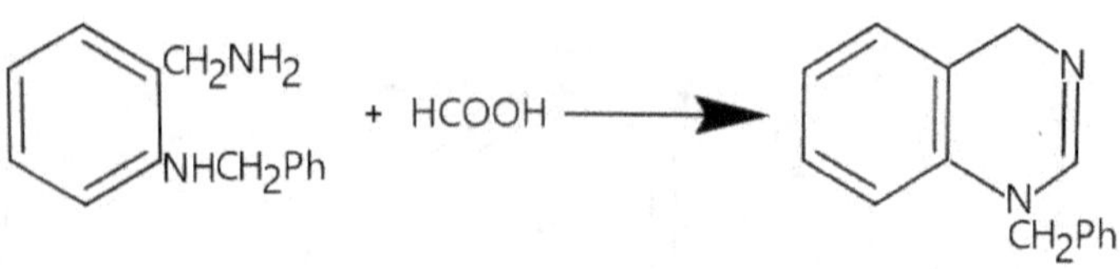

When ammonia (although better options are formamide, formamidine acetate, acetamide, etc.) at 200°C, reacts with β-acetamido vinyl phenyl ketone (A), then 2-methyl-4-phenylpyrimidine is obtained; however, when (A) reacts with guanidine carbonate, 4-phenylpyrimidin-2-amine is obtained.

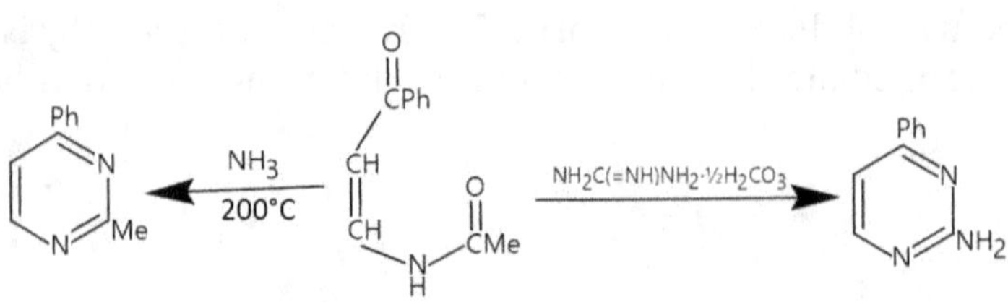

2:1 Ratio mixture of phenylacetaldehyde and urea gives 4-benzyl-5-phenyl-3,4-dihydro pyrimidin-2(1H)-one.

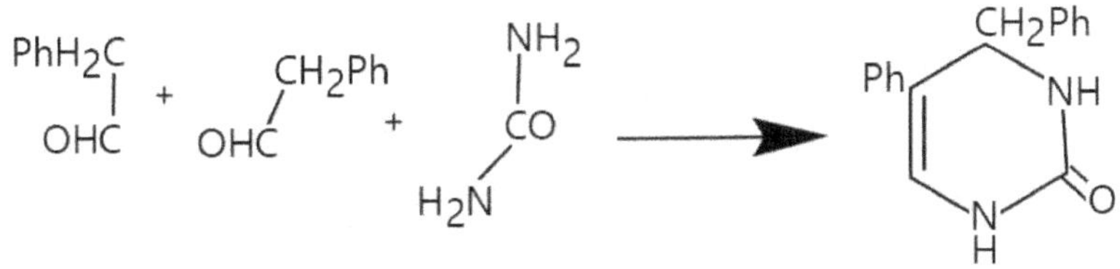

Pyrimidin-4(3H)-one **(B)** can be prepared by two ways: one is **direct synthesis** from ammonia, triethyl orthoformate and 3-ethoxyacrylate while other is **desulphurization** of 2-uracil.

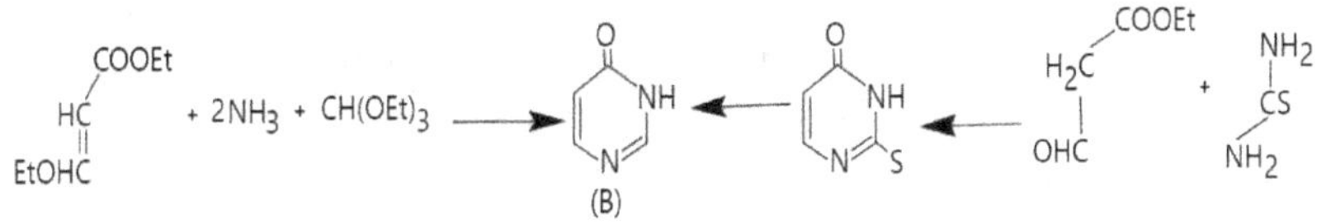

Dehydrochlorination of chloro-dihydro-cytosine gives cytosine and, in another method, 4-thiation of 2-uracil takes place giving 2-pyrimidin-2,4(1H,3H)-dithione.

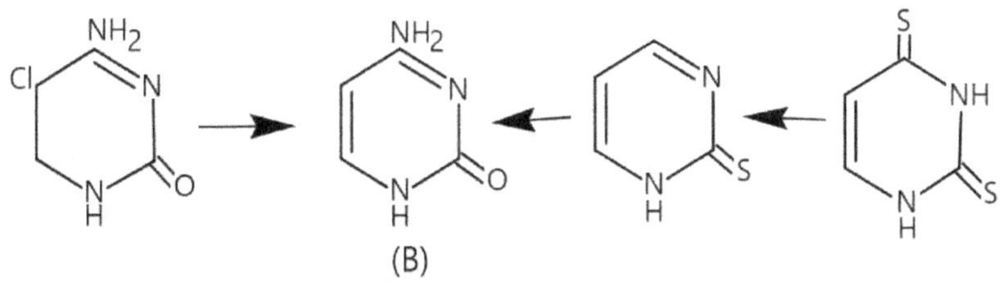

3-methyl malic acid gets converted to ethyl 2-formylpropionate which changes into thymine thereby serving as the best method of its synthesis.

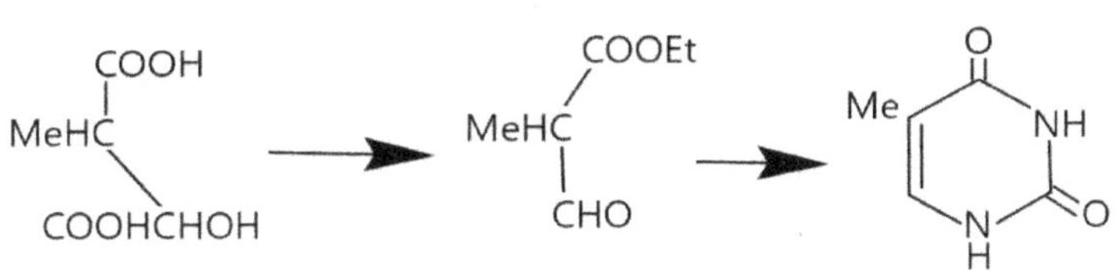

2.5 PYRAZINE AND ITS BENZO DERIVATIVES

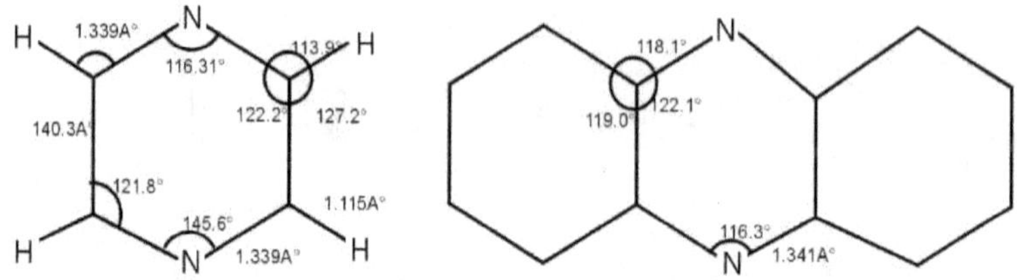

2.5.1 SYNTHESIS OF PYRAZINE

The **condensation** of 1,2-diaminopropane with 3-methylbutane-1,2-dione gives isomeric dihydro-pyrazines which on oxidation results in corresponding pyrazines. This method is one of the most ancient and applicable for synthesis of general pyrazines.

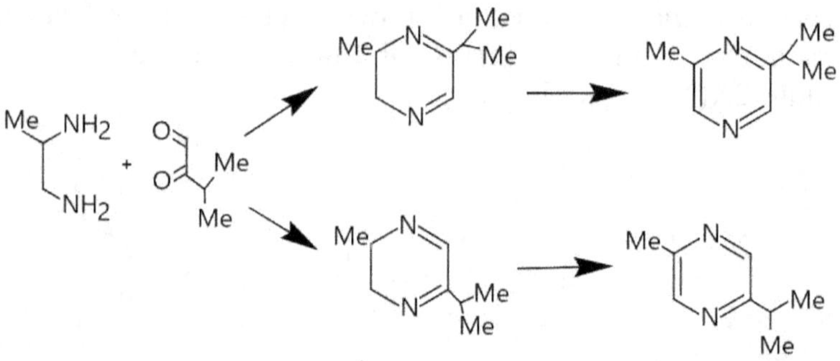

Acyl cyanides are also used and further **hydrolysed** using H_2O_2 /Na_2MoO_4 to amides which by ring closure yields pyrazine derivatives.

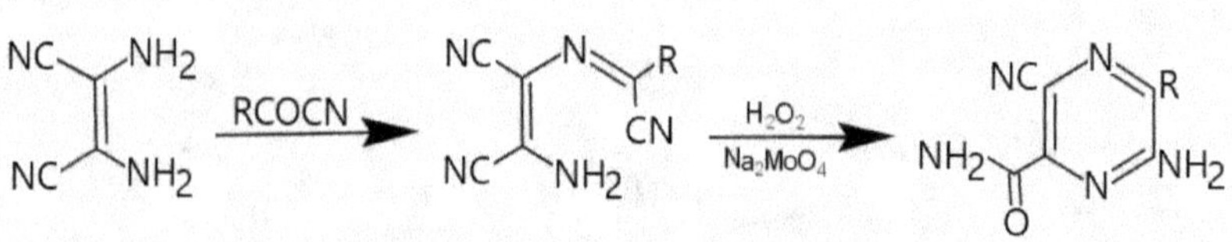

1,4-Dialkylamino-1,4-dihydropyrazines can be obtained by **condensation** of 1,1-di substituted hydrazine derivatives with α-nitro-β-ethoxy ethylene derivatives.

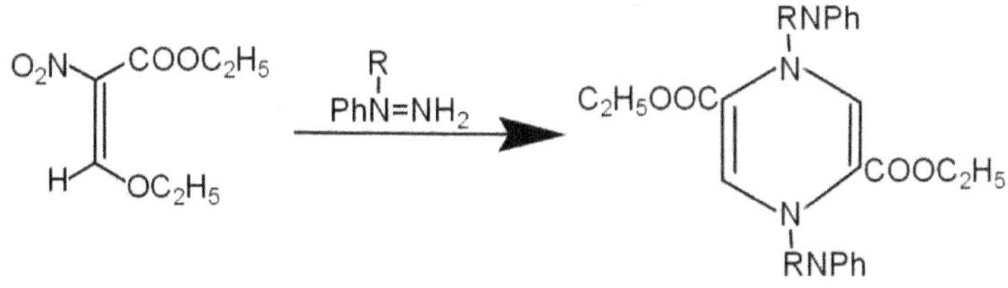

Synthesis of substituted pyrazines can be easily brought about by cleavage of fused pyrazines for example, pteridines can easily be converted to pyrazine derivatives under acidic or basic conditions.

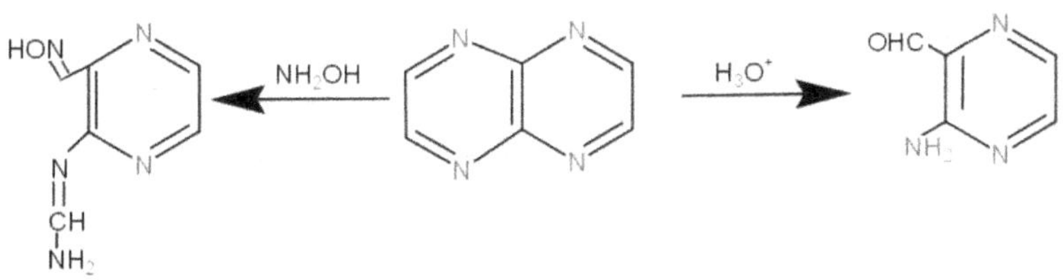

2.5.2 PROPERTIES OF PYRAZINE

Addition reaction can be understood in treatment of 2-chloro-3-dichloro methyl pyrazine with excess of sodium methoxide resulting in an introduction of methoxy group at position-6.

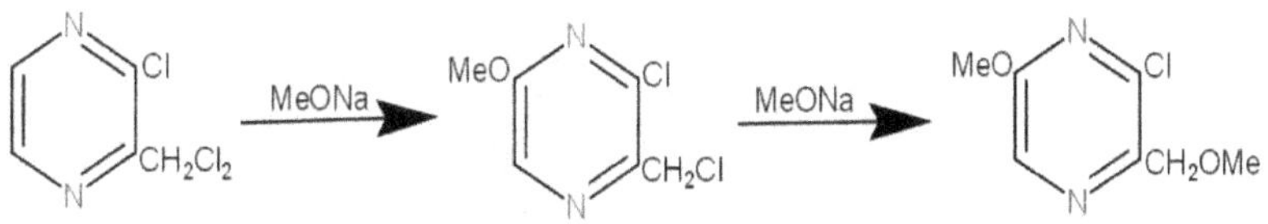

With appropriate choice of reagents, **regioselectivity** of N-oxidation reactions can be obtained. For example, in 2-chloropyrazine, N-4 is more basic; In very strong acidic medium, the pyrazine is protonated, and it is explained that although the nucleophilicity of the monoprotonated is reduced, the electrophilicity of peroxysulphuric acid is sufficient to effect oxidation of the less basic nitrogen atom.

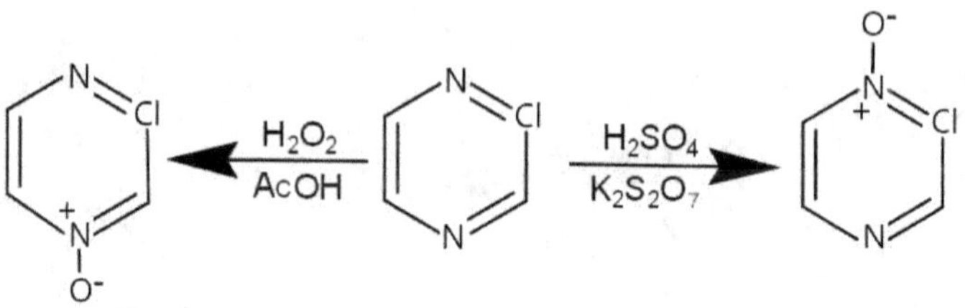

Ortho- and para- position to amino group are activated in pyrazines, quinoxalines and phenazines to electrophilic attack; hence reaction of 2-amino pyrazine with bromine is acetic acid gives 2-amino-5-dibromopyrazine.

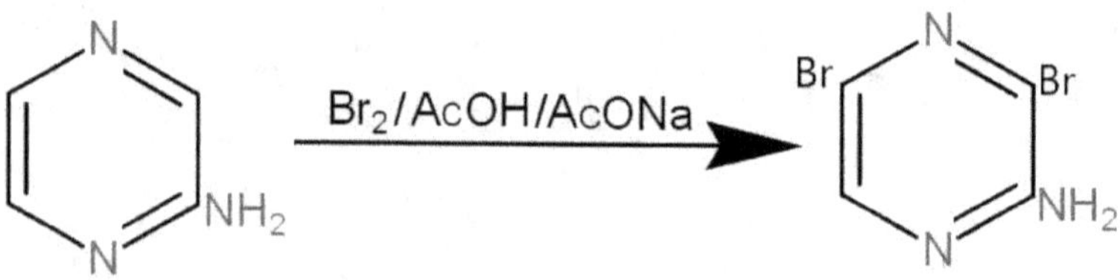

2.5.3 APPLICATIONS OF PYRAZINE

1. Phenazine rings are a component of many bacterial pigments for example, **Iodinin.** It is said to be produced in late stage of growth of *Chromobacterium iodinium.*

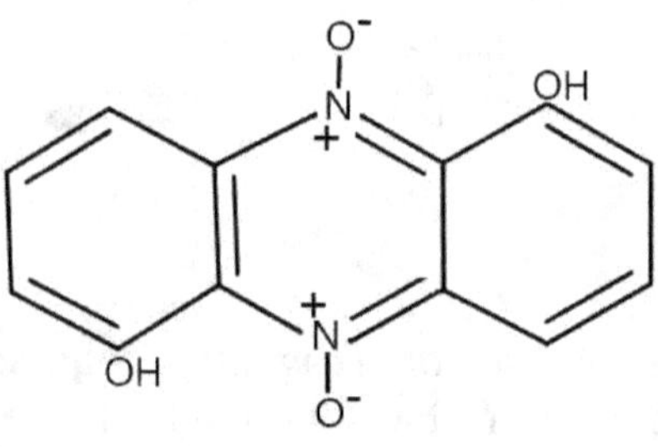

2. Simple alkylated pyrazines are part of many food products for example:

a. **Coffee** – 2,3,5,6-tetra methyl pyrazine
b. **Boiled beef** – 2,3-dimethylpyrazine
c. **Fried beef** – 2,5- dimethyl pyrazine
d. **Potato chips** – 2,6- dimethyl pyrazine
e. **Roasted barley** – 2-ethyl-3-methylpyrazine
f. **Peanuts** – 2,5-diethylpyrazine

3. Simple pyrazines are responsible for **bioluminescence** in coelenterates during enzyme–catalysed oxidation.

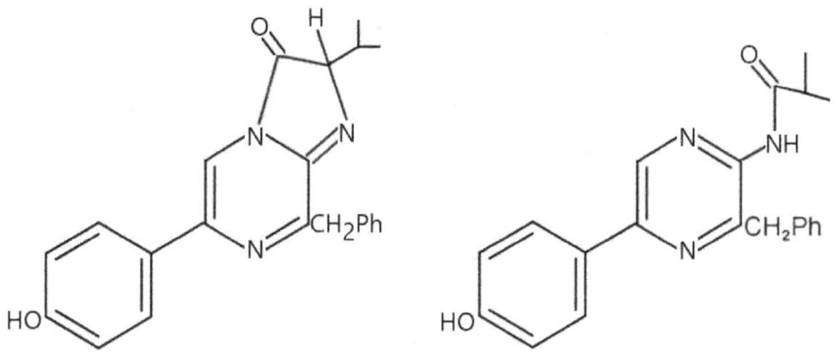

4. Before the discovery of streptomycin, **pyrazinamide** was administered in the treatment of tuberculosis.

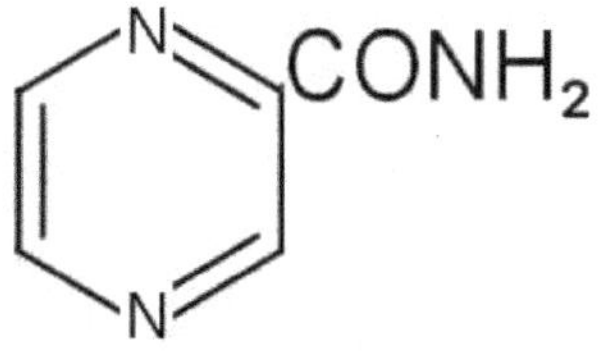

5. 1,4–Dioxide of quinoxaline–2-carboxylic acid which has been isolated from *Streptomyces ambrofaciens*_and its derived esters show **antibiotic** properties.

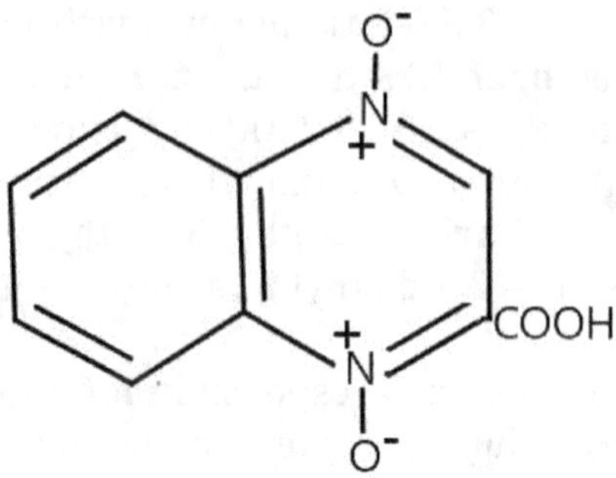

6. 5,6,7,8–Tetrachloro quinoxaline (A) and respective halogen derivatives have been found to act as **fungicide**, 6–hydroxy quinoxaline (B) as **insecticide** and 2–hydroxy quinoxaline (C) as **anthelmintic**.

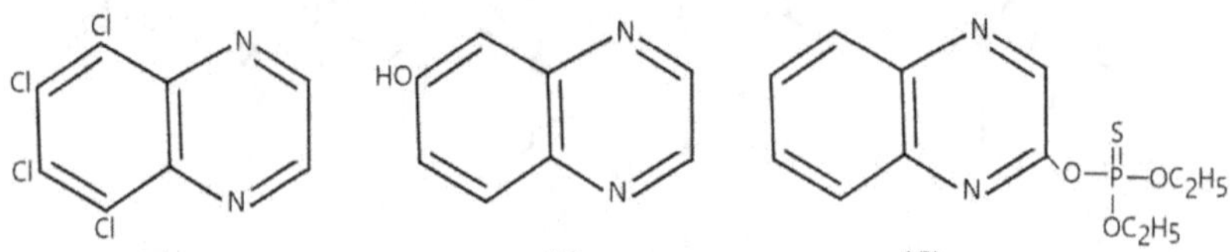

(A) (B) (C)

7. Starting diamines or monoamines result in a variety of **dyes**.

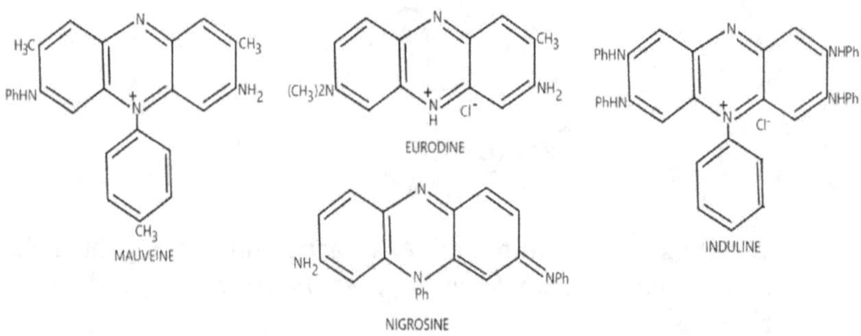

2.6 QUINOLIZINIUM ION AND ITS DERIVATIVES

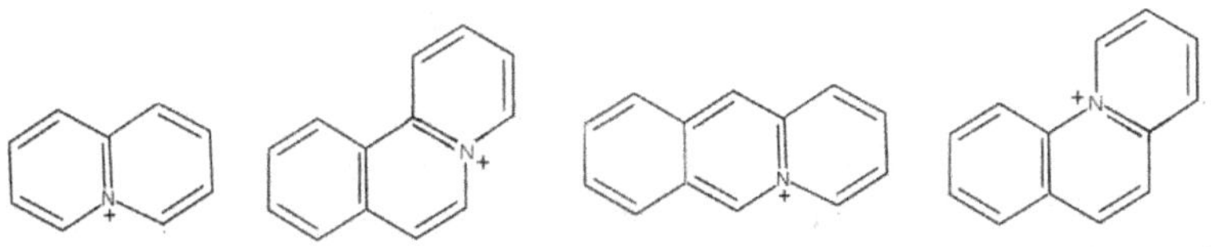

Quinolizinium ion benzo[a]acridizinium ion benzo[b]acridizinium ion benzo[c]quinolizinium ion

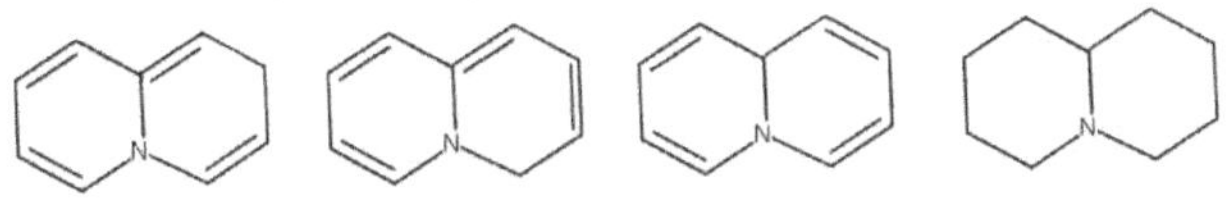

2H-Quinolizine 4H-Quinolizine 9aH-Quinolizine Quinolizidine (4a-Azadecaline)

2.6.1 SYNTHESIS OF QUINOLIZINIUM ION

2-Acetopyridine reacts with 3,3-diethoxypropyl magnesium chloride to produce 1-methyl quinolizinium ion via an intermediate.

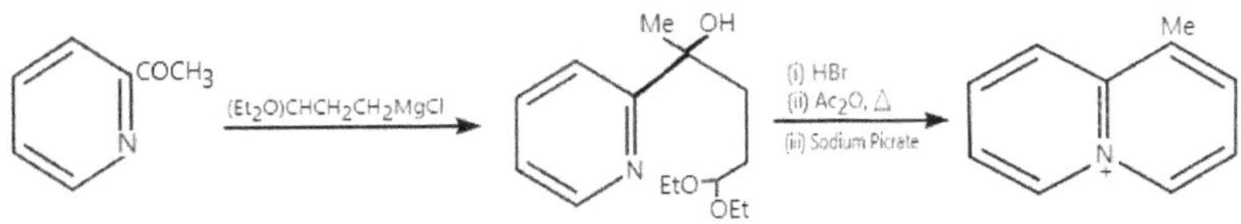

An imine results in the formation of 6-aminobenzo[c]quinolizinium chloride on heating at 200°C. Alternatively, the imine is first **hydrolysed** to ketone which, at temperature 195°C, gives 6-hydroxybenzo[c]quinolizinium chloride.

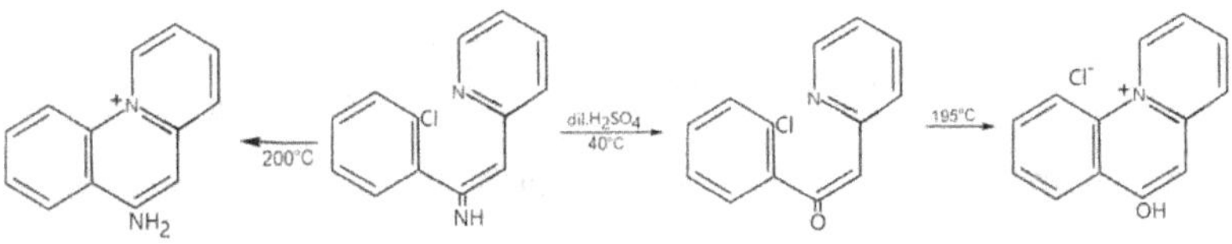

Acridizinium ion is synthesized by aromatization of quinolizinium ion. The final product obtained by this route is 4–methylacridizinium bromide.

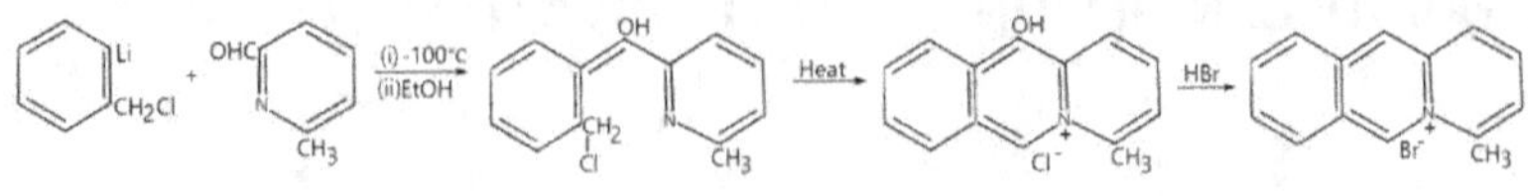

2,3-Disubstituted quinolizinium ion is obtained by **cyclization** of 2-picolinaldehyde and deoxy benzoin followed by a series of reactions.

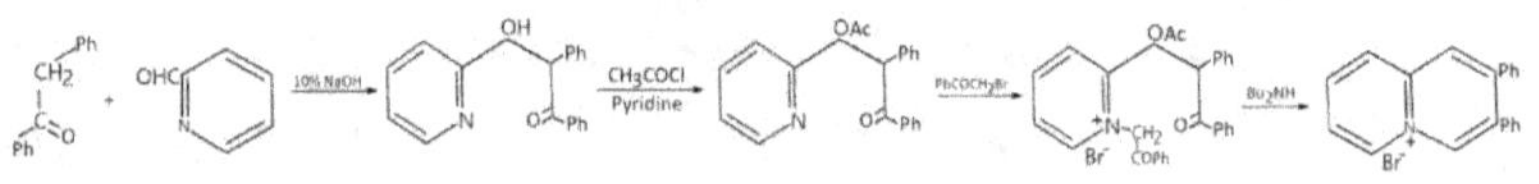

[4+2] Cycloaddition Reaction between isoquinoline perchlorate and mesityl oxide results in the formation of 2,4-dimethyl derivative of benzo[a]quinolizinium compound.

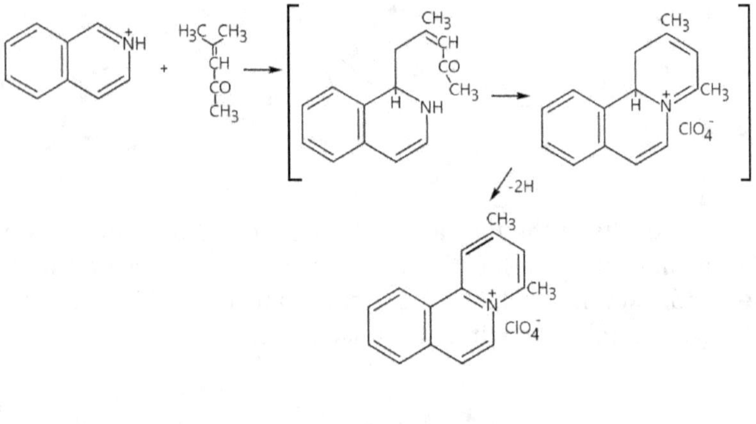

2.6.2 PROPERTIES OF QUINOLIZINIUM ION

When perbromide salt is heated at 200°C it forms 1-bromo quinolizinium bromide.

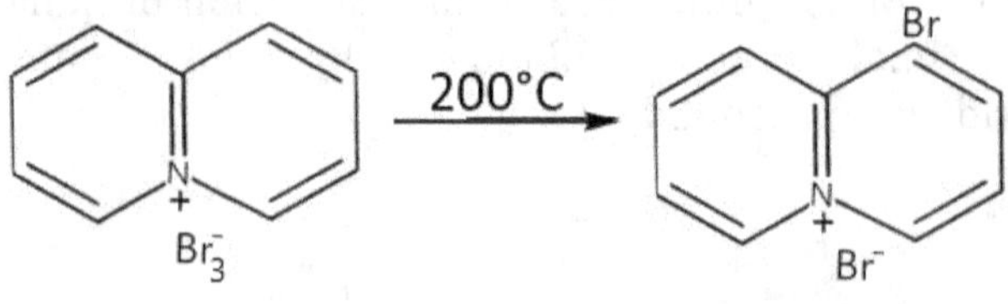

However, if position–1 is occupied by an amino group than **substitution** takes place at position-2 and if it is occupied then position-4 become prone to substitution.

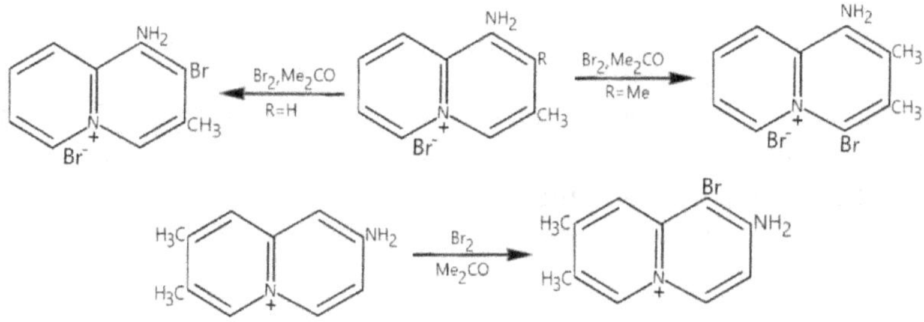

Oxidation of benzo[c]quinolizinium ion in presence of potassium permanganate results in quinoline-2-carboxylic acid.

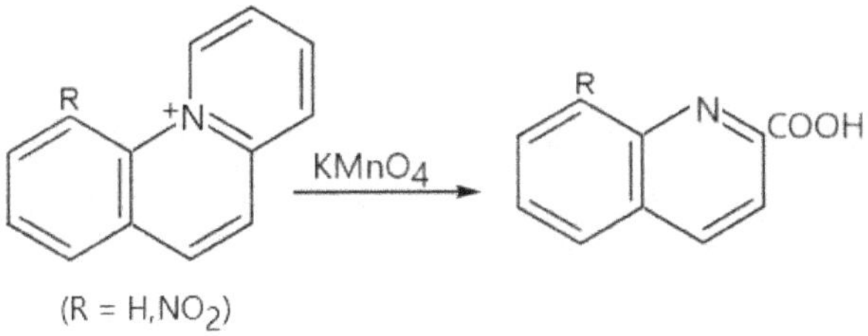

When nitrous acid attacks 1-amino-2,3-dimethyl quinolizinium bromide, then an intermediate is formed which finally forms pyrazolo quinolizinium derivative.

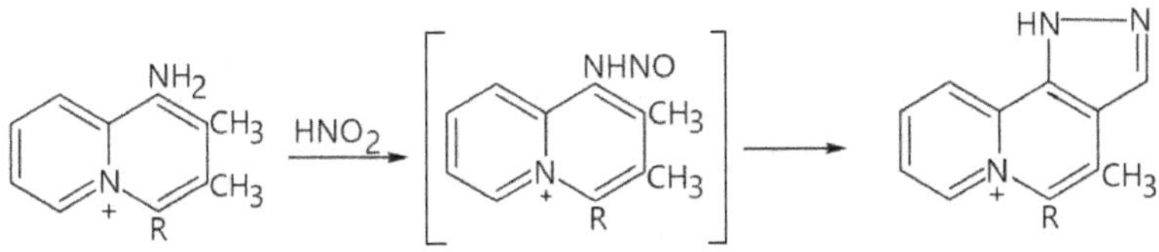

Decarboxylation of a carboxymethyl group in quinolizinium ring takes place easily and gives good results.

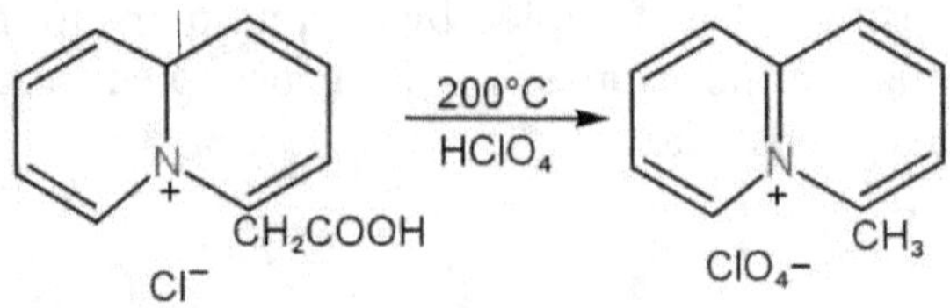

Quinolizin-2-one, which is obtained by the action of K_2CO_3 on 2- hydroxy quinolizinium ion, reacts firstly by PBr_3 to yield 2-bromoquinolizinium ion and secondly with P_2S_5 resulting in quinolizine-2-thione which gets converted to 2-methyl-thioquinolizinium iodide.

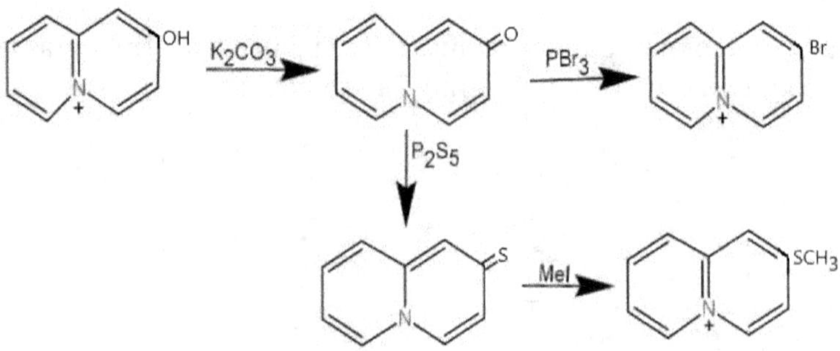

4-quinolizone results in good **conversions** to form chloro-substituted quinolizinium perchlorate, bromo-substituted quinolizinium bromide and quinolizin-4-thione.

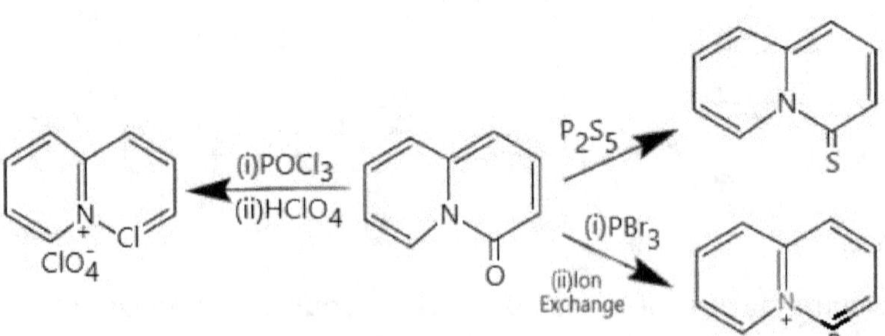

Benzo[b]quinolizinium bromide, having two hydroxyl groups on the fused ring, gives quinones on **oxidation**.

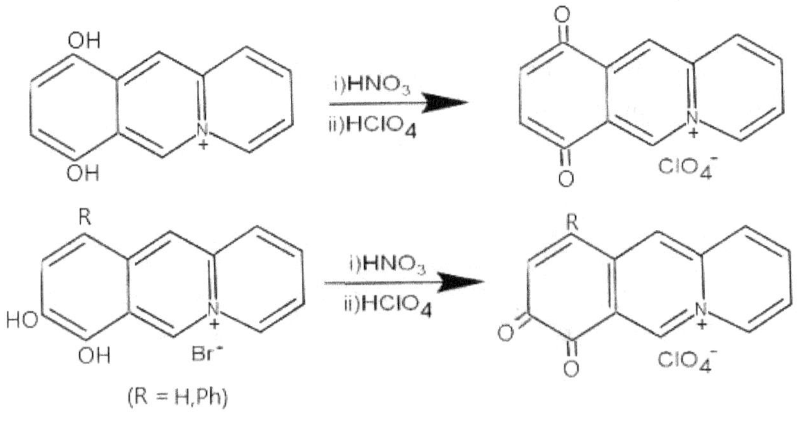

2.6.3 AZAQUINOLIZINIUM DERIVATIVES

2.6.3.1 SYNTHESIS OF AZAQUINOLIZINIUM DERIVATIVES

When 2-aminopyridine was heated with acetal of acyl acetaldehyde, an intermediate was formed which **cyclised** in hydrobromic acid to form 1-azaquinolizinium salt.

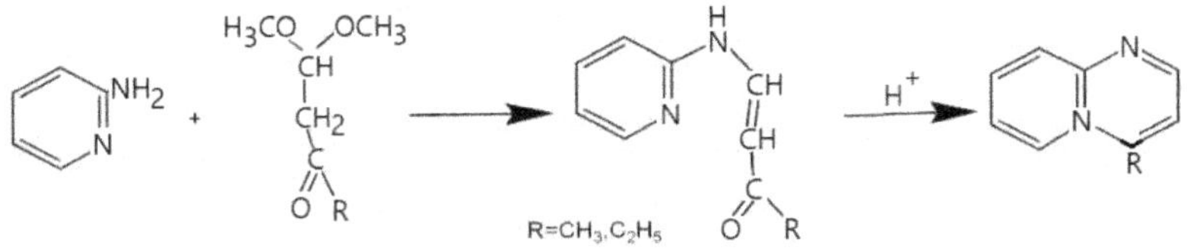

However, a pyridinium salt having an ester group at position-2 reacts with ammonium acetate, 2-azaquinolizinium salt having alkyl/aryl group at position-1.

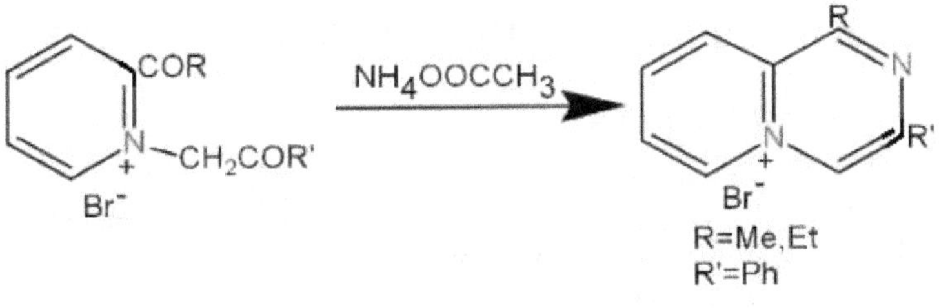

2.6.3.2. PROPERTIES OF AZAQUINOLIZINIUM DERIVATIVES

On **catalytic reduction**, 1-azaquinolizinium ion forms 1-azaquinolizidine.

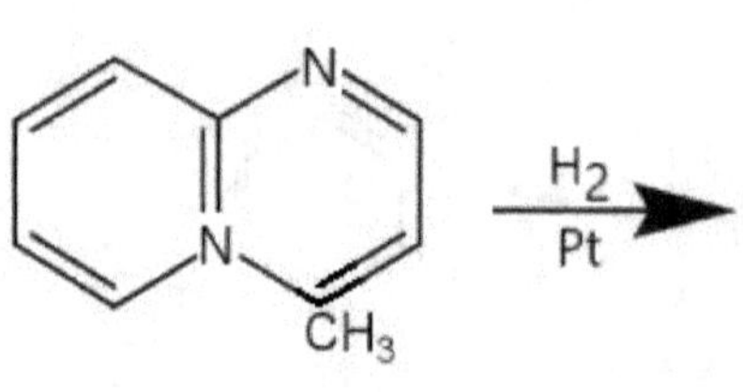

1-Azaquinolizin-4-one reacts in the same way as quinolizin-4-one; thereby 2-methyl-1-azaquinolizin-4-one reacts with P_2S_5 to form thione that reacts with methyl iodide to form 2-methyl-4-methylthio-1-azaquinolizinium iodide.

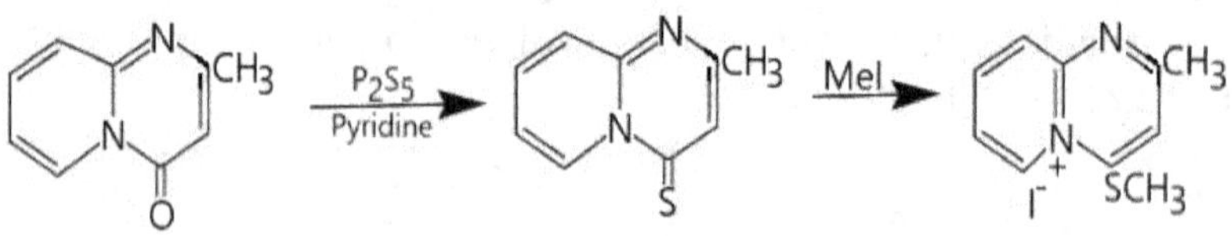

2-Azaquinolizinium ion on reduction gives 2-azaquinolizidine.

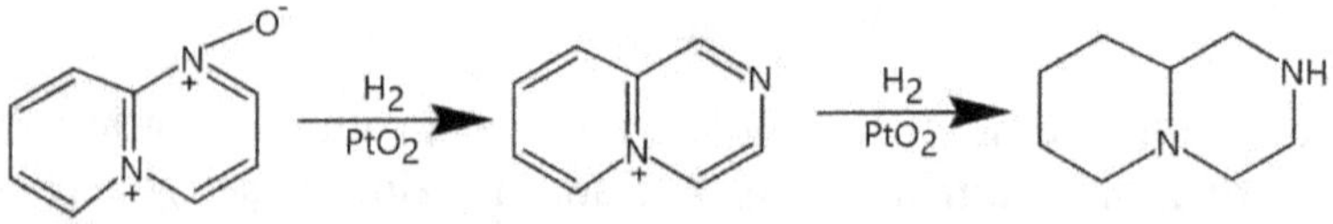

2.7. NAPHTHYRIDINES AND ANTHYRIDINES

2.7.1. NAPHTHYRIDINES

The naphthyridines consist of those compounds which are diaza-naphthalene having one nitrogen atom in both the rings but none at bridge head position.

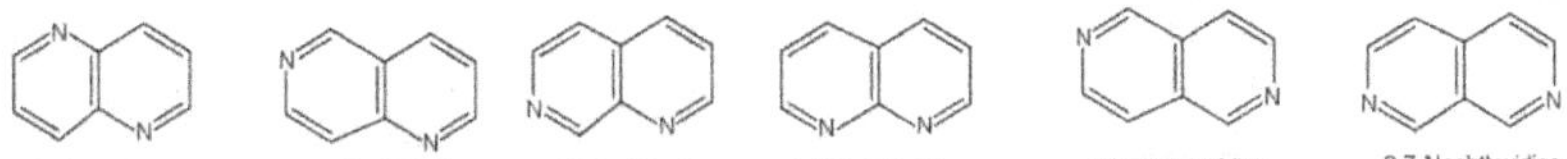

1,5-Naphthyridine 1,6-Naphthyridine 1,7-Naphthyridine 1,8-Naphthyridine 2,6-Naphthyridine 2,7-Naphthyridine

2.7.1.1 SYNTHESIS OF NAPHTHYRIDINE

Skraup method is the most widely used method of synthesis and its various modifications result in good yield. 3,5-Diaminopyridine results in the formation of 1,5-naphthyridine.

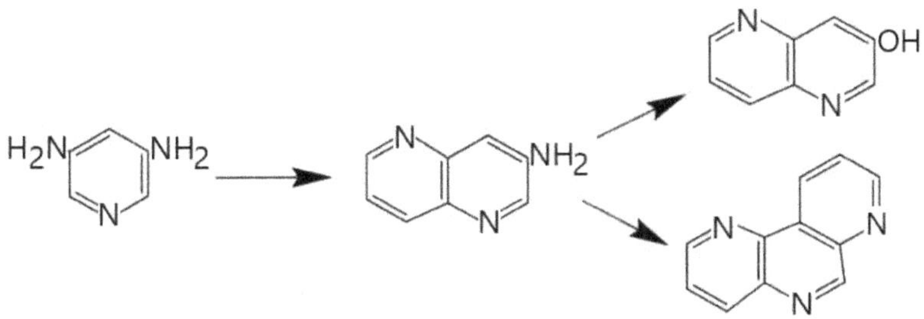

2-ethoxycarbonyl-1,5-naphthyridin-4-one is obtained when 3-aminopyridine reacts with ethyl oxalyl acetate.

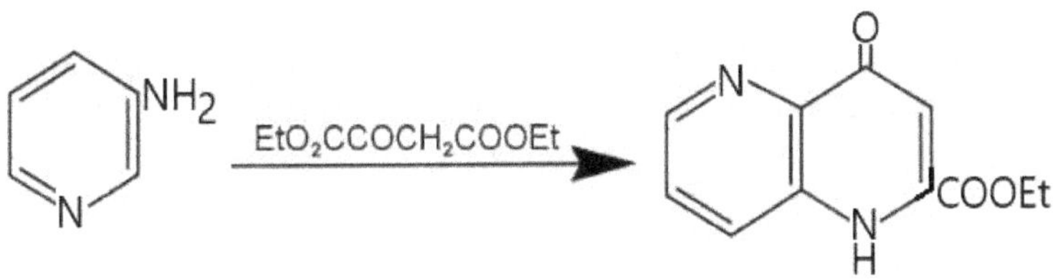

A mixture of substituted naphthyridines is obtained under **Conrad-Limpach** conditions when 3-amino pyridine reacts with ethyl acetoacetates.

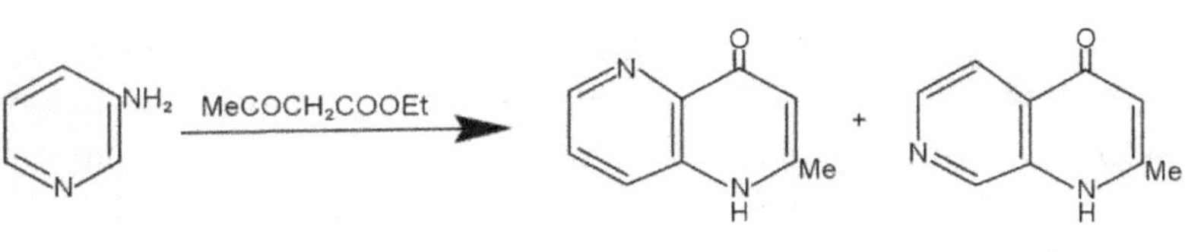

4 : 1

However, when 3-aminopyridine reacts with bis(2,4-dichlorophenyl) benzyl malonate, it gives 3-benzyl-4-hydroxy-1,5-naphthyridin-2-one.

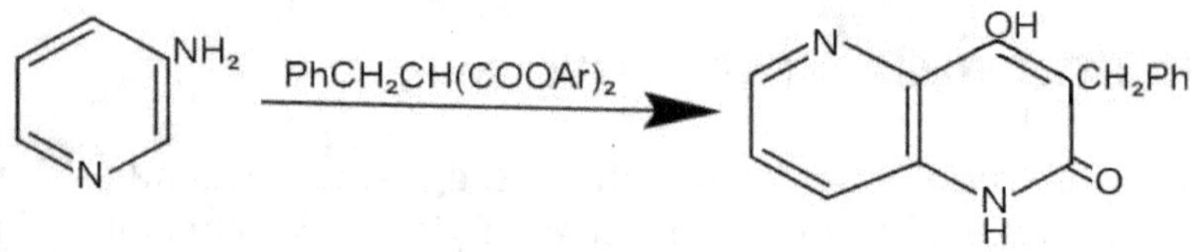

When 3-aminopyridine-2-carboxylic acid or its esters are condensed with compounds containing active methylene group, it results in the formation of 4-oxo- (**A**) and 4-hydroxy-2-oxo naphthyridine (**B**) after **hydrolysis** and **decarboxylation** of intermediates.

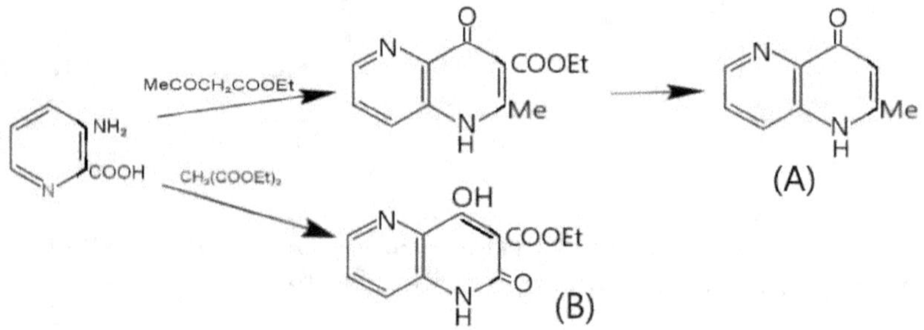

Friedlander synthesis involves condensation of 4-aminonicotinaldehyde with compounds containing active methylene group resulting in the formation of 1,6-naphthyridine.

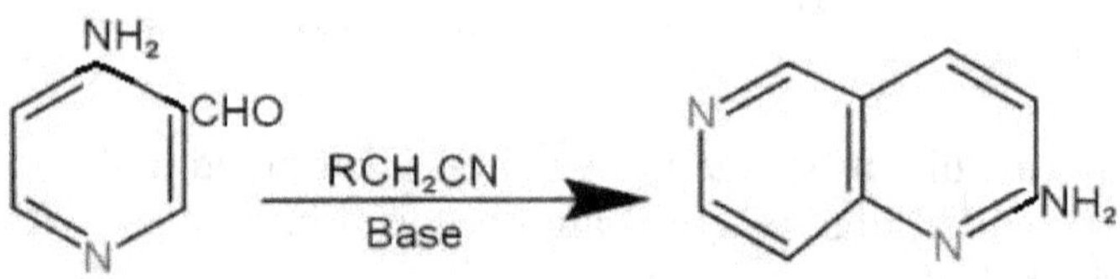

1,2,3,4-Tetrahydronaphthyridines can be obtained from enamines. The **first method** involves condensation of enamino-amide with dimethyl formamide diethyl acetal resulting in the formation of pyrido[1,2-c] pyrimidine derivative which in water rearranges to 1,6-naphthyridine derivative.

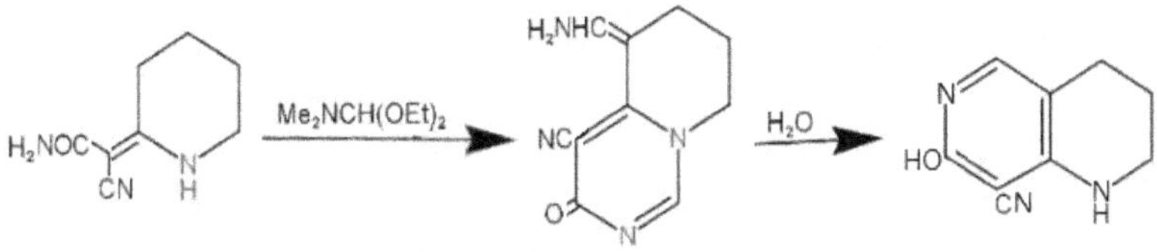

In the **second method**, enamine reacted with acetal resulting in a compound that **cyclized** with ammonia to yield 1,6-naphthyridine derivative.

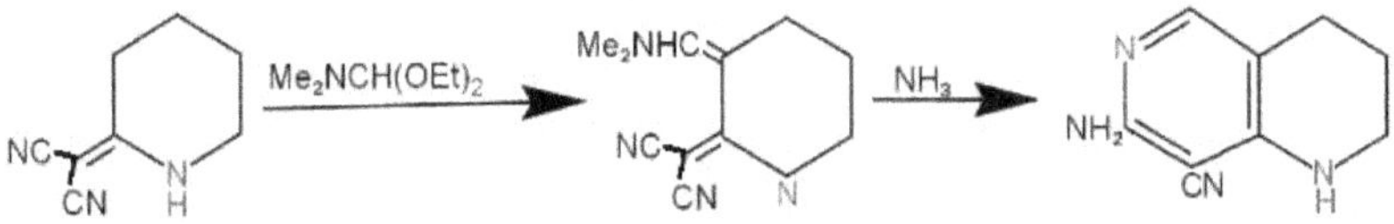

2-Cyanopyridine derivatives serve as good starting material for the synthesis of 1,7-naphthyridines when (2-cyano-3-pyridyl) acetonitrile reacts with hydrobromic acid, 6-amino-8-bromo-1,7-naphthyridine is obtained.

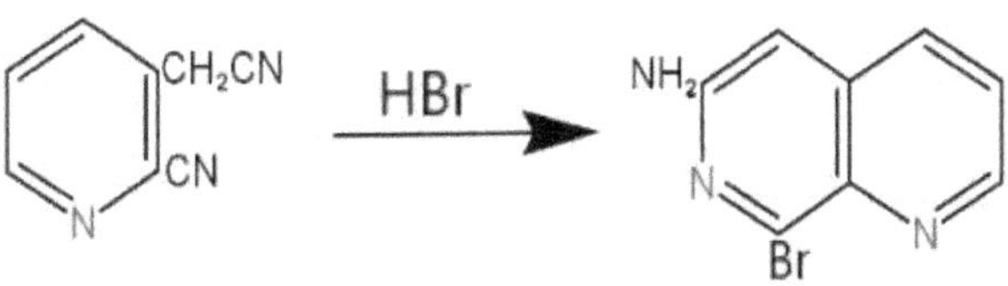

However, reaction of (2-cyano-3-pyridyl) acetonitrile with sodium methoxide results in a mixture of 1,7-naphthyridines.

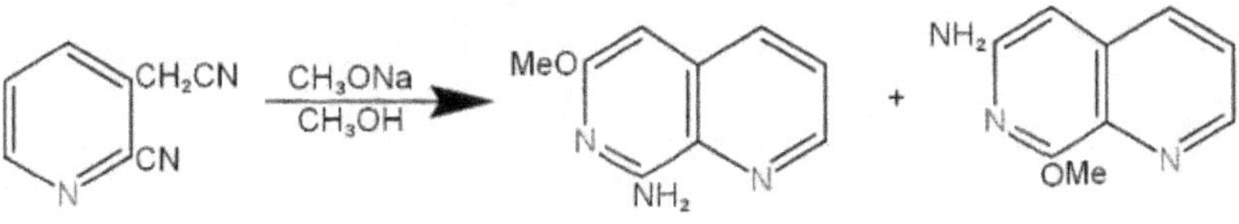

Tetrahydro-1,7-naphthylridine can also be obtained by the reaction of primary amines with lactones.

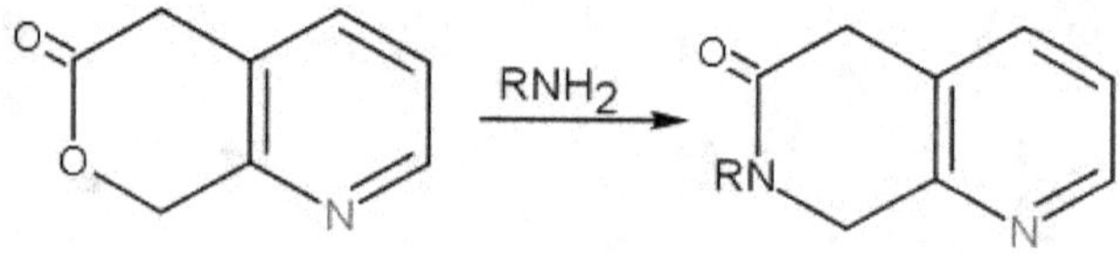

Likewise, 3-substituted amino isonicotin aldehyde undergoes **Wittig reaction** followed by reduction giving an intermediate compound that cyclizes in acid solution to yield 5-hydroxy-2-oxo-1,2,3,4-tetrahydro-1,7-naphthyridine.

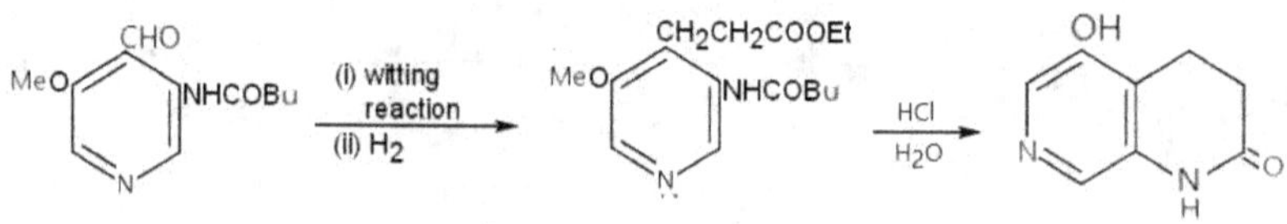

1,8-Naphthyridine-2–one is synthesised by the reaction of 6-aminopyridin-2-one and glycerol.

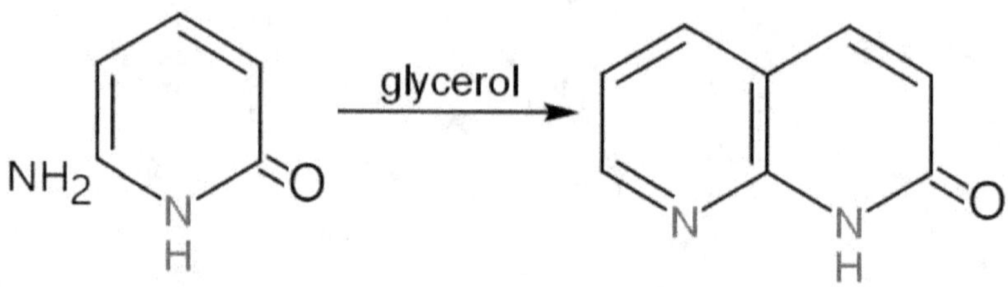

Substituted 1,8-naphthyridine is also obtained by ring expansion. 2–phenyl-7-azaindole, by carbene insertion reaction, results in ring expansion by forming 2-phenyl-3-chloro-1,8-naphthyridine.

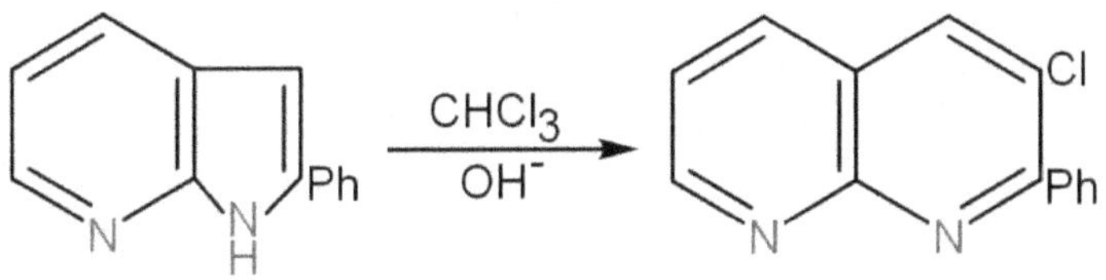

2,7-Naphthyridine can be obtained by a series of reactions with 4-methylnicotinic acid as the starting material.

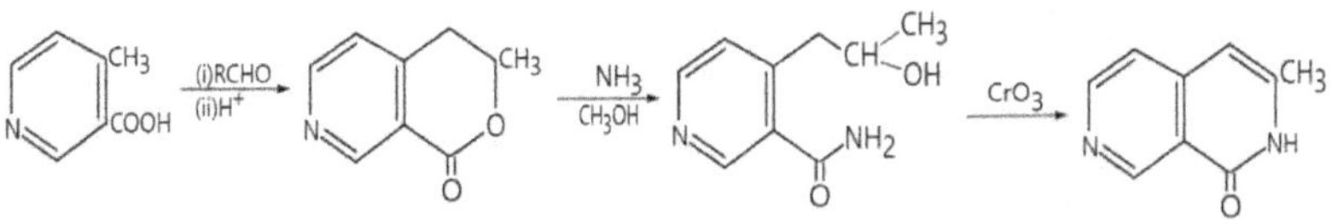

When nicotinamide meth chloride reacts with acetone in the presence of base, via an intermediate compound, it gives 2,7-naphthyridine derivatives (A); however, (A) on **sublimation** got converted to 3-methyl-2,7-naphthyridine-1-one which resulted in the final product, 2-methyl-3,7-naphtlyridine.

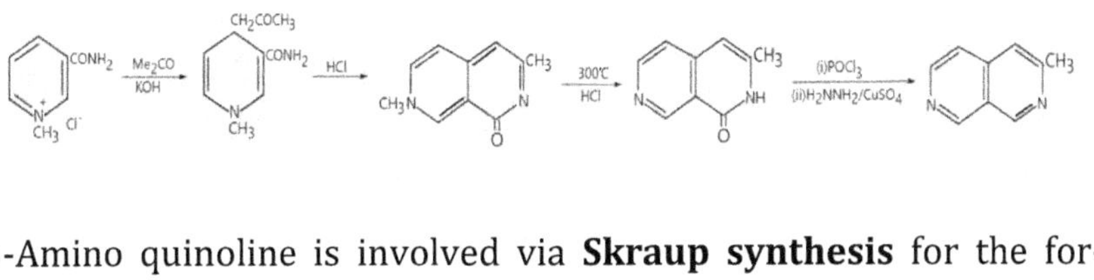

3-Amino quinoline is involved via **Skraup synthesis** for the formation of benzo[f] [1,7]naphthyridine.

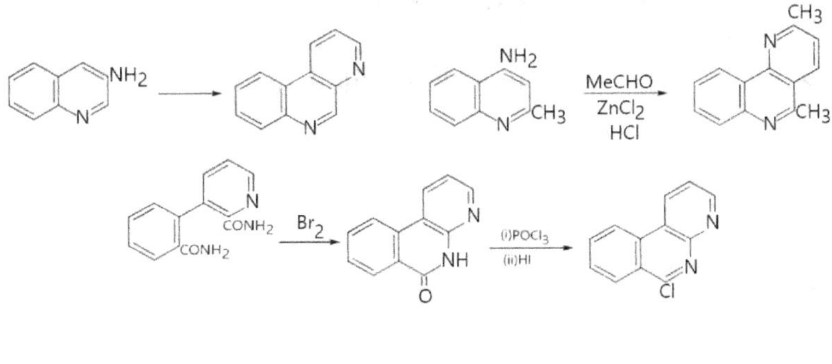

Diaza-anthracenes have been synthesized by a variety of ways. 2,6-Diaminopyridine reacts with 2-chlorobenzoic acid to result in **cyclization** product in sulphuric acid.

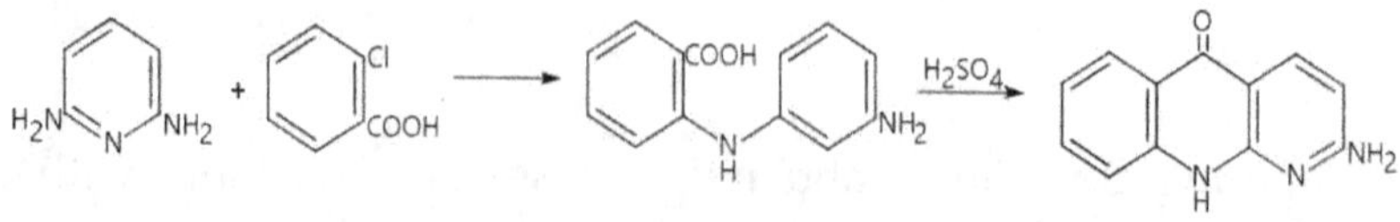

In the same way, cyclization of 2-amino nicotinic acid gives desired products in sulphuric acid/polyphosphoric acid.

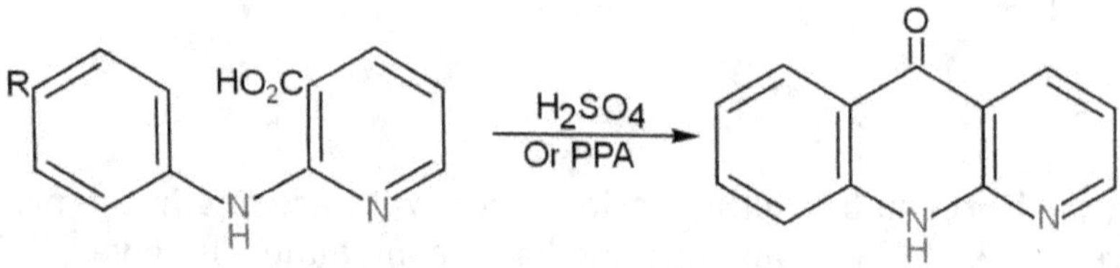

Condensation of anthranilic acid with 2,4,6-trioxopiperidine results in the formation of 3-hydroxy-2,10-diazaantracene-1,9-dione.

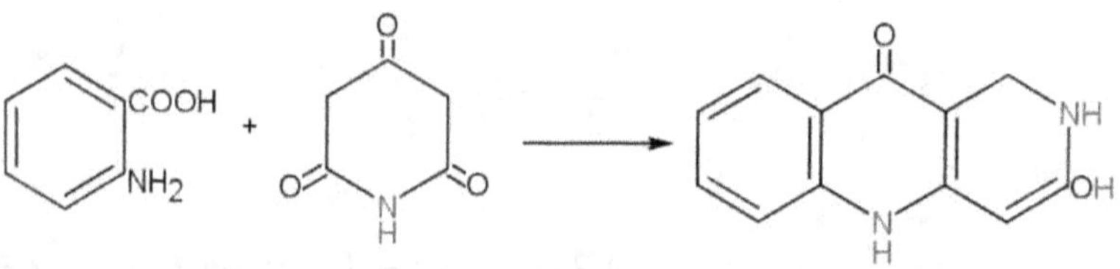

2.7.1.2 PROPERTIES OF NAPHTHYRIDINE

Various naphthyridine respond to bromine in pyridine.

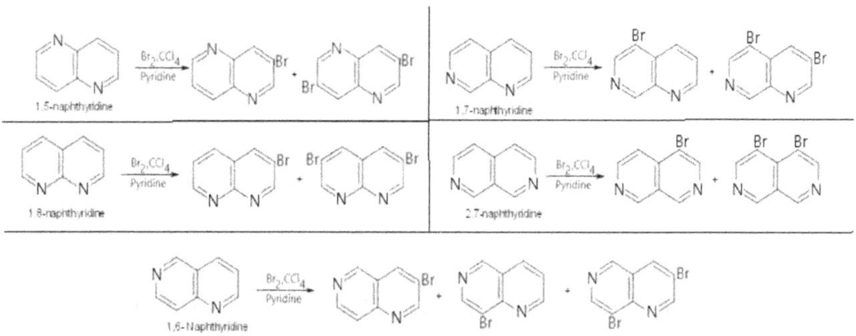

When **bromination** is brought about 1,7- and 1,8- isomers in presence of nitrobenzene then best result is obtained with naphthyridine hydro halide.

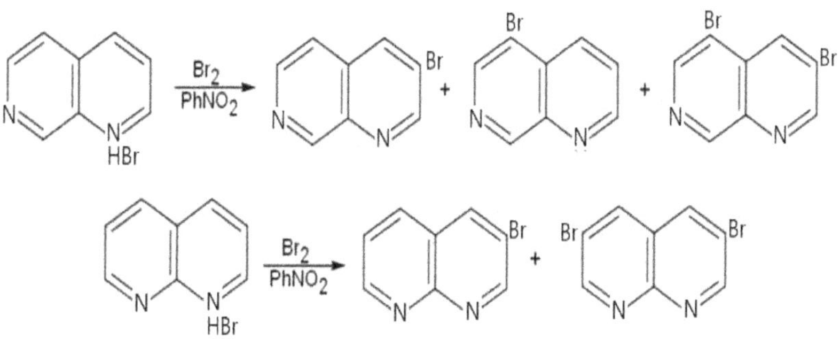

Bromination of 2- or 4-oxo-naphthyridine takes place in the presence of $POBr_3$ and PBr_5.

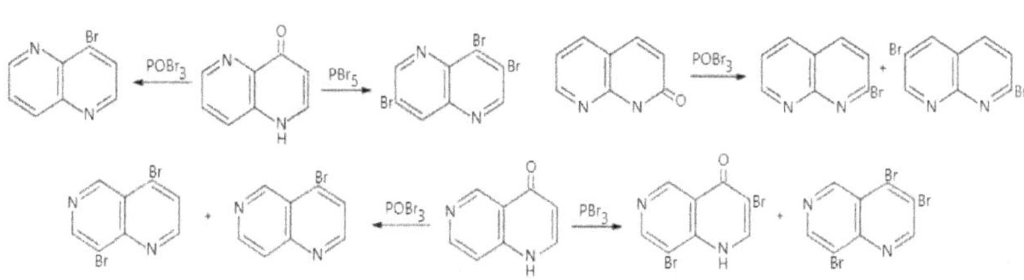

When 2- or/and 4-position are substituted with electron denoting groups only the can **nitration** be brought about. Hence, mononitration occurs in 1,7-naphthyridine-4(1H)-one at position-3.

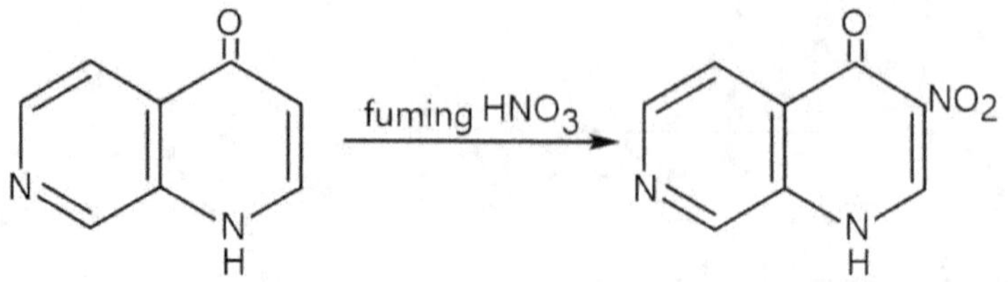

Nitration is more convenient in the presence of two-election-donating groups.

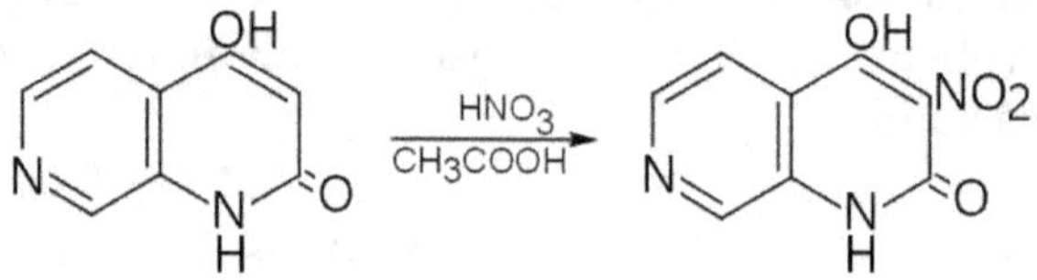

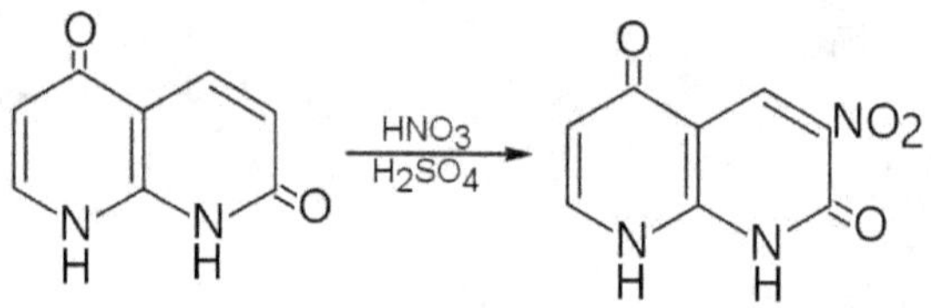

Amino-naphthyridines can easily be obtained by nitrogen via nucleophilic attack, for example, 1,5-naphthylridine results in the formation of 4-amino-1,5-naphthyridine.

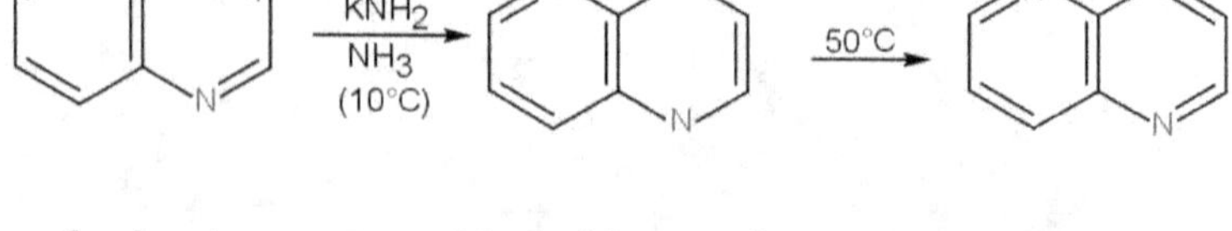

When hydrazine reacts with 4-chloro-substituted naphthyridines, then halogen-substituted ring gets broken and pyrazolyl-pyridine derivatives are formed.

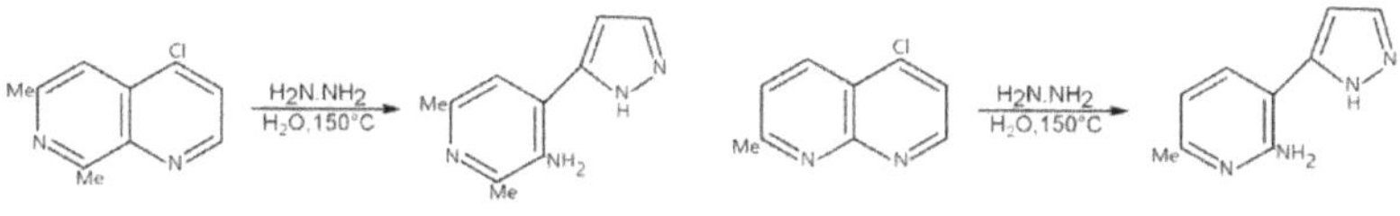

Reduction of naphthyridine is brought about by lithium aluminium hydride converting oxo compounds to 5,6,7,8-tetrahydro–1,7-naphthyridine (A) and dioxo compound to 1,2,3,4-tetrahydro-.2,6-naphthyridine (B).

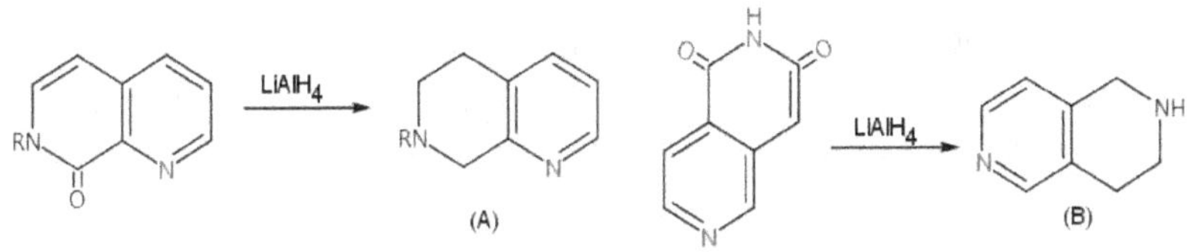

The **Reissert reaction** illustrates yet another chemical property of naphthyridine. When 1,6-naphthyridine reacts with cyanogen bromide in methanol, it yields N-cyano product that gets converted to 8-bromo-1,6-naphthyridine or 7-methoxy–1,6-naphthyridine.

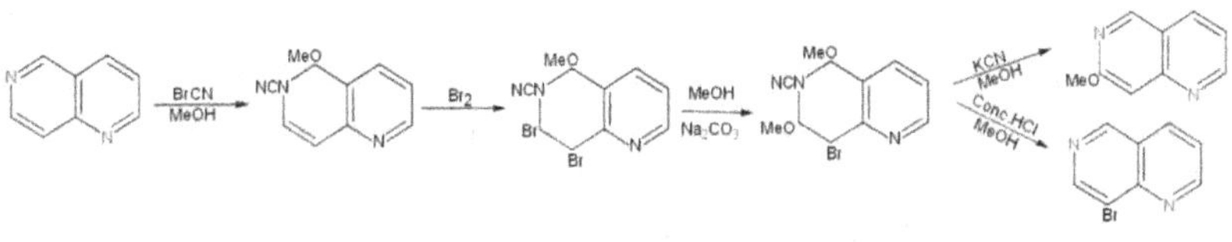

Acid hydrolysis of substituted 1,6-naphthyridine results in 1,6-naphthyridine-5-carbox amide along with carboxylic acid.

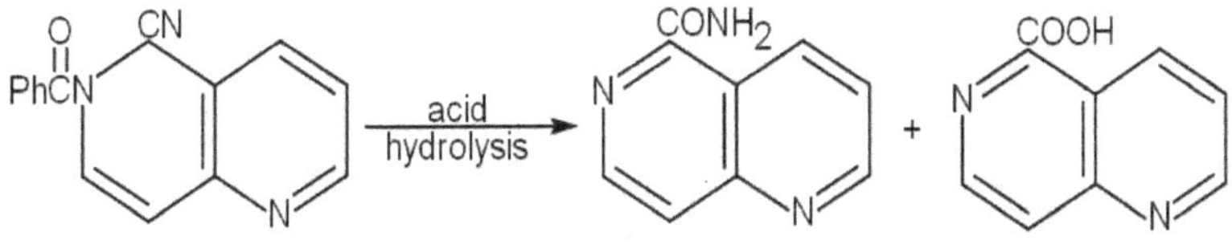

N-Alkylation of naphthyridine with methyl iodides gives N-mono methiodides which on further oxidation results in the formation of cyclic amides.

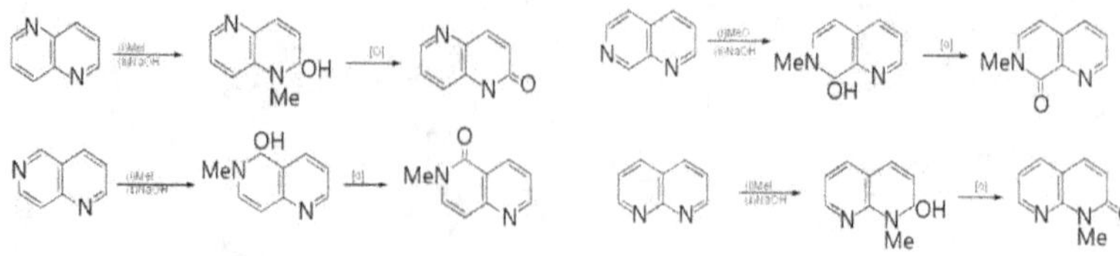

2.7.2 ANTHYRIDINES

2.7.2.1 SYNTHESIS OF ANTHYRIDINE

In **basic environment**, the condensation of 2-amino-3-ethoxy-carbonyl-1,8-naphthyridine and alkyl carboxylate results in the formation of 4-hydroxy1,9,10-anthyridines-2-ones.

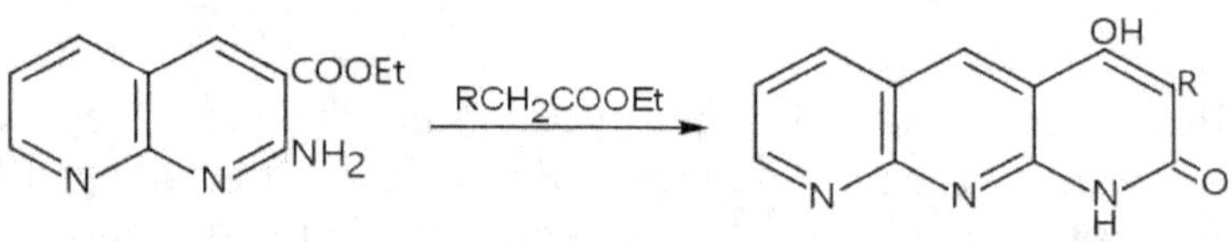

2.7.2.2 PROPERTIES OF ANTHYRIDINE

When 4,6-dihydroxy-1,9,10-anthyridine-3,7-dicarboxylic acid is tested with copper chromite in hot quinoline results in dioxo-products. This product, when exposed to phosphorus pentachloride and phosphorus oxychloride forms 4,6-dichloro-1,9,10-anthyridine, the later gets converted to 5,10-dihydro anthyridine by Pd/C in KOH/MeOH or by heating with Zn dust and $ZnCl_2$. Finally, 1,9,10-anthyridine is obtained on treatment with potassium dichromate in sulphuric acid.

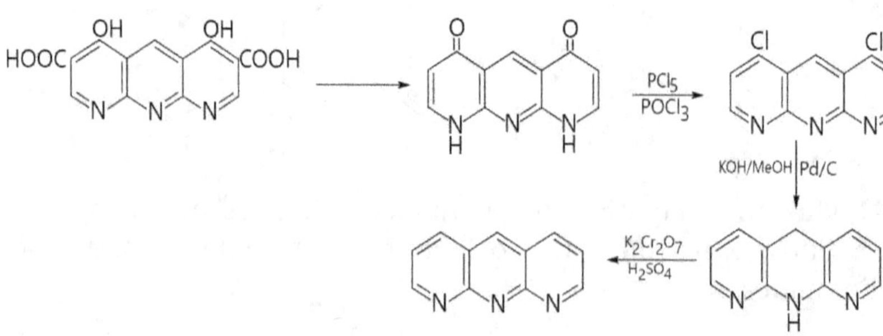

2.7.2.3 APPLICATIONS OF ANTHYRIDINE

(i) **Nalidixic acid**, 1-ethyl-3-carboxy-4-oxo-7-methyl-1,4-dihydro-1,8-naphthyridine, acts as a powerful antibacterial agent.

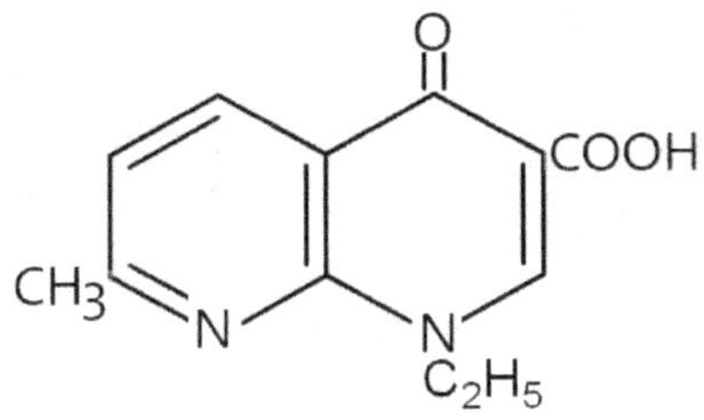

(ii) Significant activity of benzo-1,5-naphthyridine against **chloroquine–resistant** of *Plasmodium falciparum* has been reported.

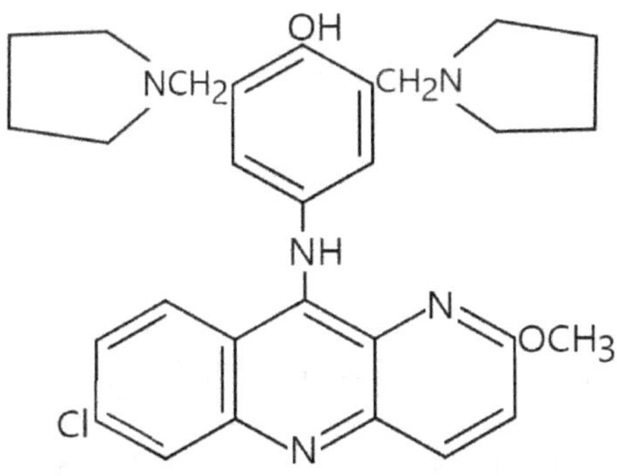

(iii) The tetrahydro derivatives of 1,6-naphthyridine have reported some muscle relaxant activity.

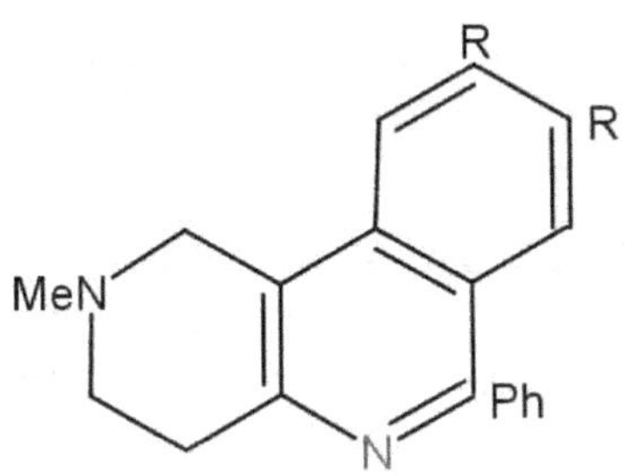

(iv) 1,5-Napthyrodine derivatives have reported **antitubercular** and **anti-dysenteric** properties.

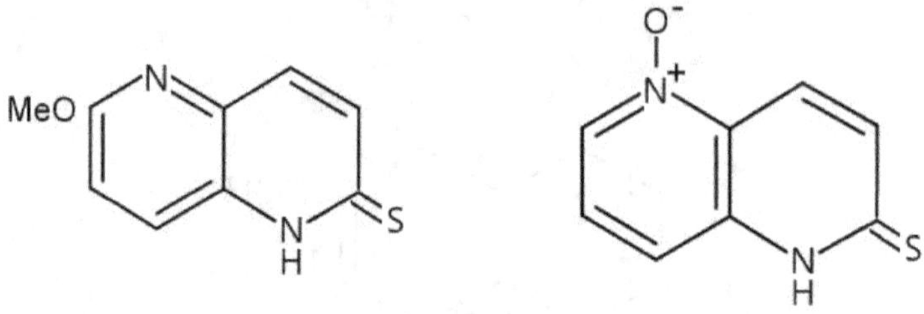

(v) Jasminine, 2,4-naphthyridine derivative is an **alkaloid** isolated from *Jasminum* species and 4-methyl-2,6-naphthyridine is found in aerial parts of *Antirrhinum majus.*

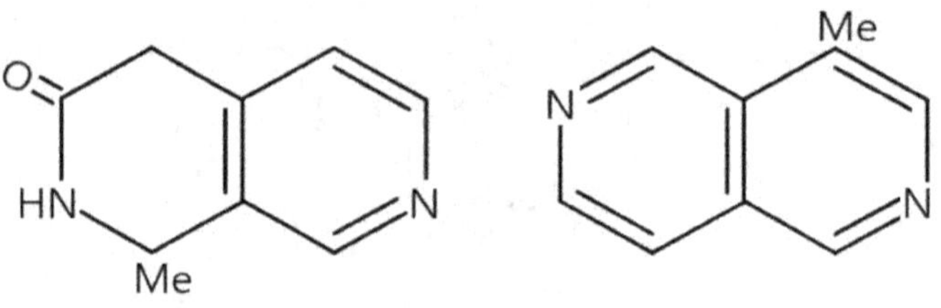

2.8 TRIAZINES, TETRAZINES AND PENTAZINES

2.8.1 1,2,3 – TRIAZINES AND ITS DERIVATIVES

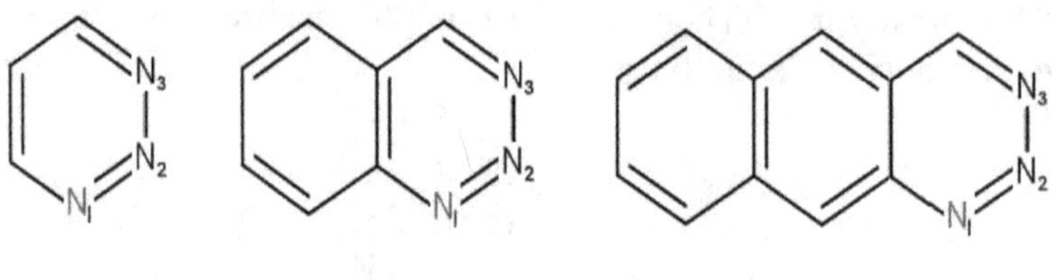

2.8.1.1 SYNTHESIS OF 1,2,3 – TRIAZINES

N-amino pyrazole (A) and N- aminoindazoles (B and C) by oxidation (mostly with nickel peroxide) offers the most convenient method for the preparation of 1,2,3–triazines and 1,2,3–benzo triazines.

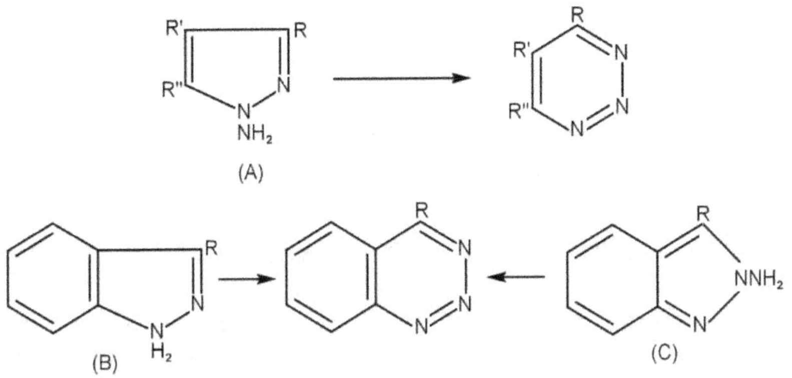

1,2,3–Benzo triazines can be easily prepared by **oxidation** of 1-amino and 3- amino- quinazolin-2-one.

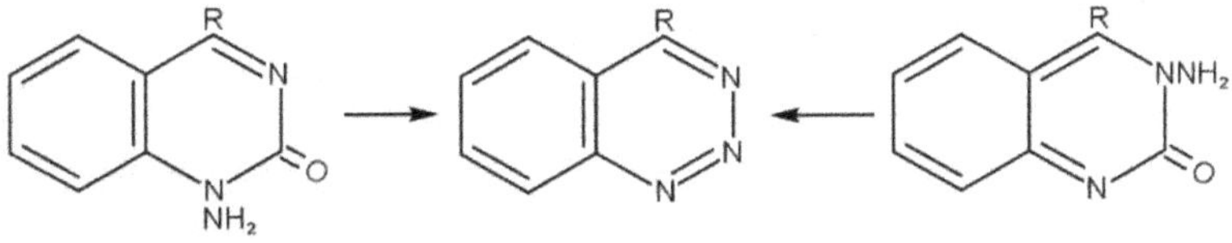

Oxidation of 3- aminoindazole results in the formation of 1,2,3 – benzo triazin – 4 – one.

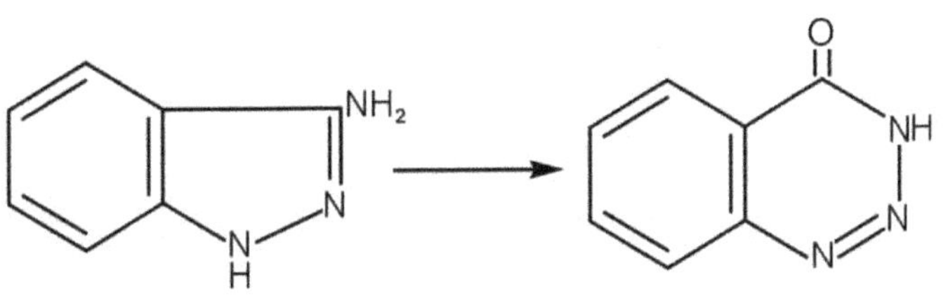

Dihydro-1,2,3-benzotriazines can be obtained by reaction of 2-aminobenzyl amines (D) and of 2-aminobenzylhydrazines (E) with nitrous acid.

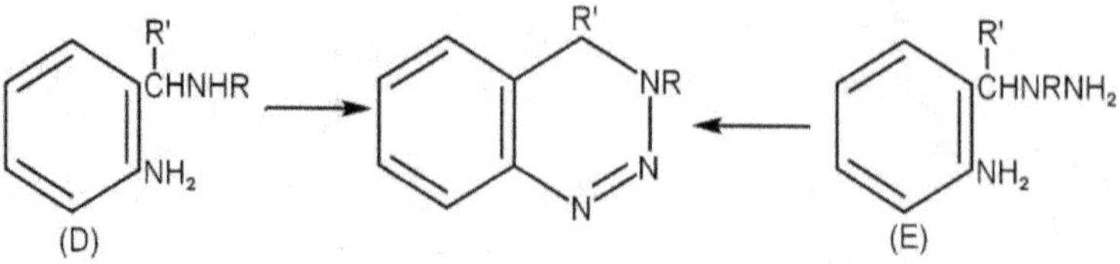

2.8.1.2 PROPERTIES OF 1,2,3 – TRIAZINES

1,2,3-Triazines and 1,2,3-benzotriazines are stable at room temperature, although some chemical properties have been reported. The hydrazino group as a substituent gets converted to azide group on treatment with nitrous acid and can be replaced by thioglycolate.

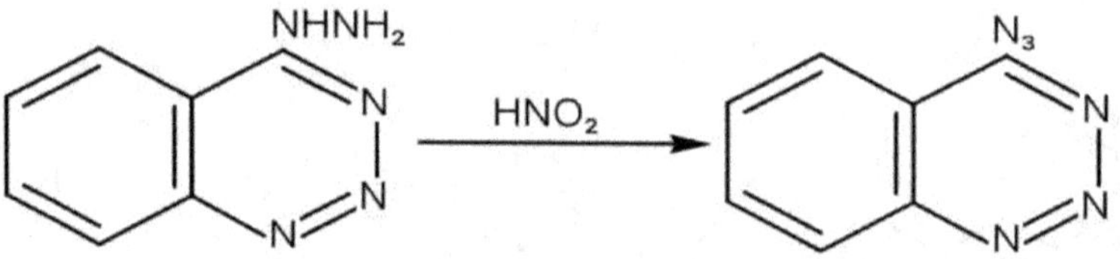

Thermal decomposition of 3-aryl-1,2,3-benzotriazin-4-one (**A**) in boiling 1 - methylnaphthalene and successive rearrangement results in the formation of 9-acridinones (**B**) and phenanthridin-6-ones (**C**); however, benzanilides are obtained in paraffin oil at 300°C.

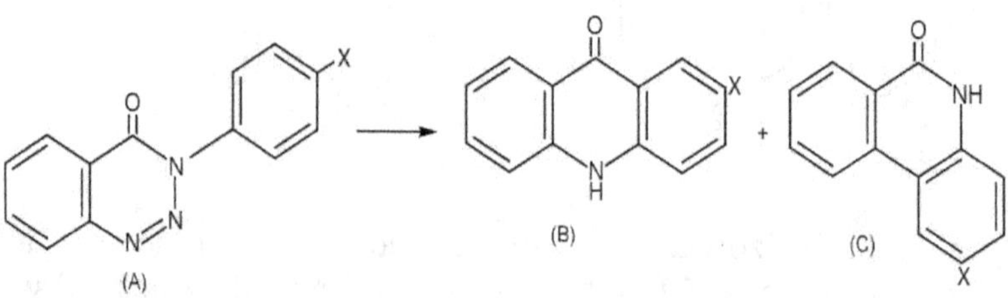

At 420°C, **pyrolysis** of 4-phenyl-1,2,3-benzotriazine-3-oxide (D) yields 3-phenylbenzisoxazole (E), 3- phenyl indazole (F) and acridin-9 -one (G).

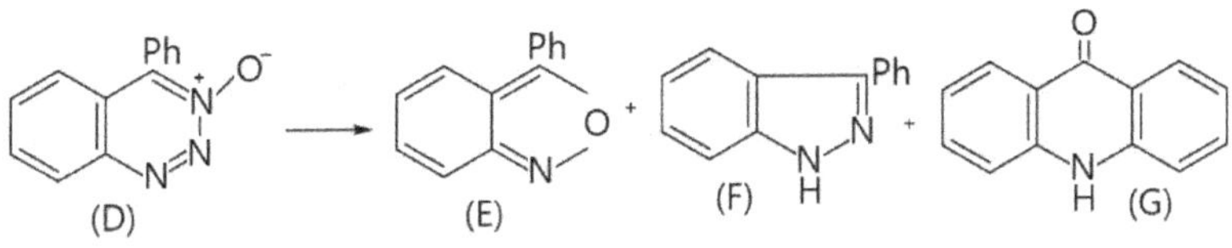

Benz-azetinones, which could not be isolated, were formed from 3-aryl-1,2,3-benzotriazin-4-one (H) by loss of nitrogen.

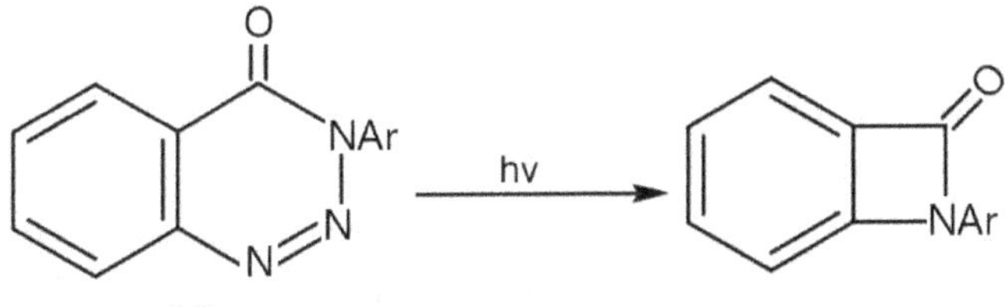

(H)

2.8.2 1,2,4-TRIAZINES AND ITS DERIVATIVES

1,2,4-trazines are well-known compounds and find a wide variety of application.

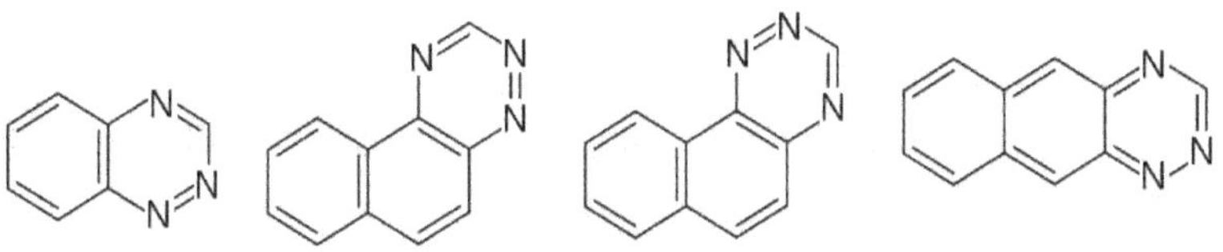

2.8.2.1 SYNTHESIS OF 1,2,4-TRIAZINES

5-Phenyl-1,2,4-triazin-3-one is obtained by the reaction of phenyl-glyoxal and semi carbazide; 6-phenyl compound is prepared from iso-nitroso acetophenone (phenylglyoxal aldoxime) through semi carbazone.

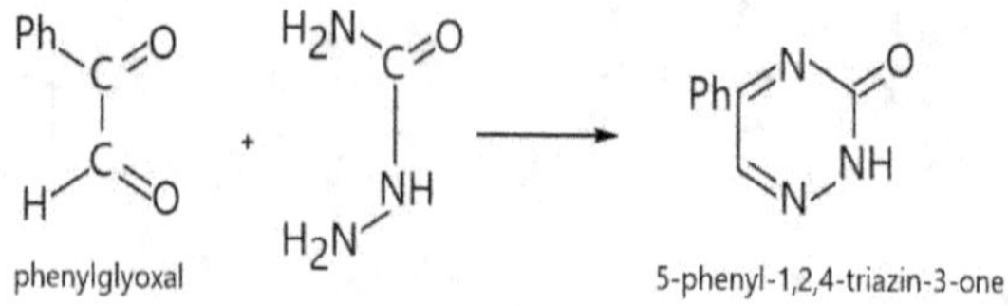

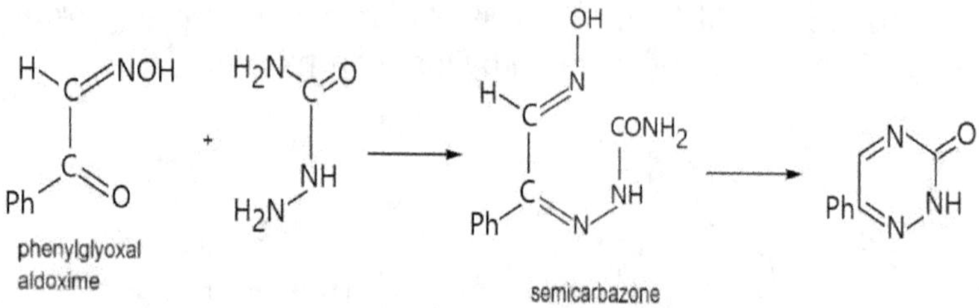

3-Hydrazino-1,2,4-triazines were obtained by the reaction of di-aminoguanidine and 1,2- di-carbonyl compounds.

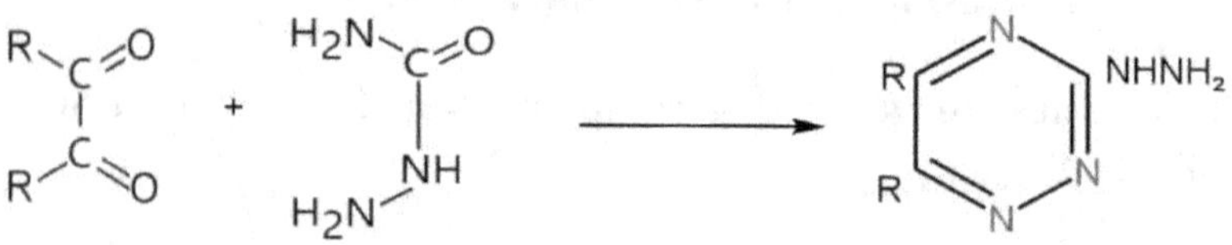

In the presence of base, semi carbazone of phenacyl hydrazine results in the formation of 4-amino-6-phenyl-4,5-dihydro-1,2,4-triazin-3-one.

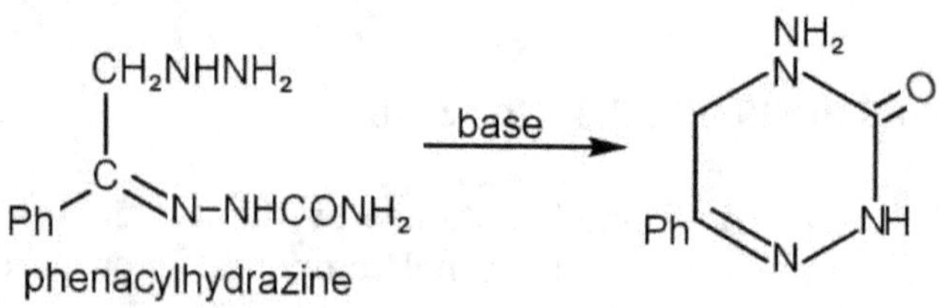

The reaction of o-phenylenediamine and methyl α-phenyldithiocarbazinate yields 1-phenyl–1,4-dihydro–1,2,4-benzotriazin-3-one.

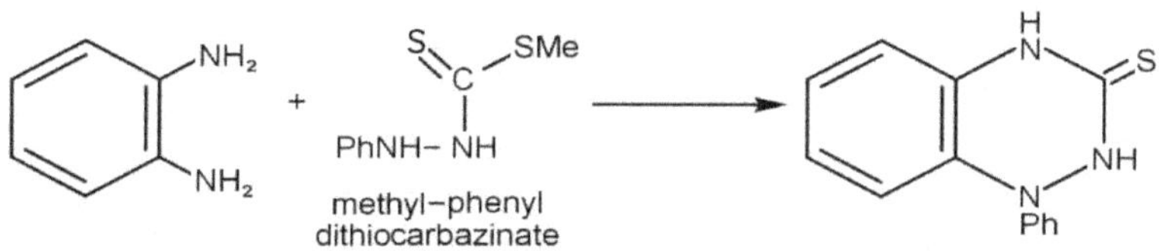

2.8.2.2 PROPERTIES OF 1,2,4-TRIAZINES

When 1,6-dihydro-1,2,4-triazine-3,5(2H,4H)-dione undergoes rearrangement with aldehydes, it forms N-aminoimidazole dione derivatives.

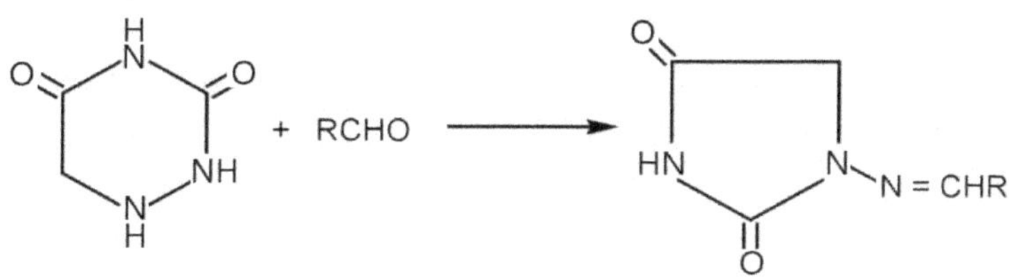

The reactions of various 1,2,4-triazine derivatives with Grignard's Reagent has been studied exhaustively. Phenyl magnesium bromide reacts with 3,5,6-triphenyl-1,2,4-triazine to give 2,4,6-triphenyl-1,3,5-tiazine (**A**) and 2,4,4,5-tetraphenyl-4H-imidazole(**B**).

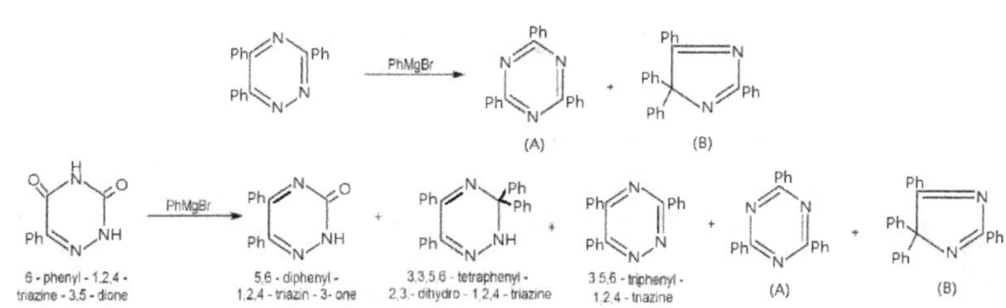

When 1,2,4-benzotriazin-3-one reacts with ethereal solution of diazomethane in two ways:

- **a.** Slow addition of diazomethane gives 3-methoxy-1,2,4-benzotriazine.
- **b.** Excess of diazomethane gives a mixture of 2-methyl and 4-methyl derivatives.

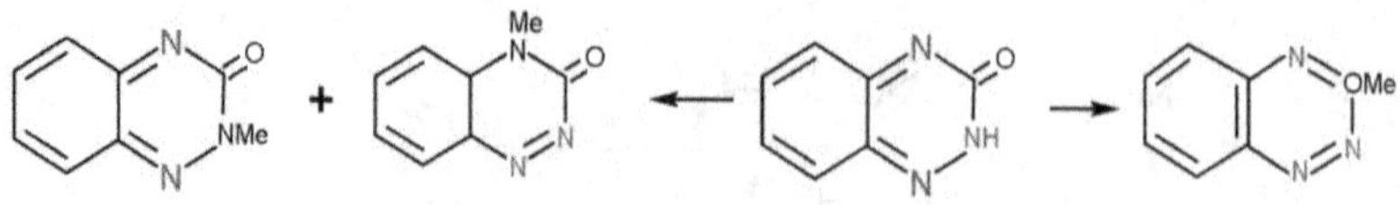

Methyl iodide reacts with 1,2,4-benzotriazin-3-one-1-oxide to give 4-methyl-1,2,4-benzotriazin-3-one-1-oxide.

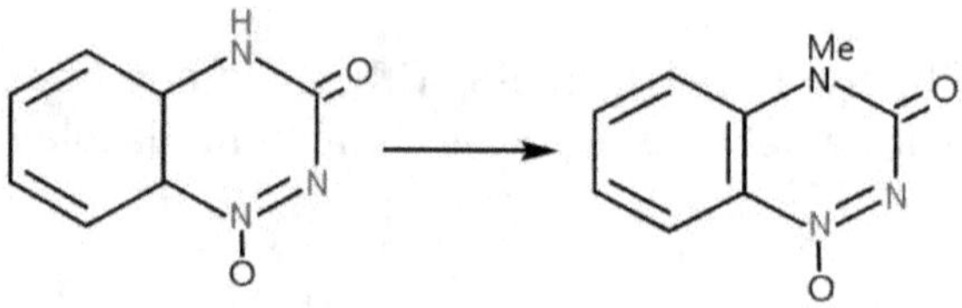

Oxidation of 3-amino-1,2,4-triazine gives 3-amino-1,2,4-triazin-5-one (C) with hydrogen peroxide and 3,5-dione (D) with permanganate.

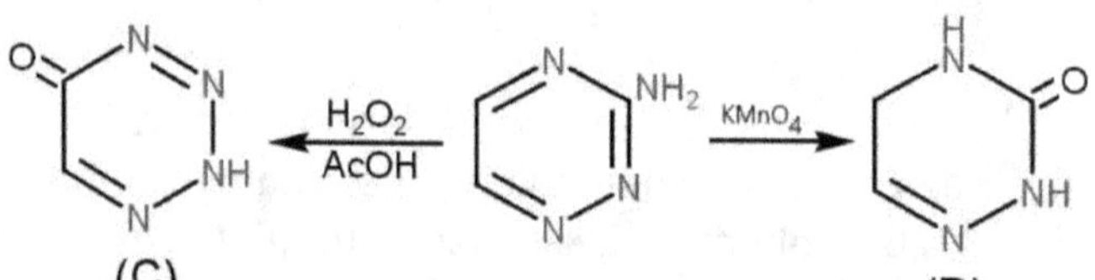

(C) (D)

2.8.3 1,3,5-TRIAZINES AND ITS DERIVATIVES

2.8.3.1 SYNTHESIS OF 1,3,5-TRIAZINES

1,3,5-triazinethione can be obtained from isothiocyanate and biguanide.

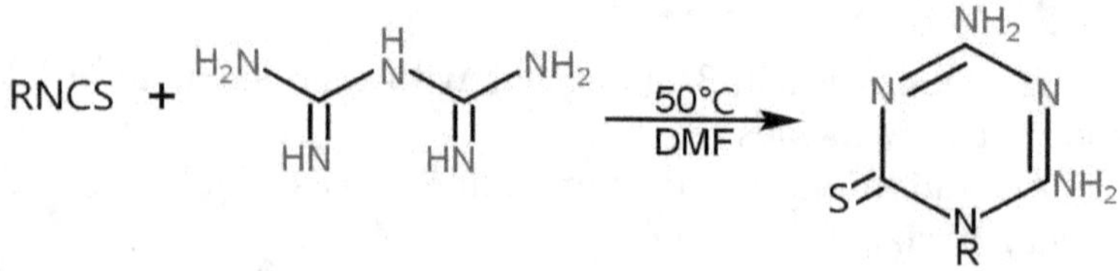

2,4-Diamino-1,3,5-triazine can be obtained by the route disordered by **Ostrogovich**.

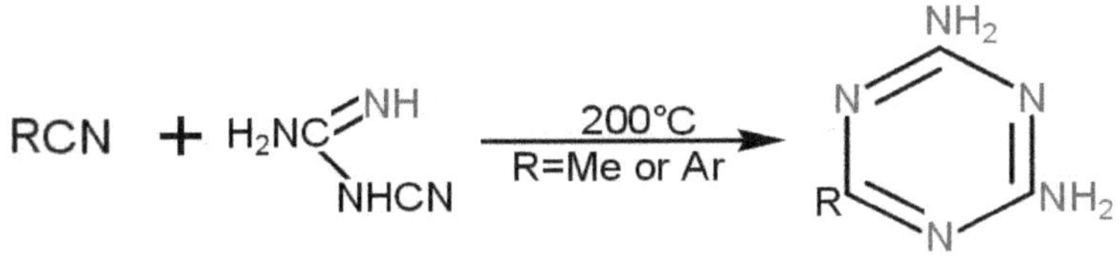

When 2-aminobenzimidazole reacts with dicyandiamide the product obtained is 2,4-diamino-1,3,5-triazino [1,2-a] benzimidazole.

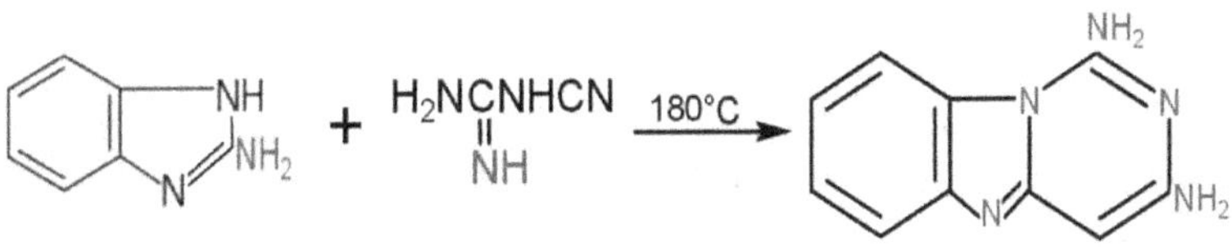

Cyanamide on being heated above its melting point results in **cyclisation** reaction.

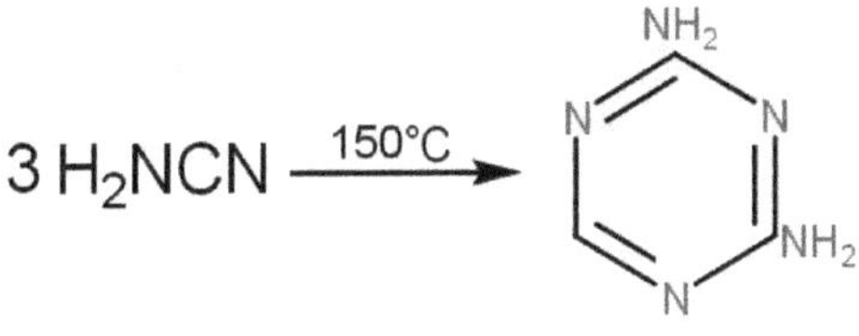

2.8.3.2 PROPERTIES OF 1,3,5-TRIAZINES

Halogenation of 1,3,5-triazine gives results in different conditions. Low yields are obtained in **chlorination** however bromination is efficient.

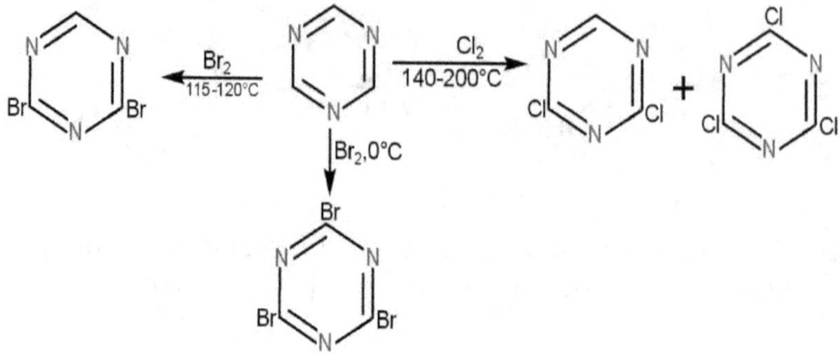

When 1,3,5-triazine reacts with compounds containing a reactive methylene group, it gives 4,5-disubstituted pyrimidine.

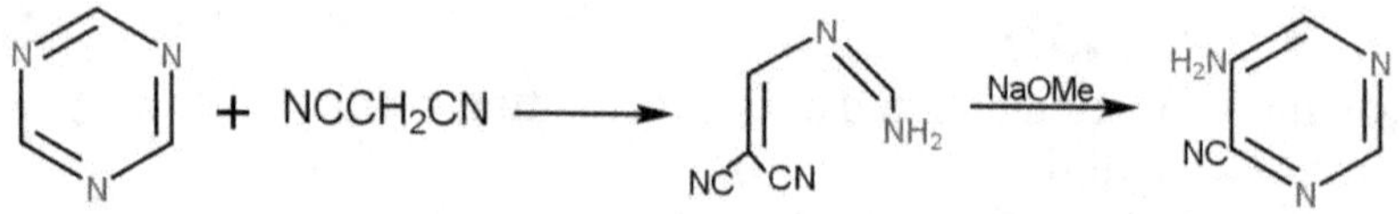

A complex is formed on reaction with boron trifluoride which gets converted to 1,2,4-oxa diazole.

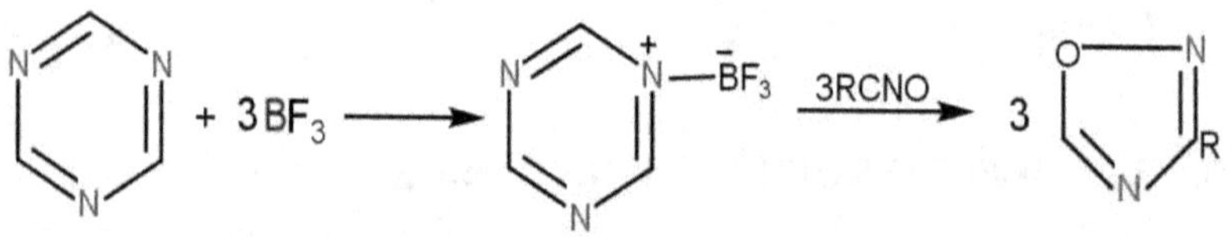

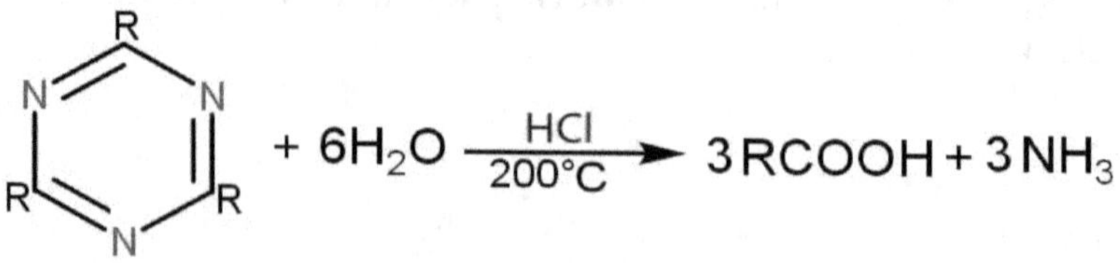

Deprotonation of allyl groups occurs in basic conductions. This take place in the allyl groups which are reactive as they are α- to two nitrogen atoms.

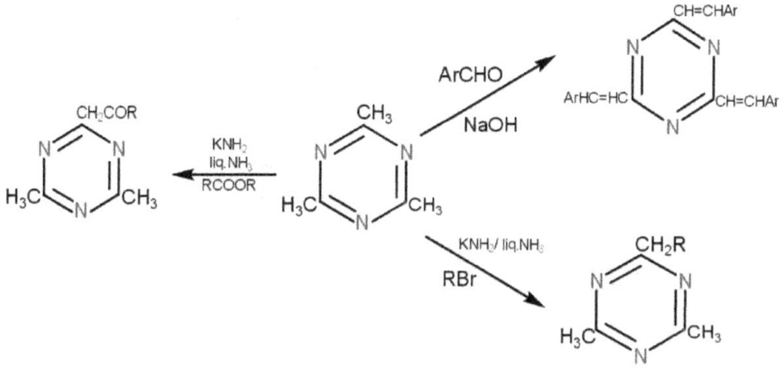

Chlorination or bromination reactions easily take place in the presence of acid.

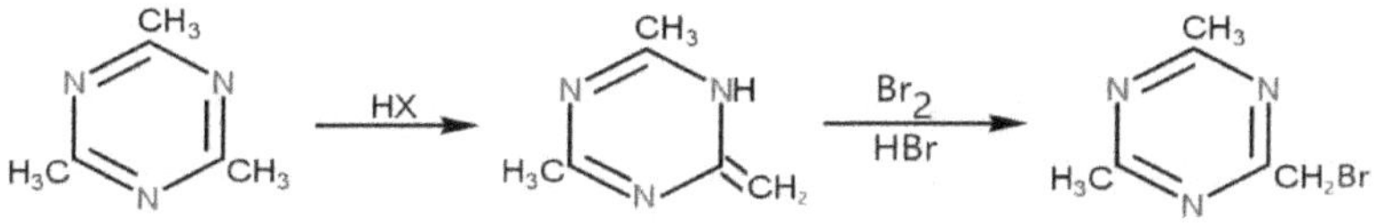

1,3,5-Triazinecarbonitriles are quite stable under the effect of heat. Cyanuric acid and hydrogen cyanide are formed when 2,4,6tricyano-1,3,5-triazine reacts with water.

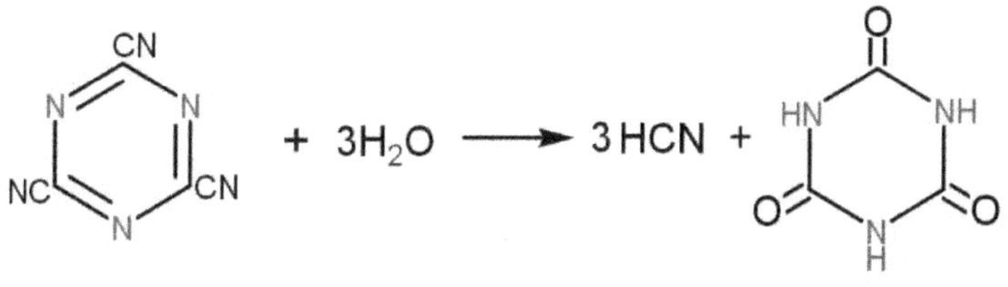

Various polycyclic compounds are formed when **self-condensation** of melamine takes place.

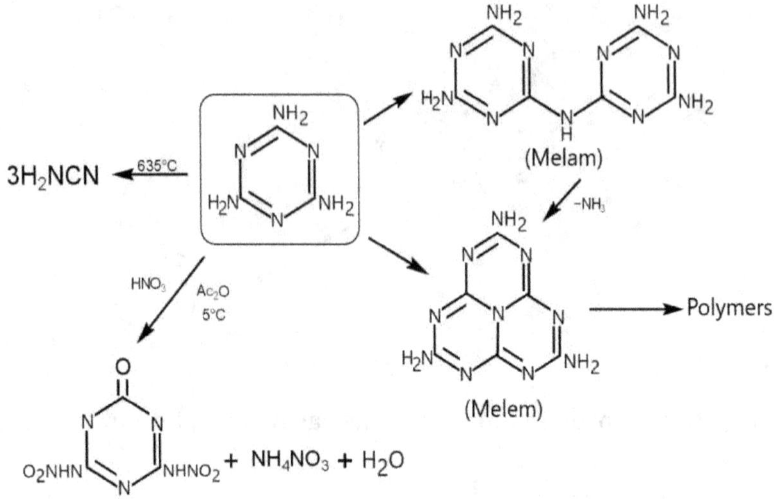

Cyanuric acid also shows activity to halogens and also exhibits **methylation** at ring nitrogen.

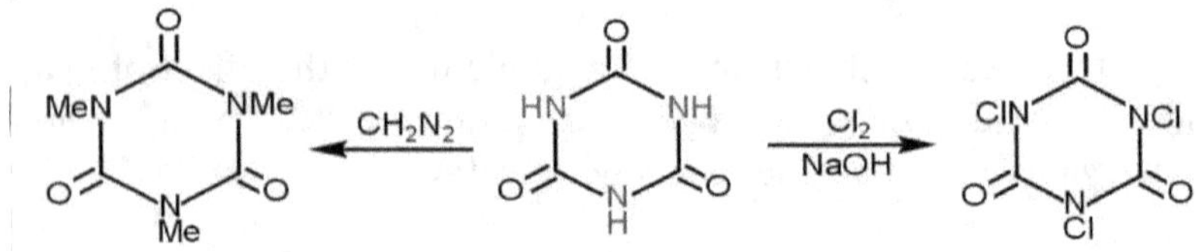

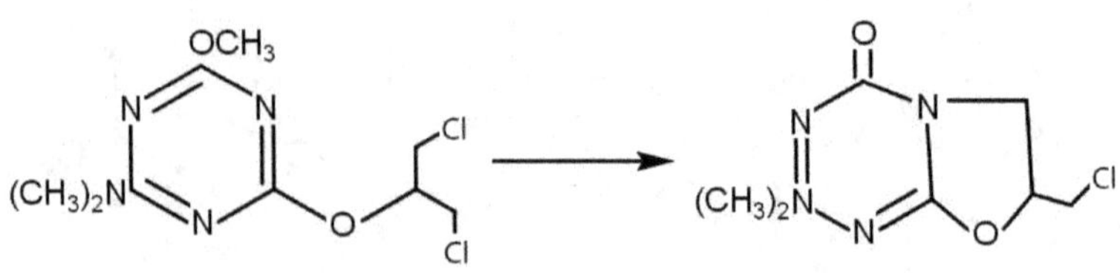

The triazine-thiones form sulphur-bridged oligomers by reaction with chloro-1,3,5-triozines, and are **oxidized** to yield the dimeric species.

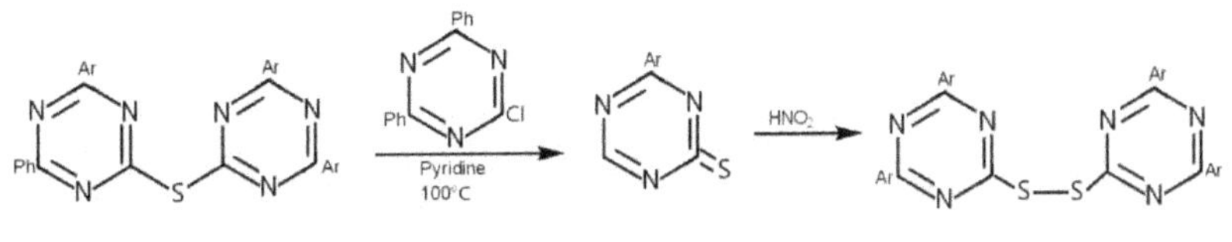

2.8.4 TETRAZINES AND PENTAZINES

2.8.4.1 SYNTHESIS OF TETRAZINES AND PENTAZINES

3-Amino-1,2,4,5-tetrazine is obtained when 4-amino-3-azido-4H-1,2,4-triazole is heated in chlorobenzene at 110°C. This reaction proceeds via intermediates like nitrene which later opens up to formazan and further gives the product by ring closure.

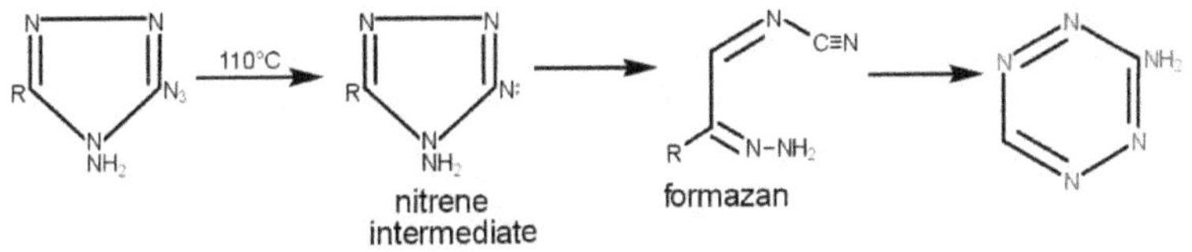

Likewise, the benzo imidoyl thiocarbodihydrazide was **cyclized** to produce 3-phenyl-1,4-dihydro-1,2,4,5-triazine-6(5H)-thione.

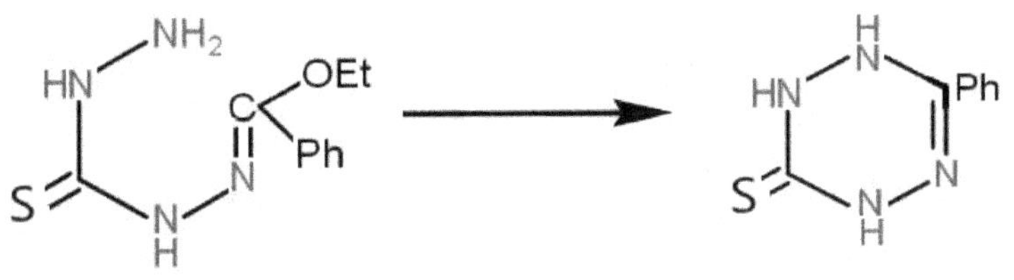

Reaction between 1,3,5-triazine (A) and hydrazine gives a light sensitive compound 1,2-diformylhydrazine dihydrazone (B) which turns pink in air due to the formation of 1,2,4,5-tetrazine (C). When (B) is heated with acetic anhydride, it gives a red volatile compound 1,2-diacetyl-1,2-dihydro-1,2,4,5-tetrazine (D).

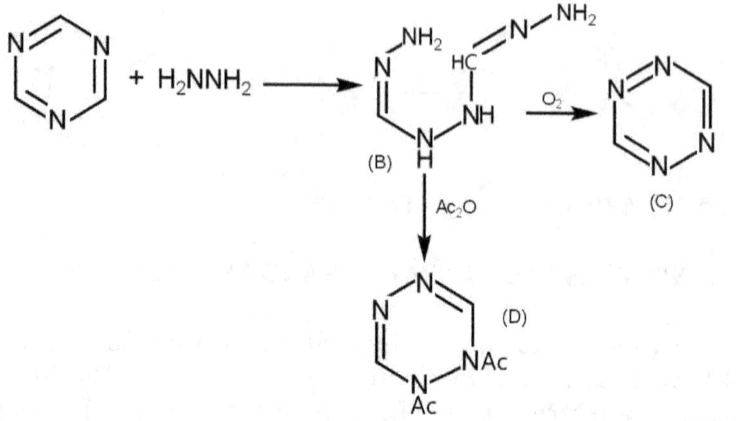

2.8.4.2 PROPERTIES OF TETRAZINES AND PENTAZINES

Hydrolysis of 3-phenyl-1,2,4,5-tetrazine, in the presence of base, results in the isolation of benzalazine. In a different case, aldehyde acyl hydrazone is obtained.

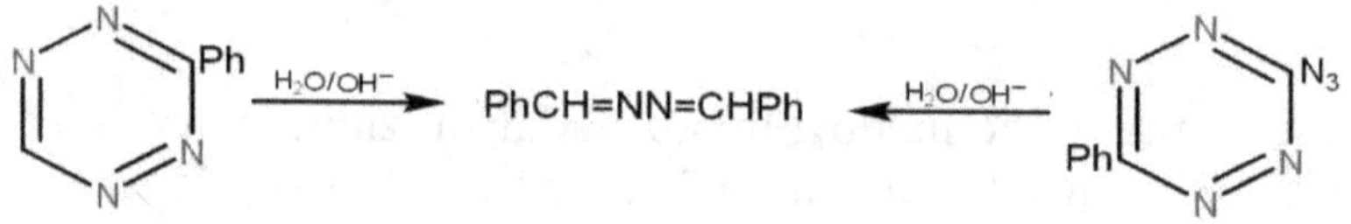

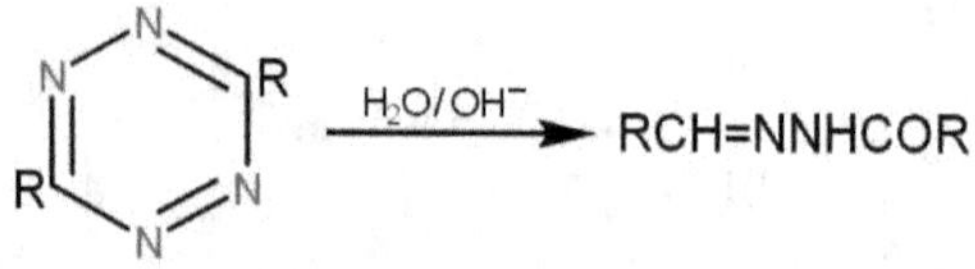

When 3,6-dibenzyl-1,2,4,5-tetrazine reacts with potassium hydroxide in methanol, it yields 3-benzyl-7-methoxy-6-phenylimidazo[1,2-b][1,2,4,5] tetrazine.

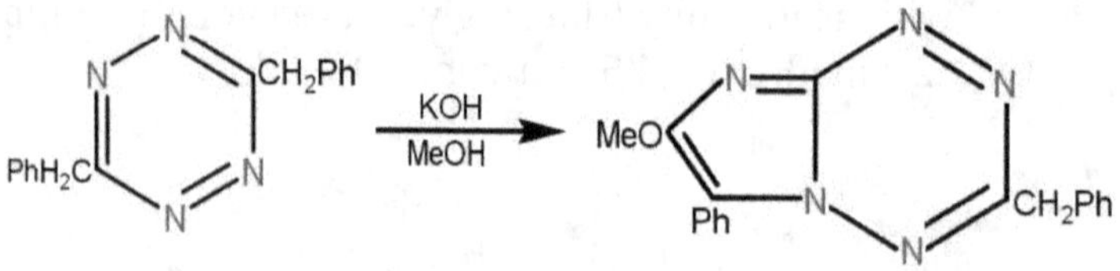

Hydrolysis of 1,4-disubstituted 3-imino-1,2,4-triazole, in the presence of base, gives 2,4-disubstituted 3-imino-1,2,4-triazoles in case-1 while 3,6-disubstituted dihydro-1,2,4,5- tetrazine got converted to 3,5-disubstituted 1,2,4-triazole in case-2.

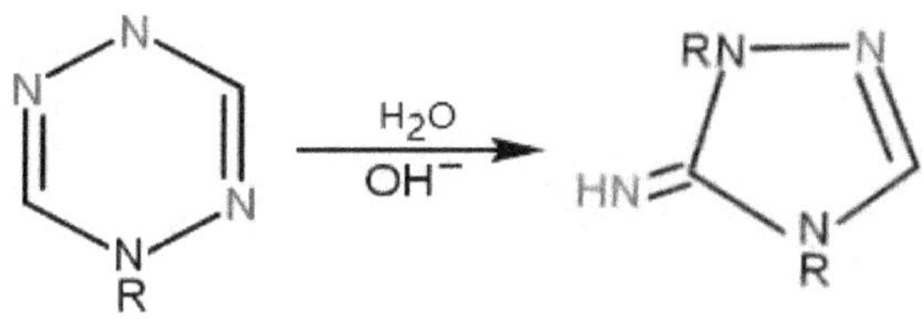

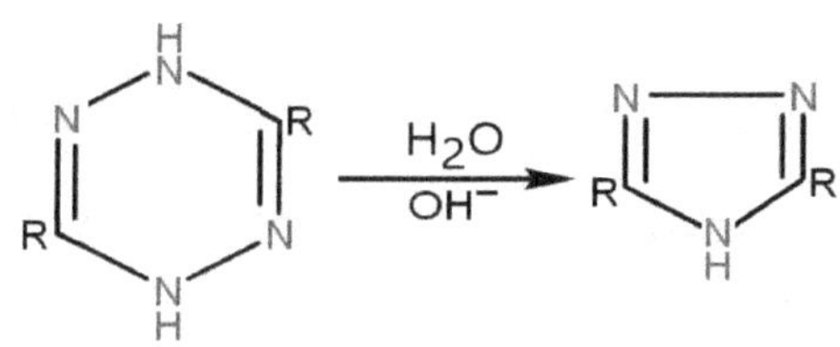

Benzalazine can be obtained on treatment of 3-phenyl-1,6-dihydro-1,2,4,5-tetrazin-6-one with base.

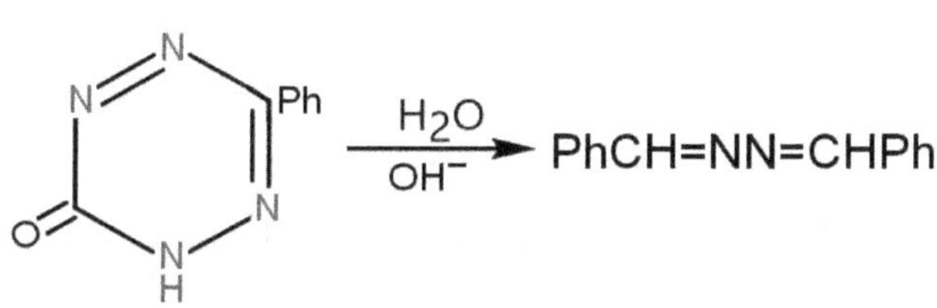

Effect of **heat** results in various formations; like 3,6-dimethyl-1,4-dihydro-1,2,4,5-tetrazine gives 4-amino-3,5-dimethyl-1,2,4-triazole and 3,6-diphenyl-1,4-dihydro-1,2,4,5-tetrazine yields 2,4,6-triphenyl-1,3,5-triazine (A), 3,5-diphenyl-1,2,4-triazole (B), benzonitrile (C) and ammonia when heated with a solvent at 190°C.

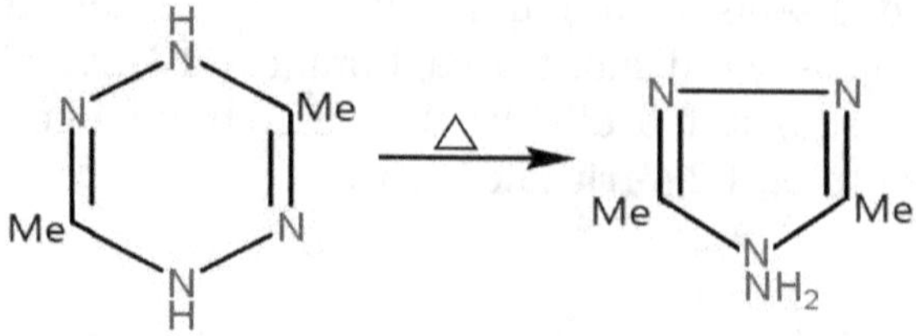

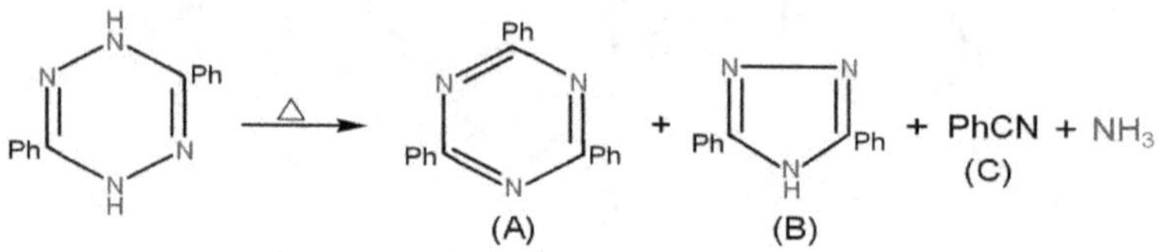

2.9 PYRIDO DIAZINES AND ITS BENZODERIVATIVES

2.9.1 PYRIDO PYRIMIDINES

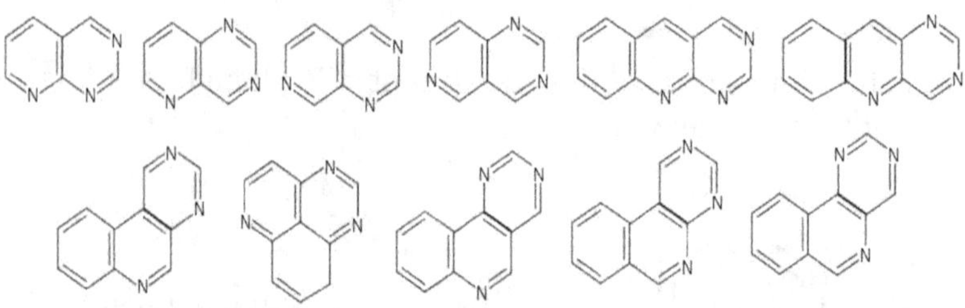

2.9.1.1 SYNTHESIS OF PYRIDO PYRIMIDINE

Pyridines have been used as the initial materials for obtaining pyridopyrimidines cyclization of aminopyridine ester and azide results is the desired product.

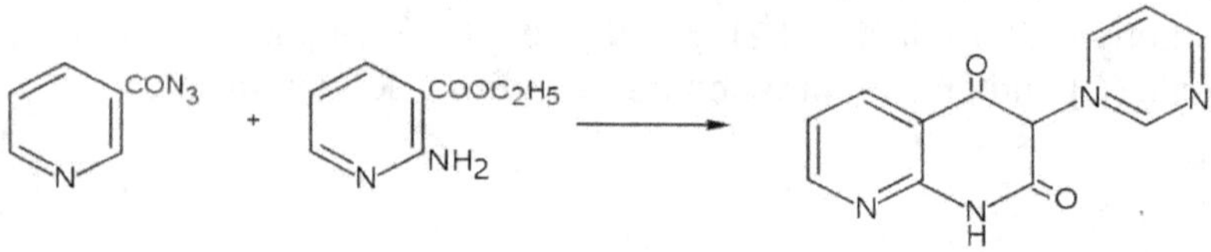

Pyridine ring can be obtained from a fused pyran intermediate, however if pyrylium salt is employed, quaternary pyrido-pyrimidinium salt is formed.

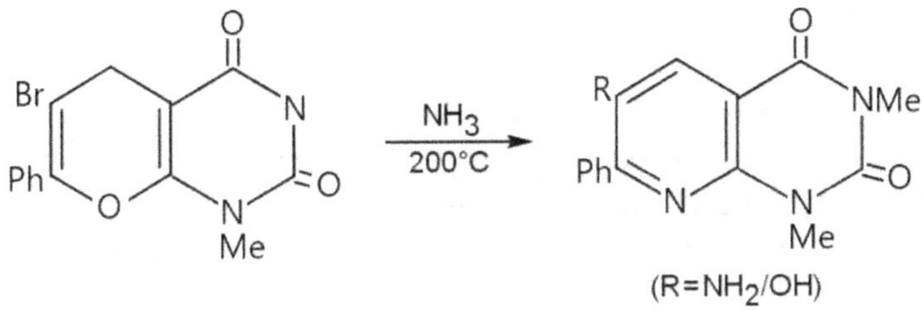

There are certain examples of reverse behaviour in the form of opposite **regioselectivity**.

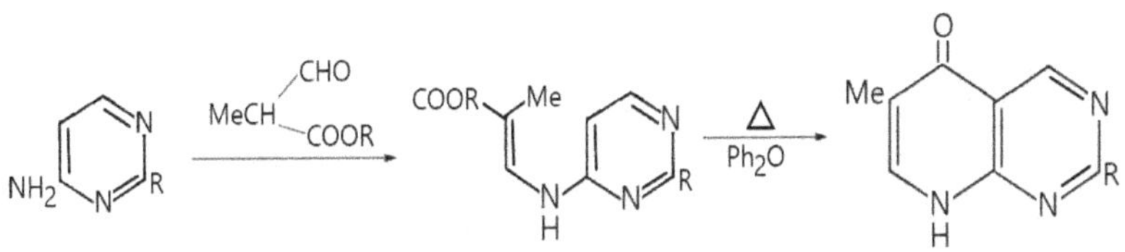

Reaction of urea with 2-aminopyridine-3-carboxylic acid results in 2,4-diones.

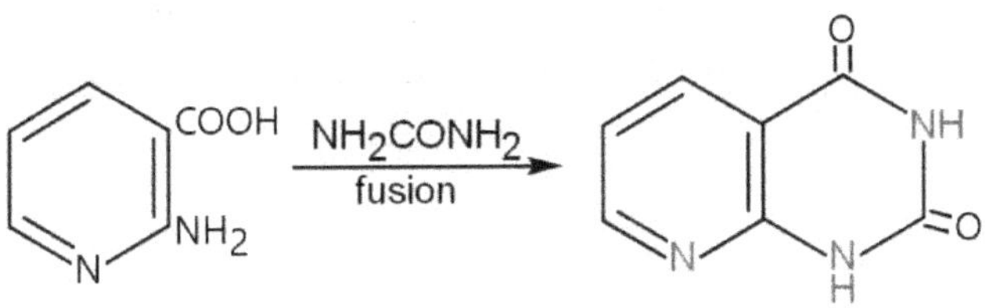

2.9.1.2 PROPERTIES OF PYRIDO PYRIMIDINE

Electrophilic attack at ring nitrogen is easily brought about. Alkylating agents as alkyl halides/alkali, alkyl sulphate/alkali, alkyl halide results in N-alkylation of oxo-di- and tetra-hydro derivatives.

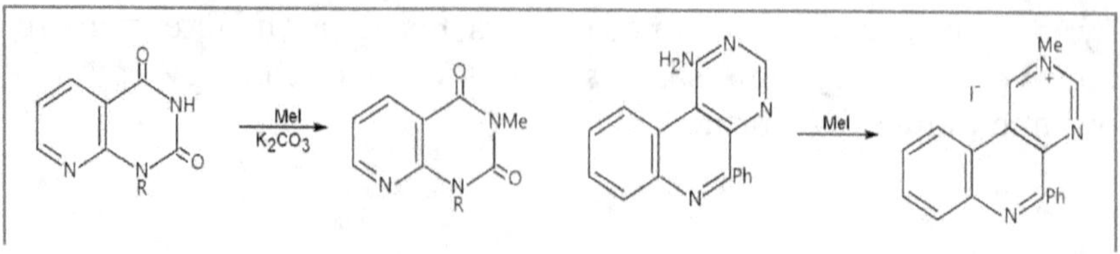

Nucleophilic attack at ring carbon atom occurs quite readily. Grignard's reagent reacts with pyrido[2,3-d] pyrimidine to result in the formation of 4-substituted product. However, in the presence of methyl lithium, both addition as well as substitution reactions occur.

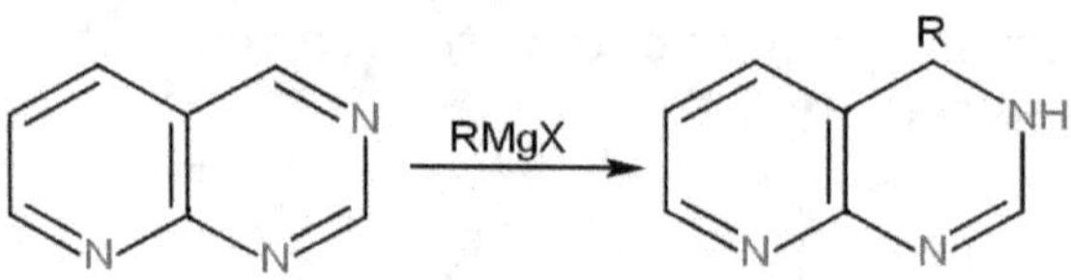

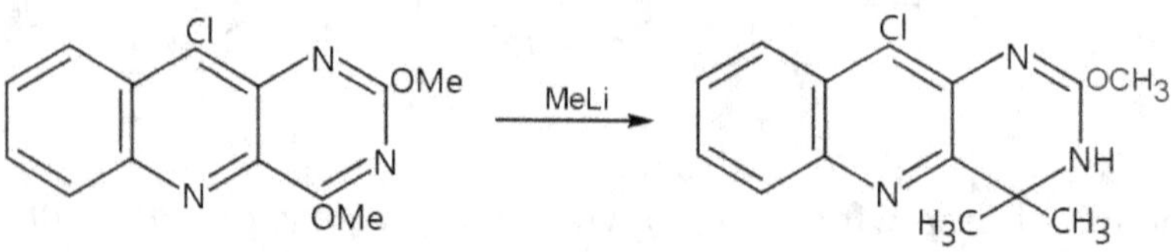

Ring opening reactions can also occur from nucleophilic attack at ring carbon in the presence of a base or conc. H_2SO_4 to give formyl amino nitrile.

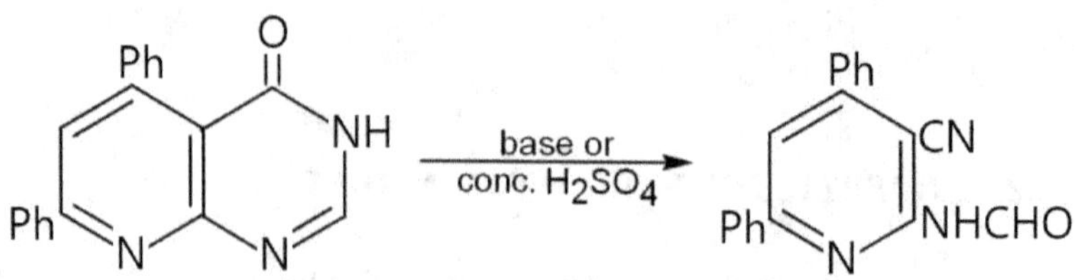

Pyrolyzed polyacrylonitrile can be obtained from 2-aminonicotinaldehyde.

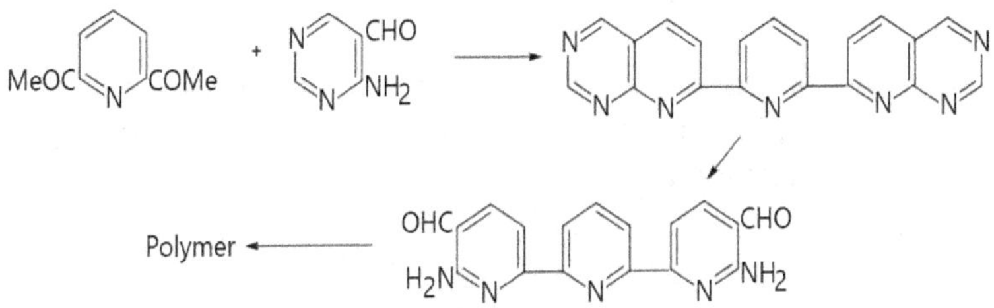

Fused heterocyclic ringed compounds can be obtained in presence of formic acid.

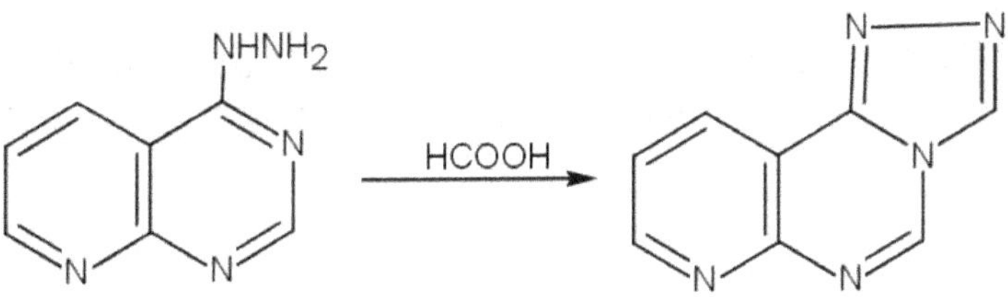

2.9.2 PYRIDO PYRIDAZINES

2.9.2.1 SYNTHESIS OF PYRIDO PYRIDAZINES

N-amino pyrido [2,3-c] pyridazine (A) is obtained from the reaction of hydrazine on acrylic ester. This ester is obtained from pyran[2,3-d] pyridazinone which is further yielded by **ring expansion** of furo-pyridazine in presence of KCN/DMSO.

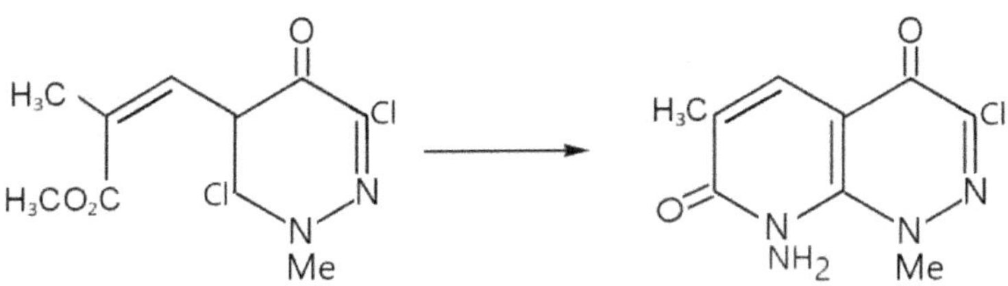

2.9.2.2 PROPERTIES OF PYRIDO PYRIDAZINES

Electrophilic substitution is brought about with great difficulty. In the presence of phosphorous pentachloride on strong heating, such substitution is obtained.

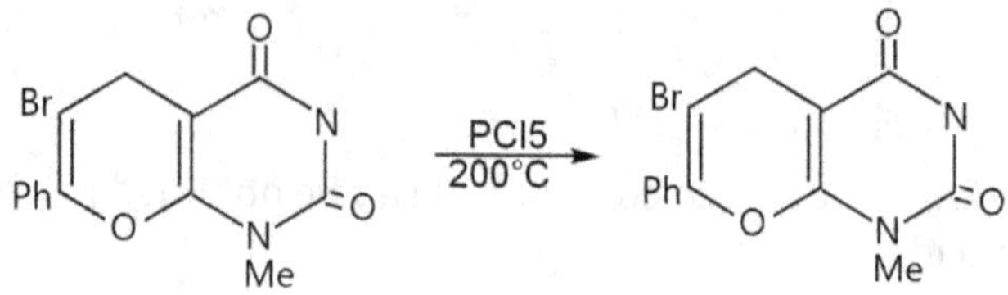

[4+2] Cycloaddition with butadiene results in the addition of a fused ring and hence electrophilic attack at ring nitrogen is brought about.

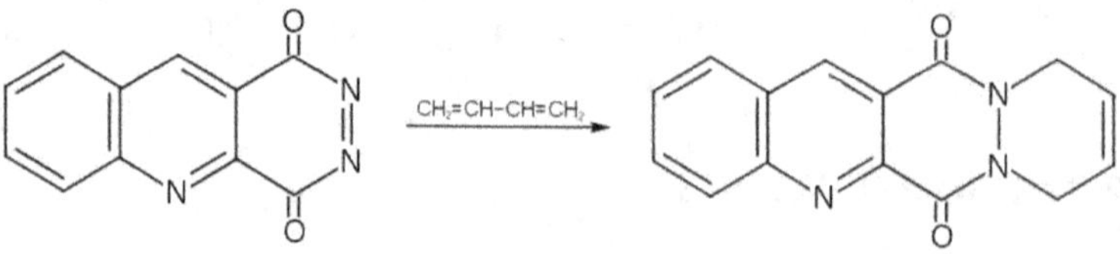

Ring opening reactions are easy to occur. N-aryl cyclic hydrazides (A) give N-aminoimides (B) with alkali. Alkaline hydrolysis of ester gives hydrazone.

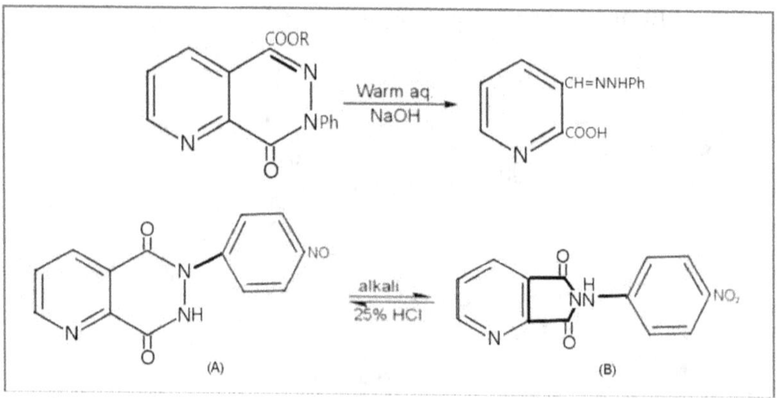

2.9.3 PYRIDO PYRAZINES

2.9.3.1 SYNTHESIS OF PYRIDO PYRAZINES

Cyclization of amino alcohol in HBr results in 3-methyl-1,2,3,4-tetrahydro derivative.

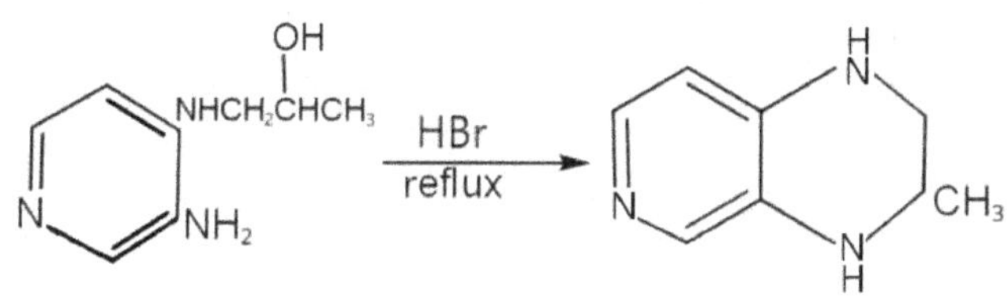

2.9.3.2 PROPERTIES OF PYRIDO PYRAZINES

Nucleophilic attack on substituents can be carried out. The reaction of bases on N-aryl-2,3-dimethyl pyrido[2,3-b] pyrazinium salts gives anhydro-bases.

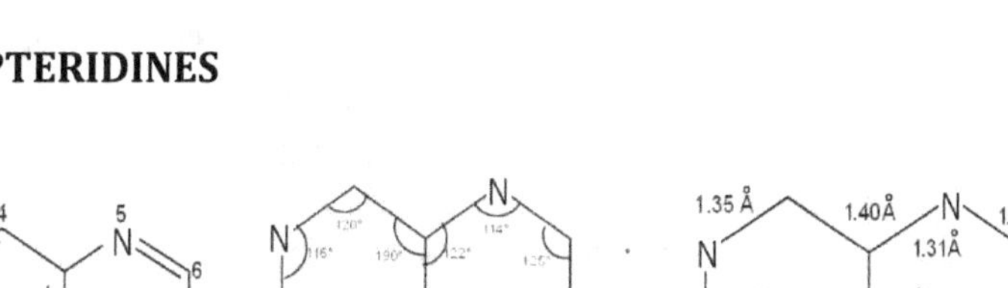

2.10 PTERIDINES

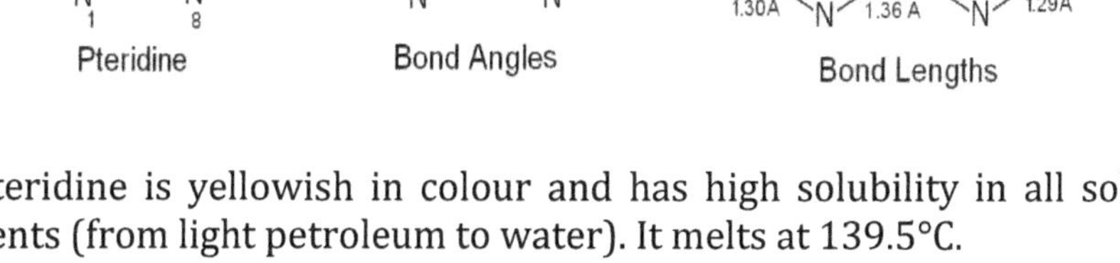

Pteridine is yellowish in colour and has high solubility in all solvents (from light petroleum to water). It melts at 139.5°C.

2.10.1 SYNTHESIS OF PTERIDINES

The **Taylor method** of synthesis is a new route to produce pteridines by the reaction of an oxyimino-carbonyl compound and ethyl-α-amino cyanoacetate or amino malononitrile thereby resulting in the formation of substituted pyrazine oxides that cyclize in the presence of guanidine to give pterin-8-oxides or 2, 4-diamiinopteridine-8-oxide. Reduction of these oxides give 7, 8-dihydropteridines.

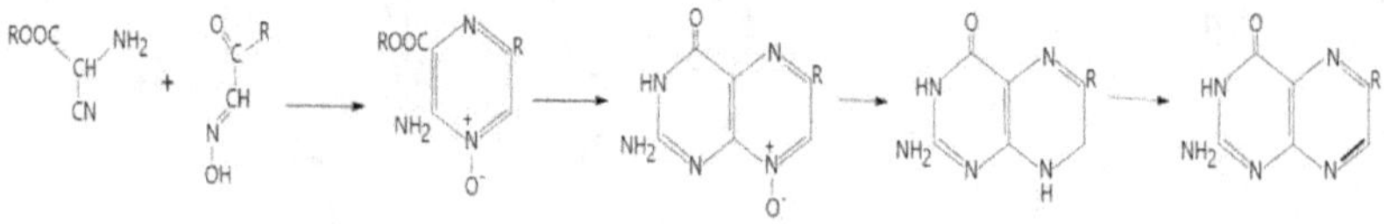

Gabriel-Isay reaction gives the simplest synthesis of pteridines by **condensation** of a 5,6-diaminopyrimidine with a 1,2-dicarbonyl compound via Gabriel's method to prepare 4-methyl pteridine and Isay's method to obtain 5,6-diphenyl pteridine.

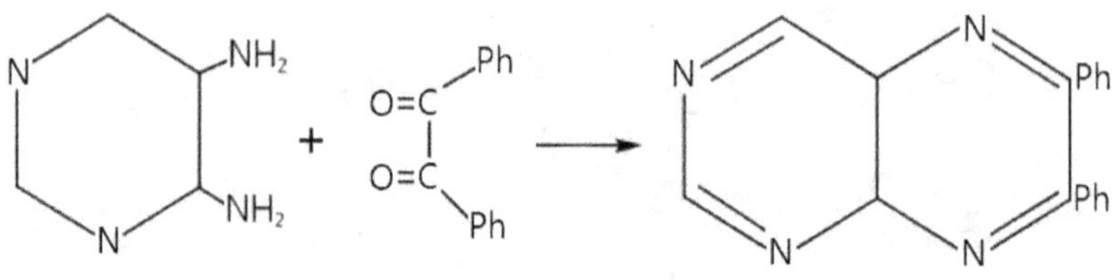

6- And 7-substituted pteridines are obtained via **Timmis reaction** when keto-methylene compounds react with 6-amino-5-nitrosopyrimidines. It is not so that the required product is obtained with carbonyl compounds, but compounds like nitriles, esters, and acyl halides, that have activated methylene group next to the relative functional groups, also give good results.

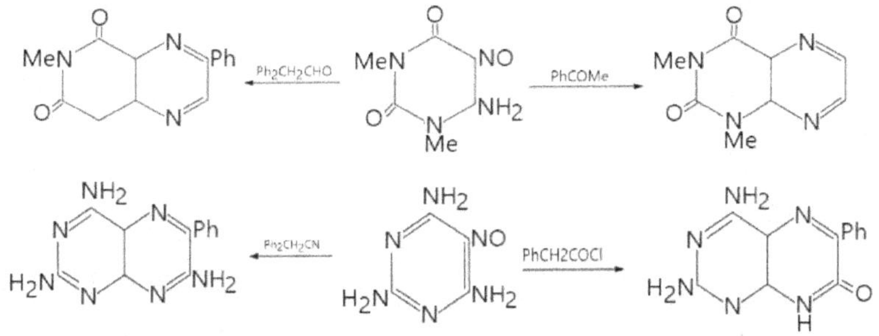

An advanced version of Timmis condensation is **Pachter reaction** used to obtain 4-amino-7-methyl-2,6- diphenyl pteridine (**A**) and 4,7-diamino-2,6-diphenylpteridine (**B**).

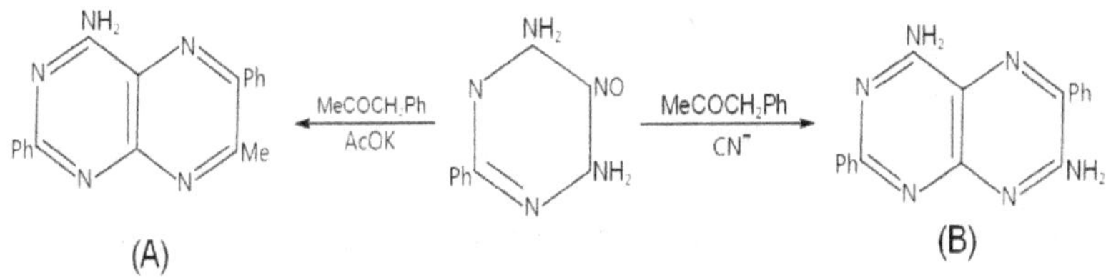

Polonovski and **Boon** proposed a nucleophilic substitution of 6-chloro-5-nitropyrimidne and α-amino carbonyl compounds to yield 7,8-dihydropteridine on reduction. Thus, ethyl-α-aminoisobutyrate and 2-amino-6-chloro-5-nitropyrimidin-4-one (**A'**) result in the production of 7,7-diethyl-7,8-dihydroxanthopterin (**B'**).

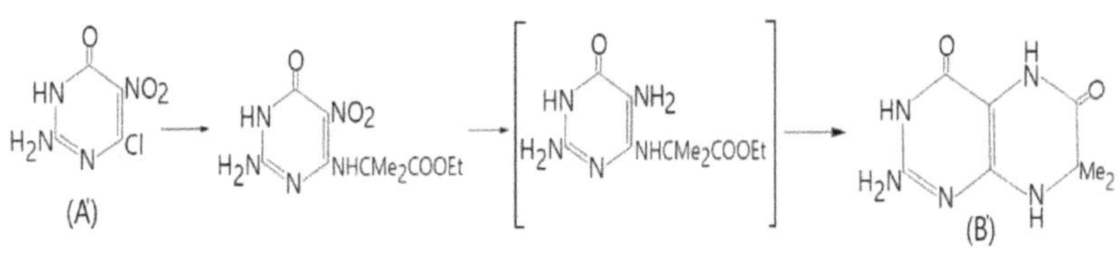

2.10.2 PROPERTIES OF PTERIDINES

Since the pteridine nucleus is highly π-deficient, it shows interest in nucleophilic attack. Pteridin-6-one (**C**) offers for addition of carbanions in low temperature alkali solutions, thereby final degraded product is 7-cyanoethyl-7,8-dihydropteridin-6-one (**D**).

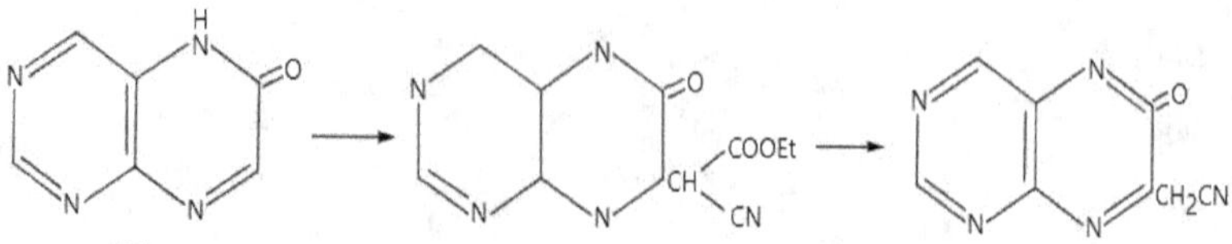

(C) (D)

Xanthopterin (**C'**) facilitates acid- catalysed addition at 7- and 8- positions in the presence of oxalo-acetic acid.

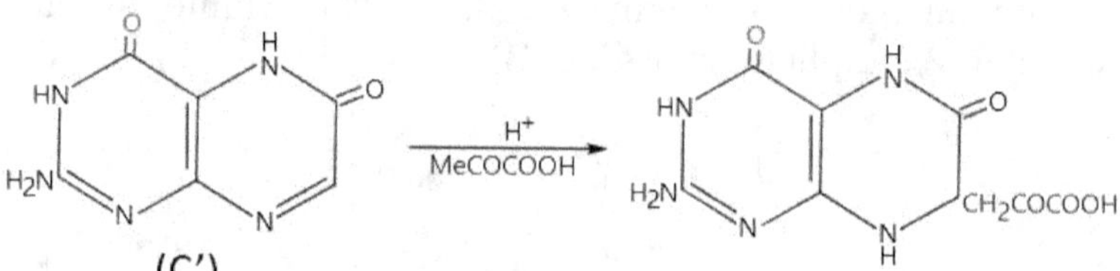

(C')

The activity of halogen atoms can be well understood by lumazine series.

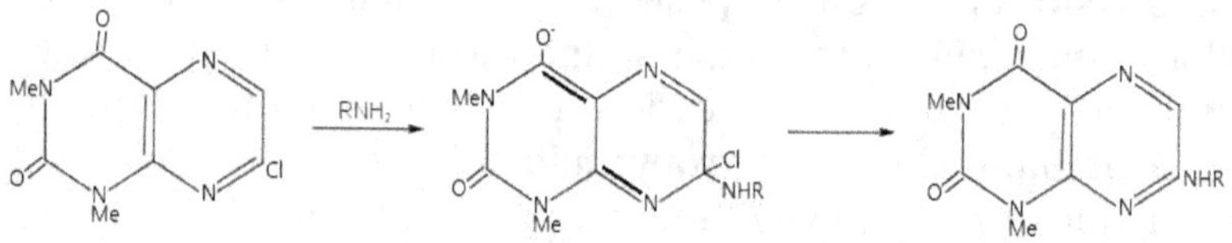

7 - chloro - 1,3 - dimethyl lumazine

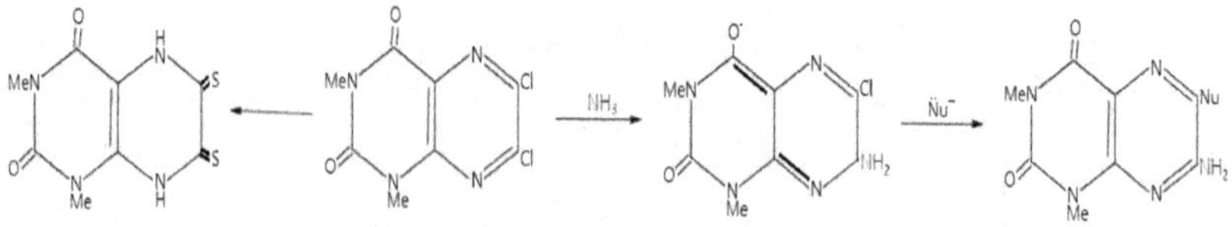

6,7 - dichloro - 1,3 - dimethyl lumazine

7 - amino - 6 - chloro - 1,3 - dimethyl lumazine

Methyl-substituted pteridine is oxidized to pterin-6-carbonylic acid in the presence of potassium permanganate, while demethylation occurs in the presence of light.

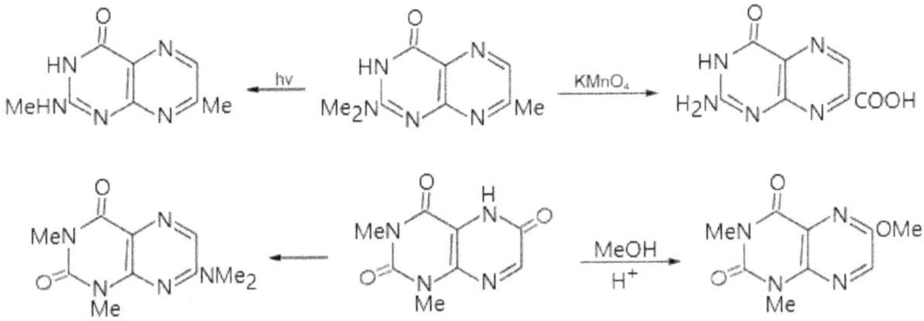

Alkylation takes place on sulphur atom of pteridine thiol forming corresponding alkylthio derivatives.

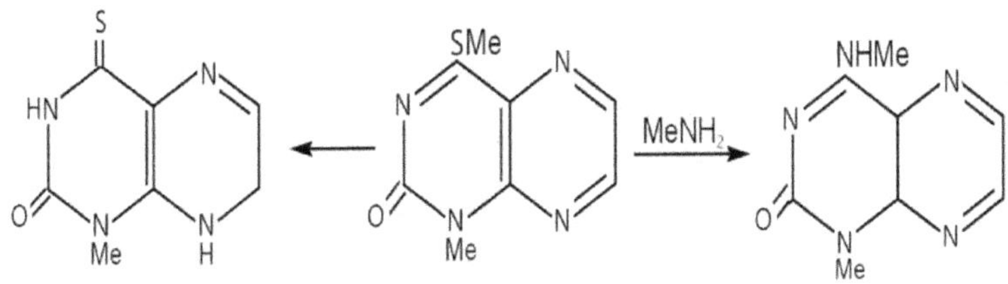

1-methyl-4-methylthio
ptredin-2-one

Alkylation of amide functional groups is the most prominent substitution reactions involving ring nitrogen atom in pteridine.

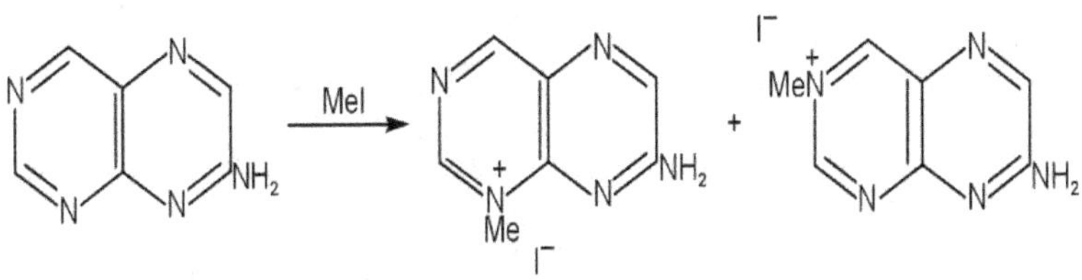

2.11 PYRANS AND ITS DERIVATIVES

The heterocyclic compounds that are six membered and has oxygen as the heteroatom are widely spread in plants.

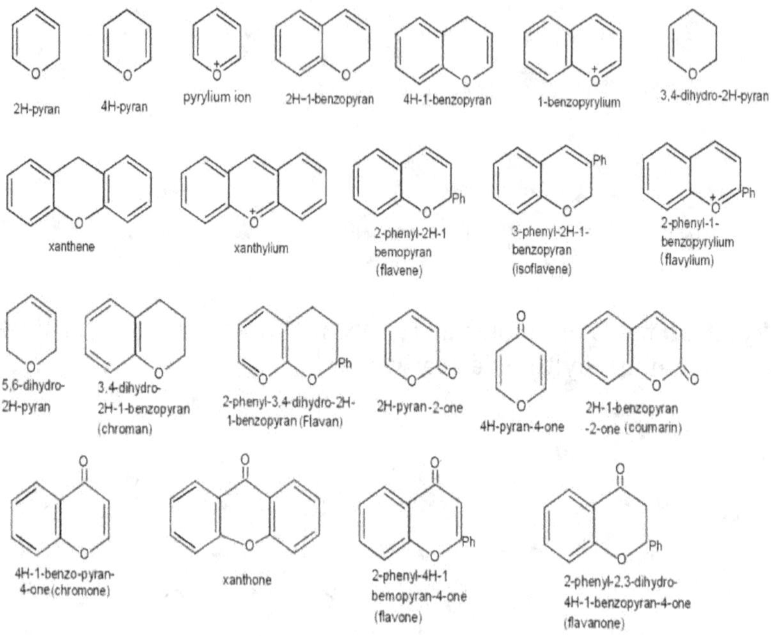

2.11.1 SYNTHESIS OF PYRANS

Dienone, on **irradiation**, gives a mixture of products including pyran.

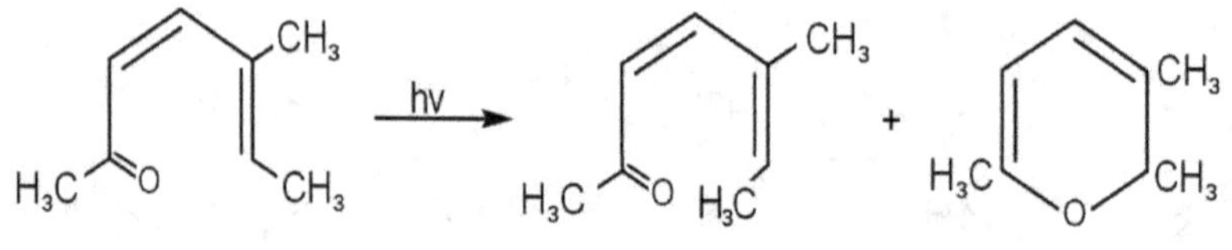

When arynes are trapped with α, β – unsaturated aldehyde, chromenes are obtained. Likewise, the reaction of cinnamaldehyde and tetrachlorobenzyne results in formation of benzoxete (**A**) by [2+2] cycloaddition which on ring opening yielded (**B**). Finally, by electrocyclic ring closure 5,6,7,8 – tetrachloro flav-3-ene (**C**) is obtained.

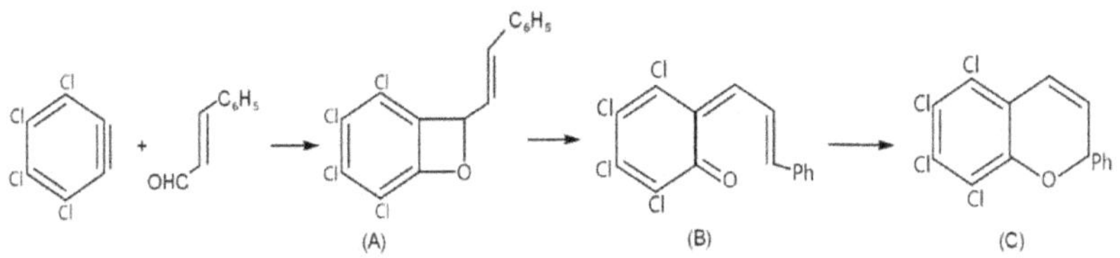

Propynal in piperdine acetate leads to the synthesis of 4-ethynyl - 4H - pyran - 3,5 - dicarbaldehyde whose structure was corrected. If the reaction is conducted in the absence of moisture, then no results is obtained. Thus, the **hydration** of alkyne forms di vinyl ethers which cyclizes to form products.

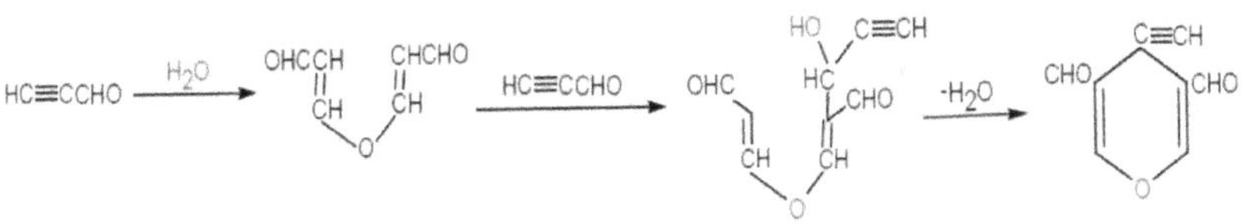

In yet another method of synthesis, luteolinidin tetramethyl ether, in the presence of lithium aluminium hydride gives flavene.

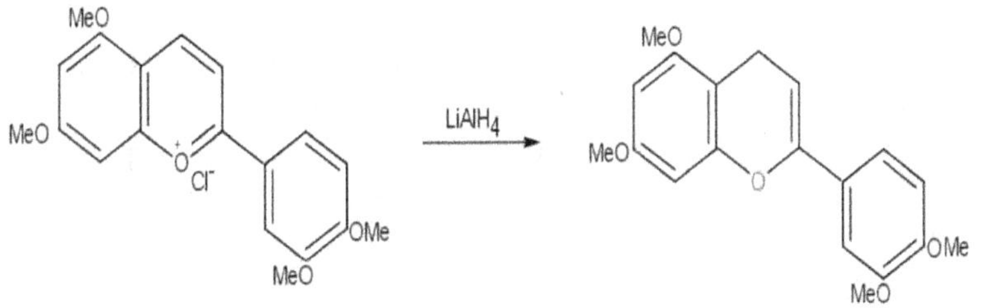

3-Formyl derivative (**D**), that is formed by the base catalysed reaction of dihydrocoumann and ethyl formate, rearranges to chroman easter(**E**) in alcoholic hydrogen chloride. Finally, Chrom-2-ene(**F**) is obtained by heading(**E**) with polyphosphoric acid.

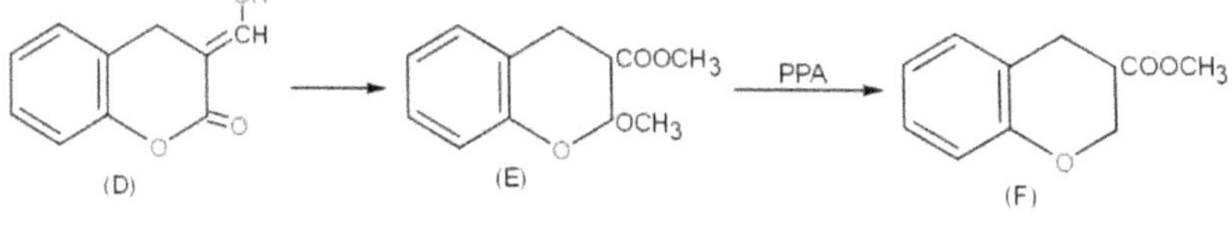

o-Quinone methides on **dimerization** lead to the formation of spiro-annealed chromans. Of the two molecules, one acts as hetero diene while the other as dienophile.

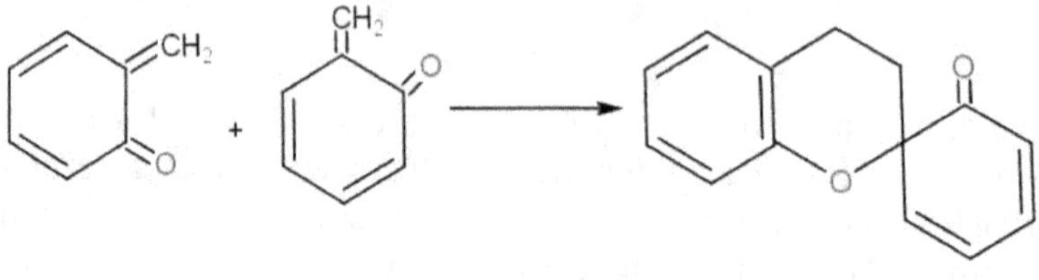

Pyranopyrandiones and dipyranopyrantriones are synthesized from mannonate derivatives.

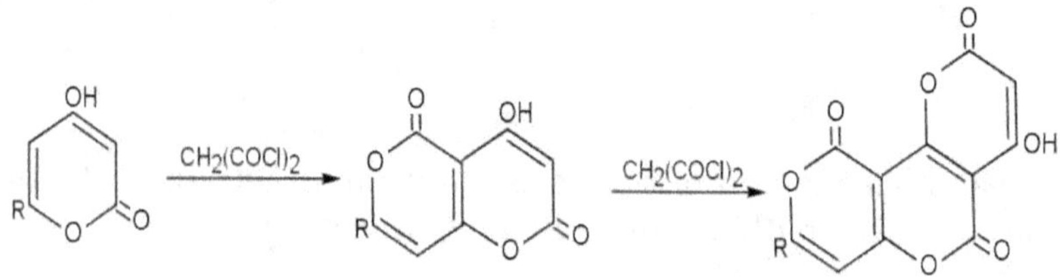

Chromans and chromenes are obtained via **Claisen Rearrangement** of allyl and propargyl aryl ethers and finally to coumarin. For instance, allyl aryl ether (prepared from o- methoxyphenol and methoxy allene) rearranges and cyclizes into 2-methoxychroman (R=CH_3) and chroman-2-ol (R=H) in boiling dimethylaniline. The second step of oxidation converts chromanol to coumarin.

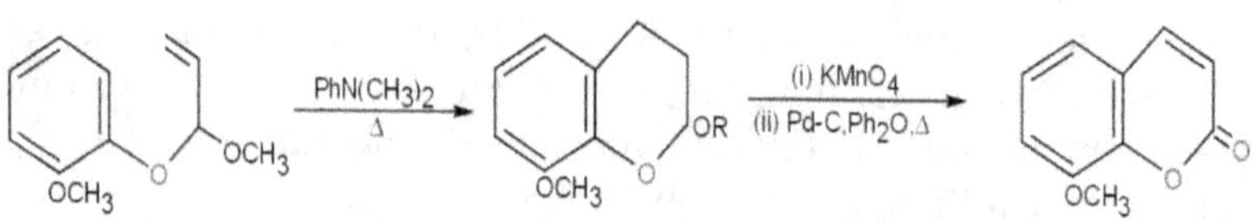

C- and **O- acylation** is involved when 1,3- diketones are treated with malonyl dichloride to obtain pyranone.

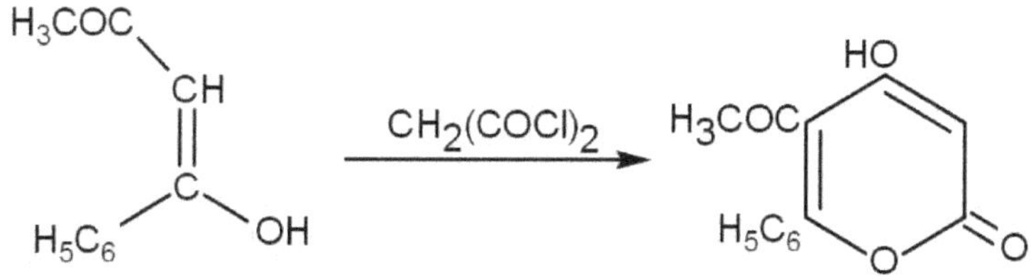

Photo-Fries rearrangement of ester to obtain a ketone is an interesting method of preparation. Cyclization is obtained on heating under pressure with water to obtain dihydroxy-xanthone.

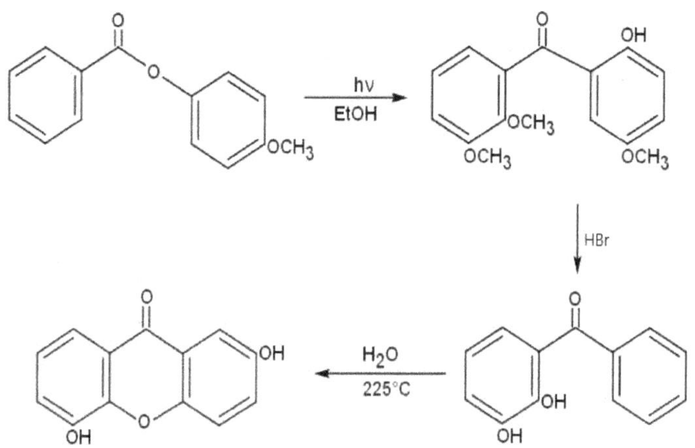

In the same way, **photochemical Fries rearrangement** of methyl 4-(3-methoxy benzoyloxy) benzoate leads to preparation of benzophenone and further by **para-coupling** during alkaline ferricyanide oxidation of 4-hydroxy-3-(3-hydroxybenzoyl) benzoic acid thereby forming 7-hydroxy-9-oxoxanthene-2-carboxylic acid.

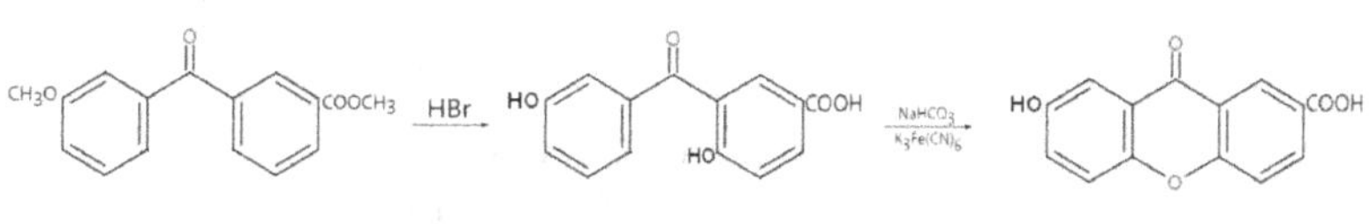

The trans-isomer of 2-hydroxybenzoyl acrylic acid forms on cyclization, 4-oxochroman-2-carboxylic acid **(G)** while the cis-isomer forms coumaranones. This cyclization reaction is facilitated by barium hydroxide or even weaker bases.

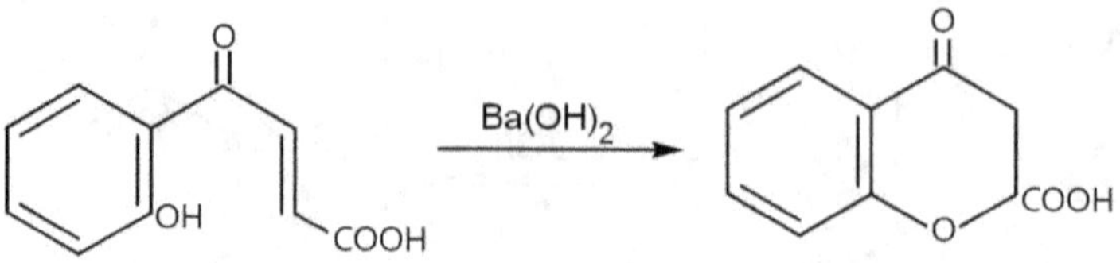

(G)

Diesters, via **Dickman cyclization,** gives chroman-3-ones. The products obtained are 3-oxochroman-4-carboxylate (H) and 2-carboxylic ester.

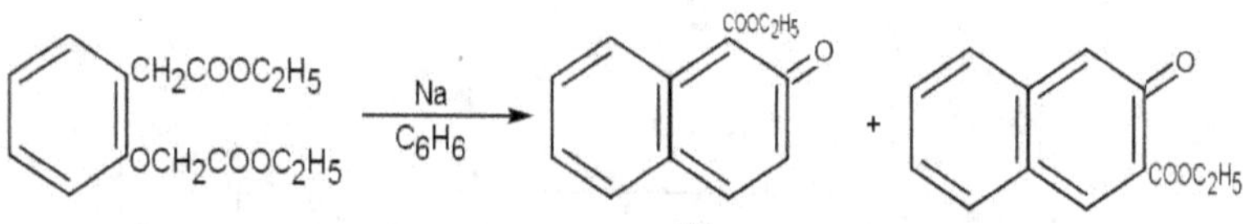

(H)

2,4,6-Trimethylpyrylium perchlorate is obtained by treating 4-methyl pent-4-en-2-one with acetic anhydride and perchloric acid.

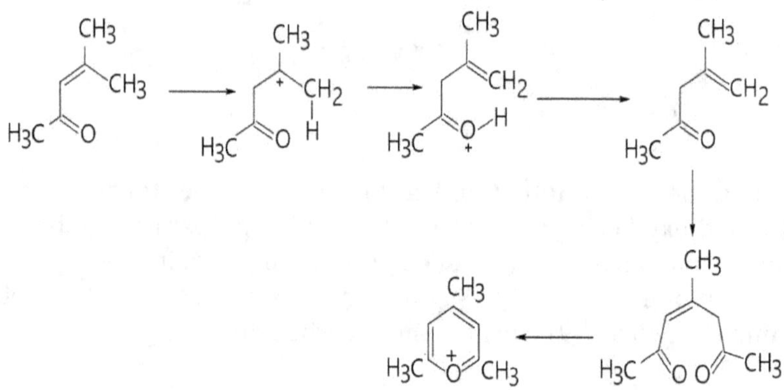

Under mild conditions, acetophenone gets converted to pent-2-ene-1,5-dione (I) that gets cyclized to 2,6-diphenylpyrylium salt.

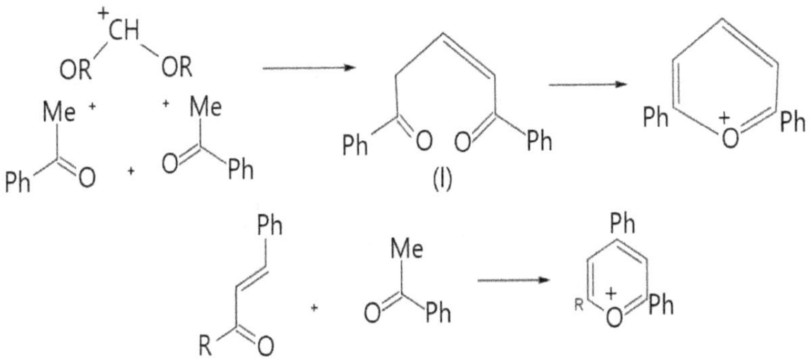

When isoflorene reacts with triphenylmethyl perchlorate, isoflavylium perchlorate (J) while with perchloric acid, di(isoflavenyl) ether(K) is formed.

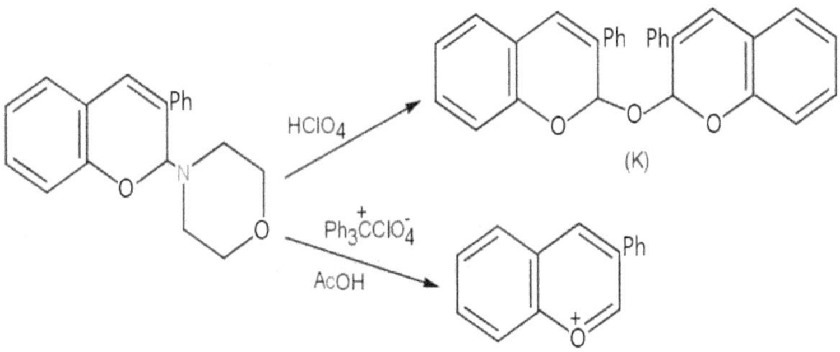

2.11.2 PROPERTIES OF PYRANS

The pyrylium ring has a positive charge on the heteroatom and hence does not support the attack of electrophiles. 2,4,6-Triphenylpyrylium perchlorate is nitrated at m- and p- positions of benzene rings.

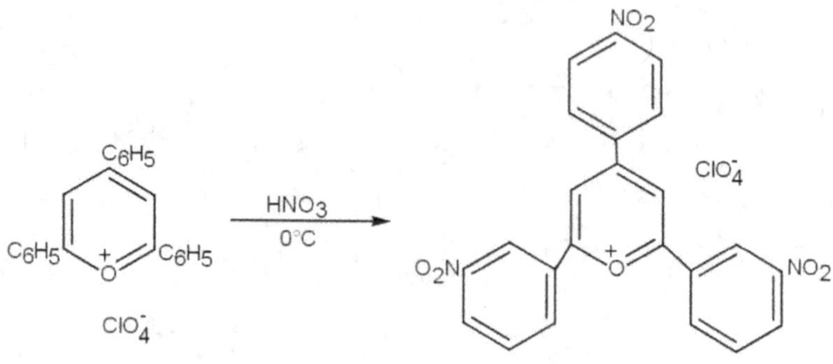

Hydrogen peroxide acts as a good oxidising agent for 2,4,6-trialkylpyrylium salt and gives 2-acyl furans.

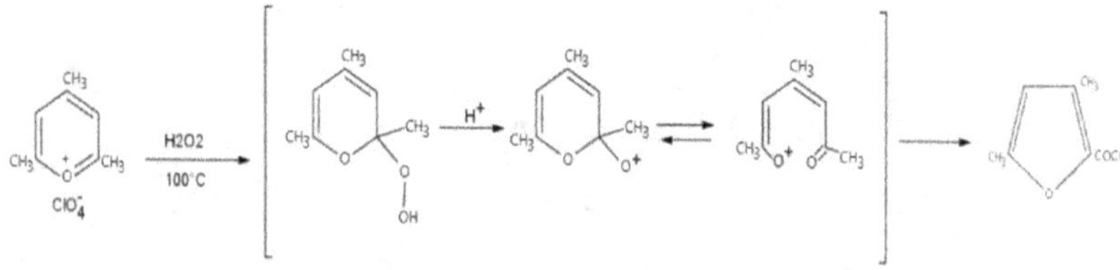

Ozonisation of pyrylium salts results in oxo acids or salicylaldehyde after treatment with peroxide or zinc-ethanol couple.

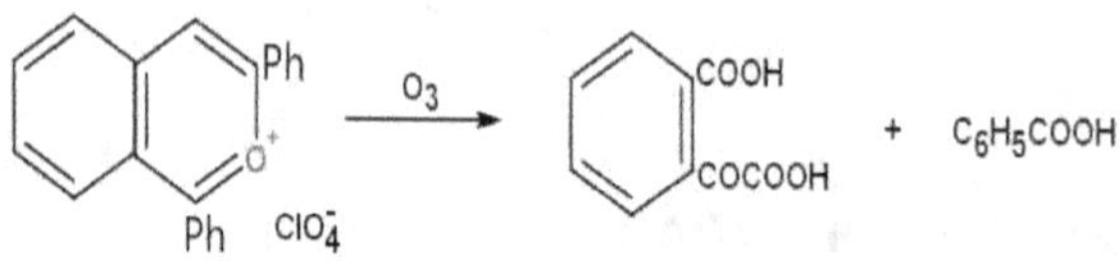

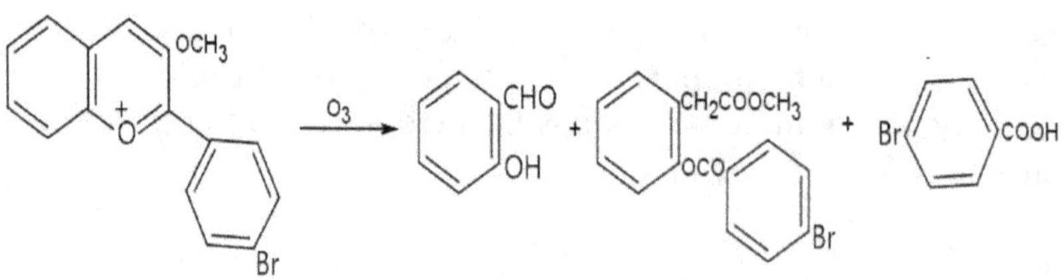

Nucleophiles attack the pyrylium salts easily and in the presence of hot strong alkali, peonidin gives phloroglucinol and vanillic acid.

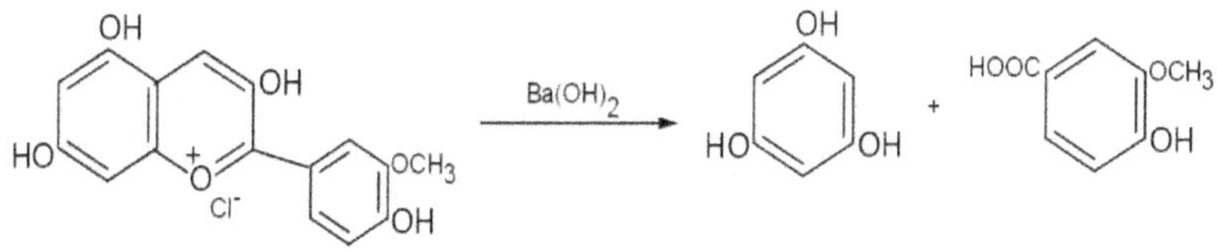

4-Ethoxybenzopyrylium perchlorate, if reacts with mole of Grignard's reagent, results in addition at position-2 but leads to ring opening on reaction with 2 moles of Grignard's reagent. With methylamine, 4-methoxy-2,6-dimethylpyrylium perchlorate reacts in two ways.

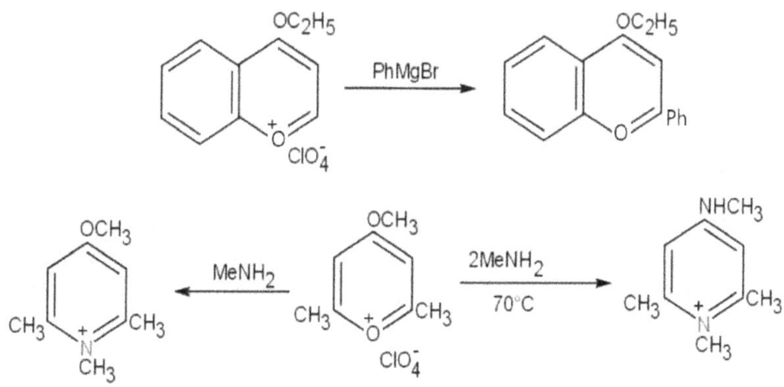

In the same way, pyrylium ring reacts with sodium sulphide forming a blue coloured dienone that results in thiopyrylium salt in the presence of an acid.

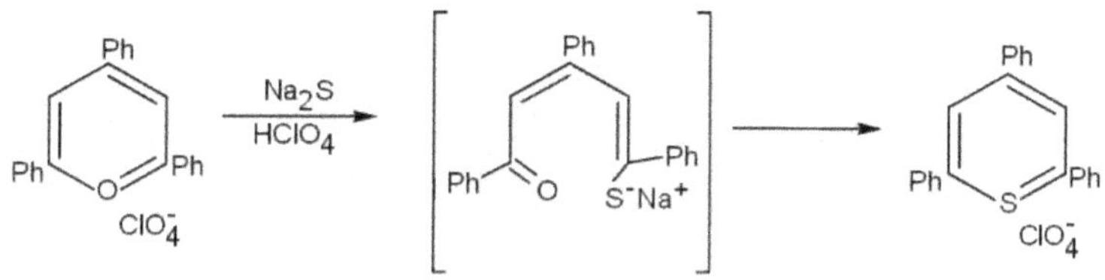

Dyes like tri-pyran are formed when pyrylium salts are condensed in the presence of phosphorus oxychloride.

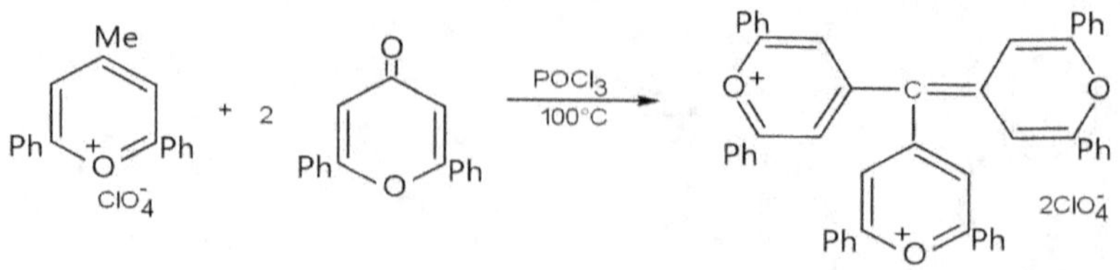

Pyrylium salts to be substituted at position-4 require alkyl thio groups at the same position to serve as useful precursors.

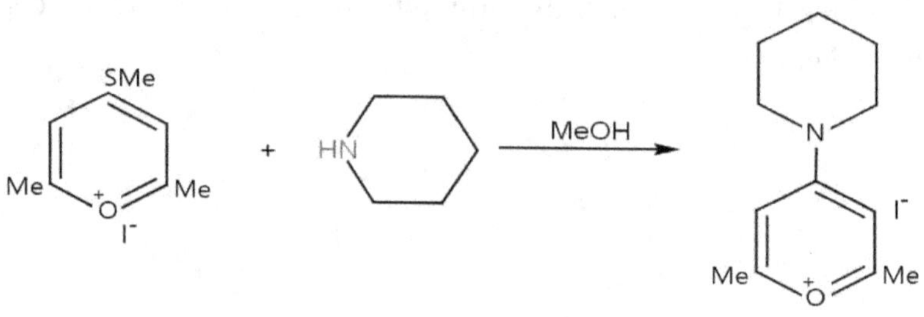

Chromans are formed on **catalytic hydrogenation** of chromenes. Likewise, simple chroman is obtained from ethyl-3-hydroxy-7-methoxy-2H-chromene-4- carboxylate in ethanol but 3-substituent is removed in acetic acid.

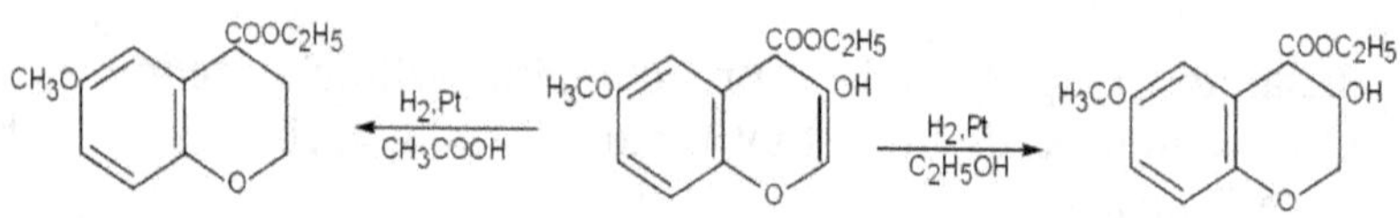

Oxidation of lonchocarpin (a naturally occurring compound) does not disturb the ring. Thus, the oxidation reaction serves as a great help in determining the structure of lonchocarpin.

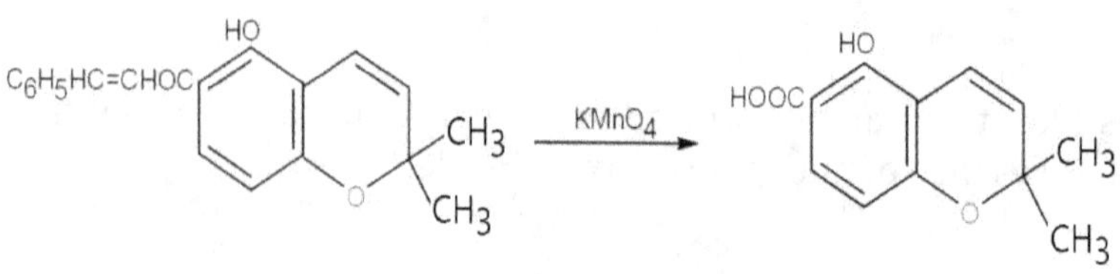

Reactive lactones are obtained when 4,6-dimethyl pyran-2-one is irradiated in methanol and these lactones gives ester with acids.

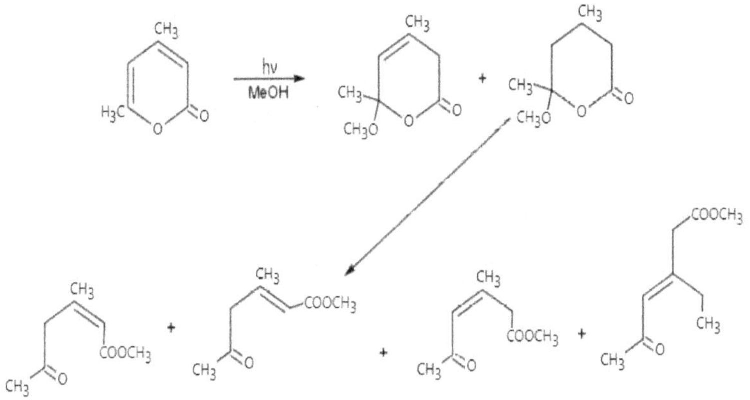

The **Mannich reaction** in chloromethylation of pyran-2-ones like 5,6-dimethylpyran-2-one occurs at C-3, however aminomethylation of 3-hydroxy coumarin occurs at position-3.

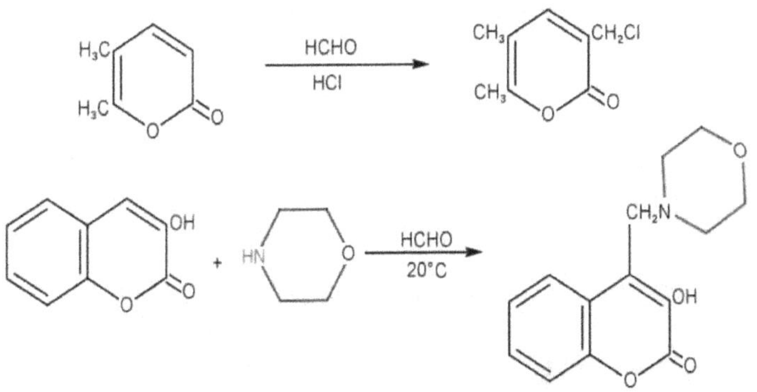

When isocoumarin is treated with Grignard's reagent in the presence of perchloric acid, 2-benzopyrylium salt is prepared. Methyl-2-oxo-pyran-5-carboxylate is methylated by diazomethane at position-6 and position-4.

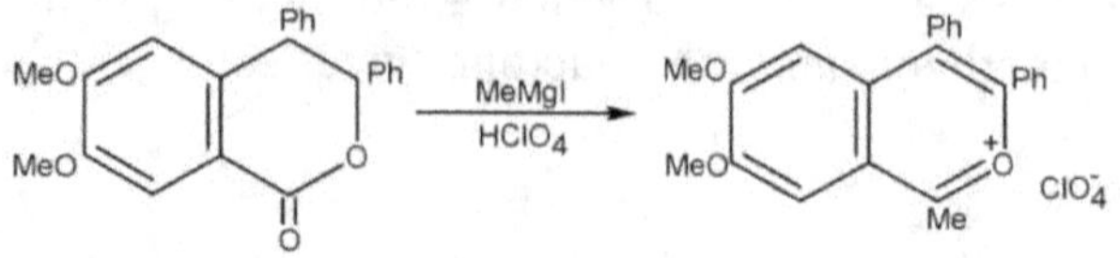

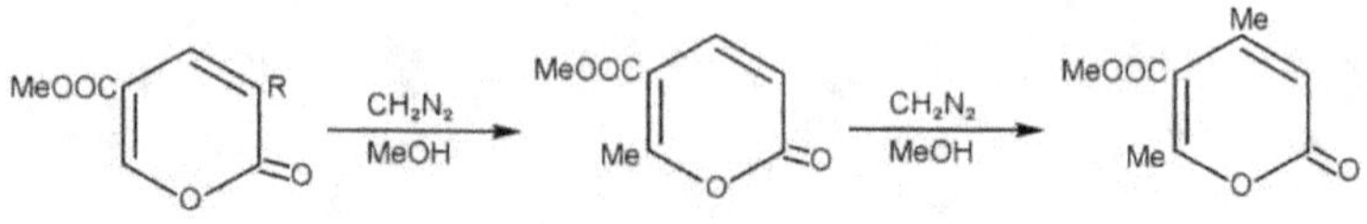

Amines tend to open the ring of pyran-2-one and further on acidification, the cyclisation to pyridone occurs like at high temperature and longer exposure time the former is formed.

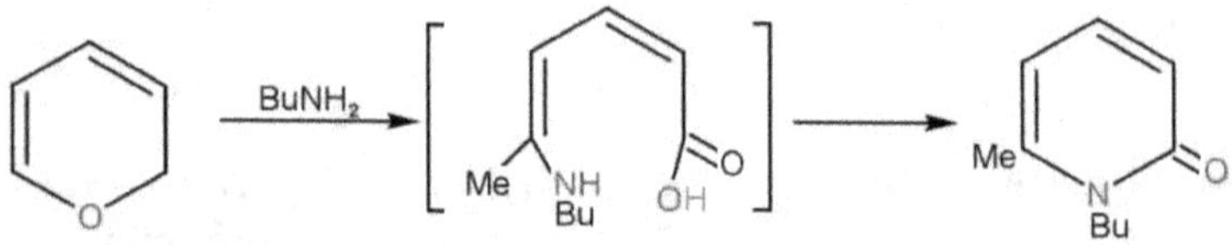

Furan-2.4-dicarboxylic acid are formed on reaction of methyl-3-bromocoumalate with hot aqueous alkali.

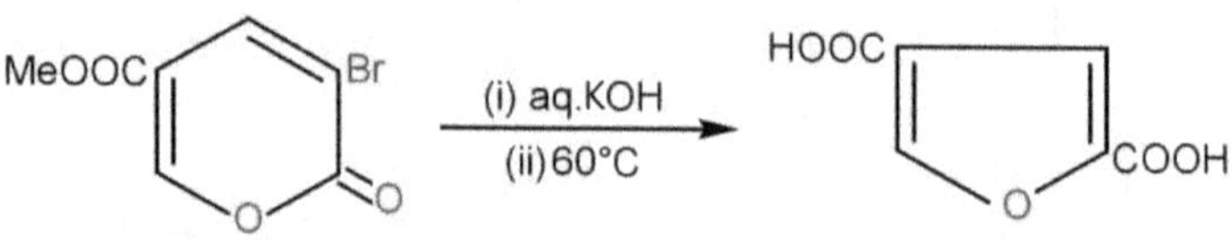

Coumarin exposed to diborane and then treated with hydrogen peroxide results in reduction of carbonyl group to methylene and addition of hydroxyl group across double bond of heterocyclic ring.

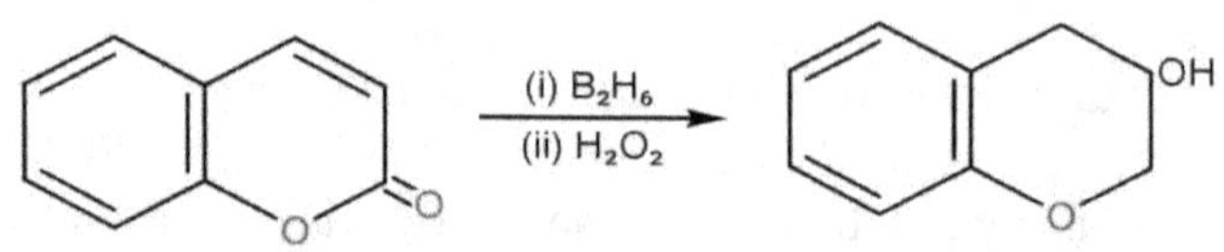

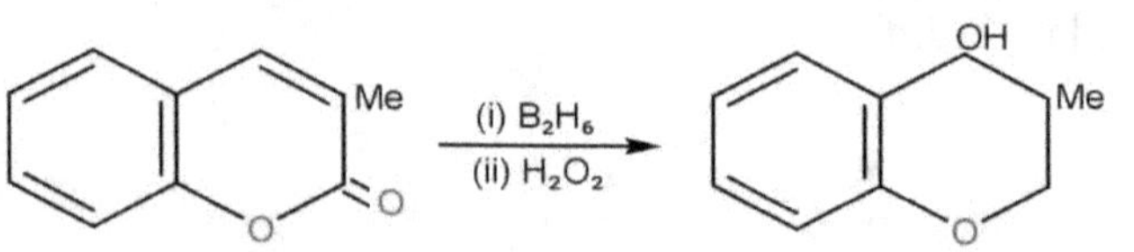

Pyran-4-one gets converted to pyrylium salt on reaction with Grignard's reagent followed by acid. In the presence of Grignard's reagent, pyran-4-ol is obtained that converts to salt.

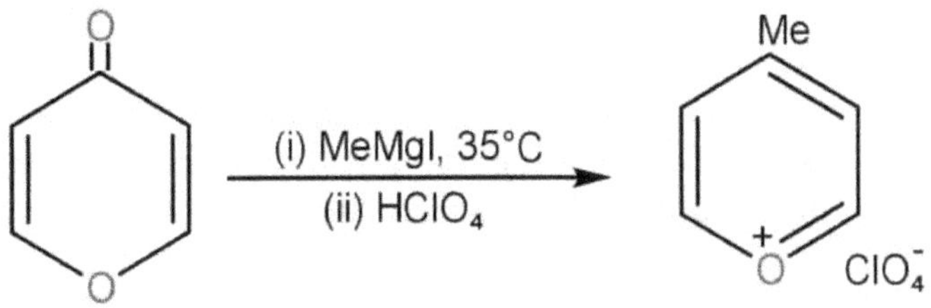

In the same way, benzopyrylium salt is obtained finally on treatment of 3,7-dimethoxy-2-methylchromone first with Grignard's reagent and then with acid.

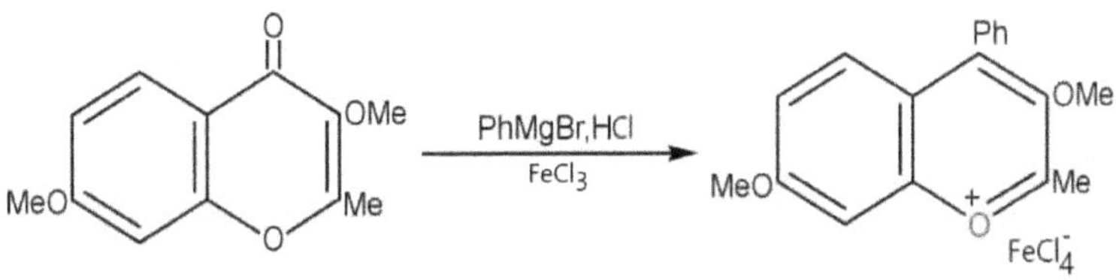

Degradation of chelidonic acid occurs in strong basic conditions thereby yielding 2,4,6-trioxoheptanedioic acid which further degrades to simple acids.

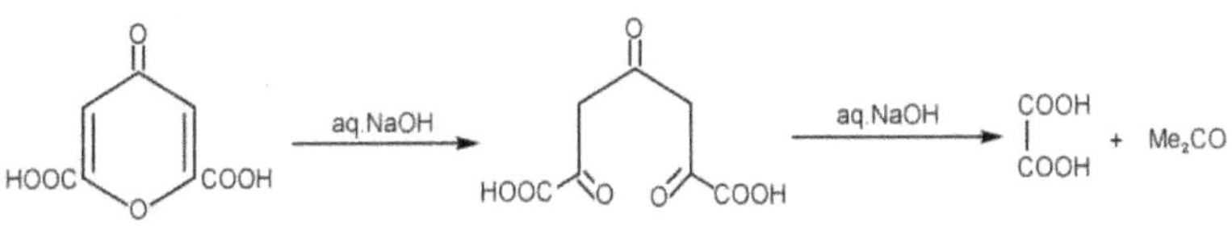

Chelidonic (R=H) acid is widely distributed in flora and hence its **decarboxylation** is useful in structural determination and giving NMR signals. Thus, double decarboxylation takes places in presence of heated Cu powder giving pyran-4-one.

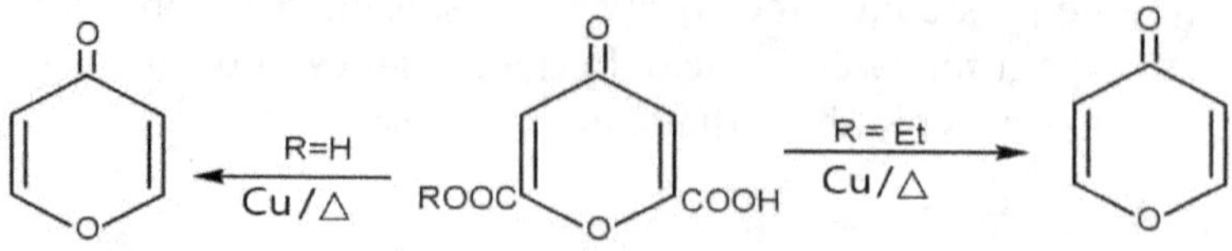

When chromone-3-carboxaldehyde reacts with arylamine, whose another amino group is at the ortho position, then dihydrotetraaza[14]annulene is obtained which can easily form complexes with Ni, Cu and Zn.

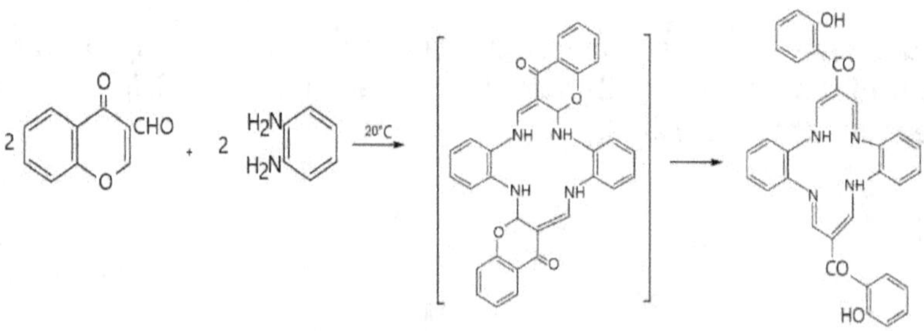

A new ring is spontaneously formed when hydrogenation of nitro group in presence of a side chain take place, like in nitro ketone, at atmospheric pressure and room temperature, dihydropyran[3,2-g] benzoxazine is formed on reduction with Raney nickel. Otherwise, under pressure and in presence of palladium charcoal, pyran[2,3-f] [1,4] benzoxazine is obtained.

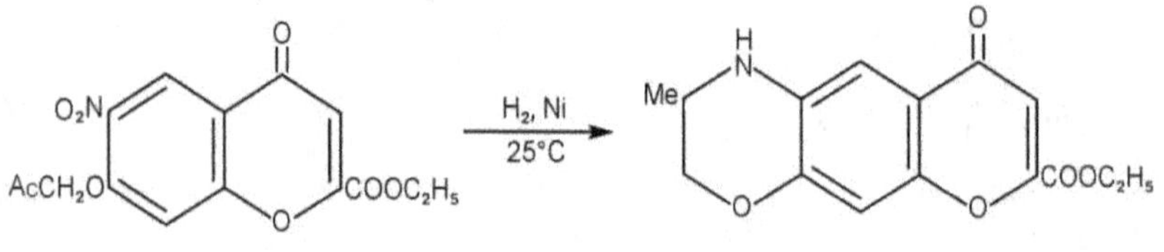

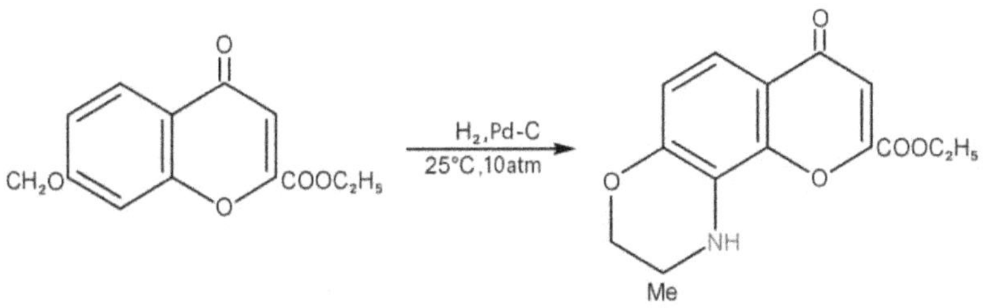

Photodegradation of dipyran occurs on irradiation however, with increase in substituents, photostability increases and hence 2,3-dihydro-2,2,5-trimethyl pyran-4-one gets slightly affected by irradiation and hence only a small amount of dimer is formed.

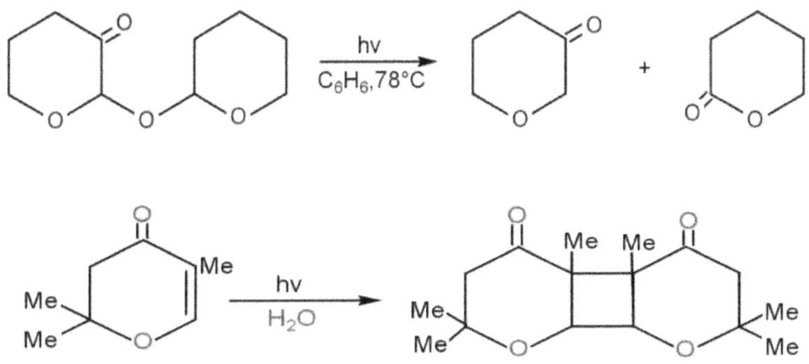

Addition across double bond takes places in dihydropyran when halogens and hydro halides are present. Hence, 3,4-dihydro-2H-pyran yields 2,3-dichlorotetrahydropyran. Formaldehyde in acid solution gives spiran.

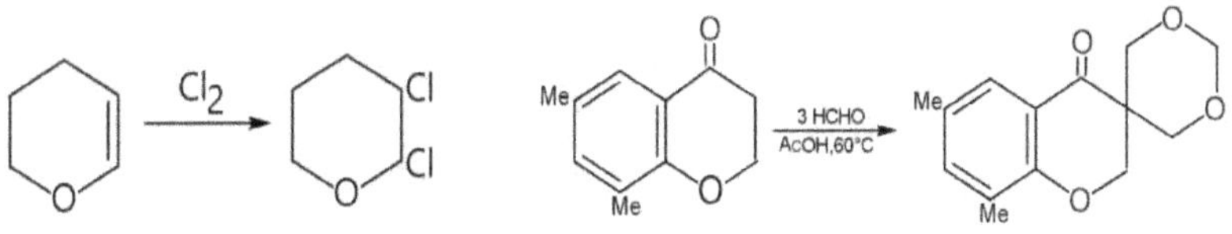

Oxidation of 2,3- dihydro-4H-pyran can be obtained by several reagents.

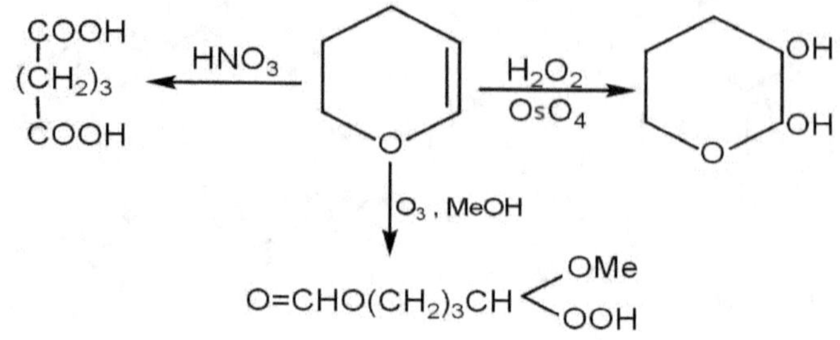

Unsymmetrical diene like 2-methylprop-1-ene, when irradiated with 3,3-dimethylpyran-2,4-dione, gives a mixture of cis- and trans-fused pyran-4-one.

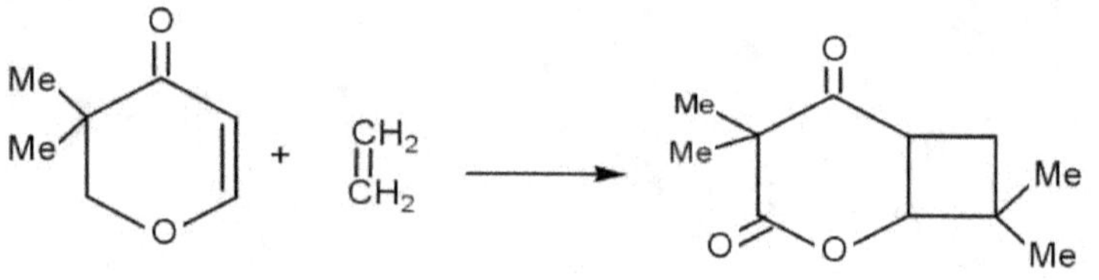

2.11.3 APPLICATIONS OF PYRANS

1. The flavone morin gives yellow shade to wool and is also used as a spot test reagent for salts of aluminium, beryllium, zinc, and other metals.

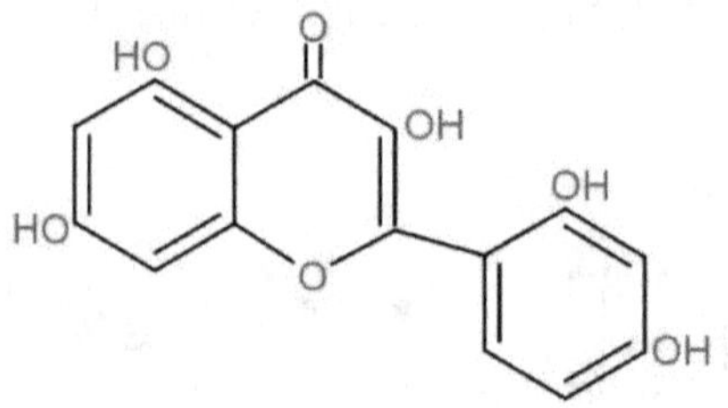

2. Eosin is used in inks, lipsticks, and nail varnish in addition to be used as a dye for silk and paper.

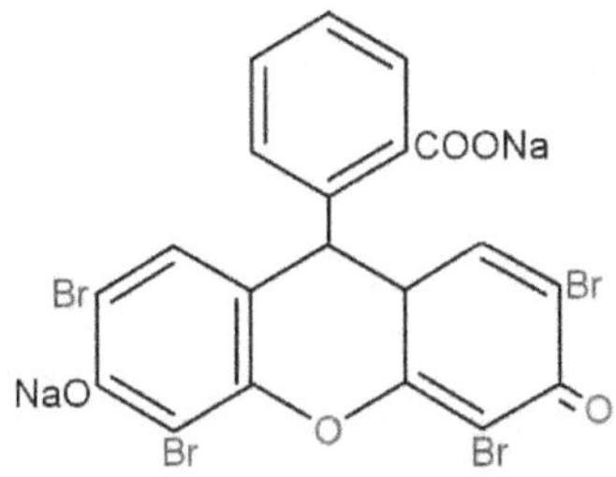

3. 2,2'-Spirobichromans are used as stabilizers in colour photographic emulsions.

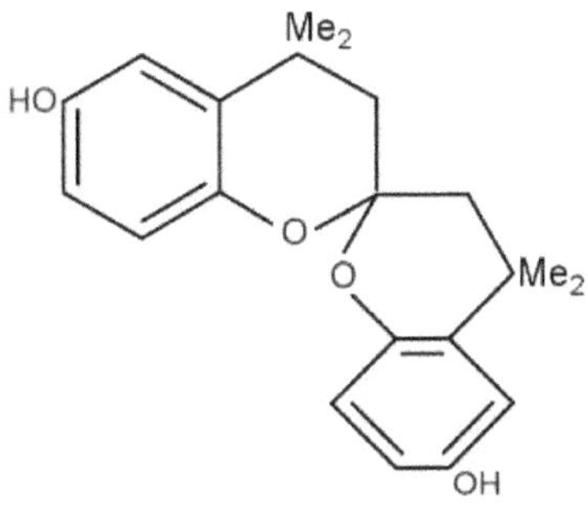

4. **Warfarin** (4-(4-hydroxy coumarin -3-yl)-4-phenylbutan-2-one) is used as rodenticide.

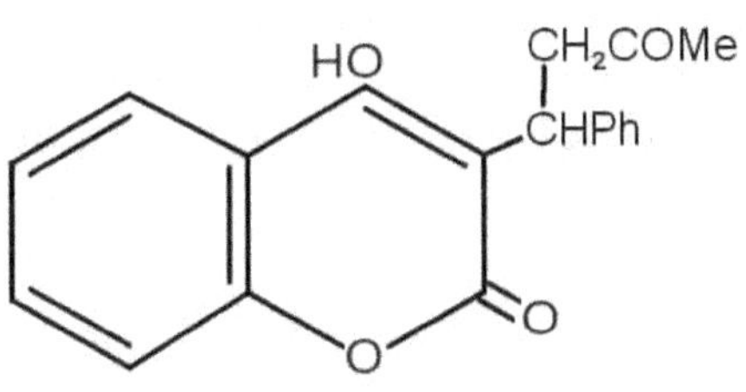

5. i. Acenocourmarol (**A**) is used as an anticoagulant.
   ii. Chromonar (**B**) and Visnadin (**C**) are used as vasodilators
   iii. Haloxon (**D**) is used as anthelmintic.
   iv. Mercumallylic acid(**E**) act as diuretic.

v. Hymecromone (**F**) acts as an insecticide especially for Colorado beetle.

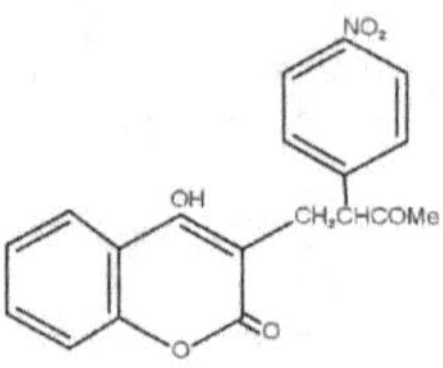

(A)

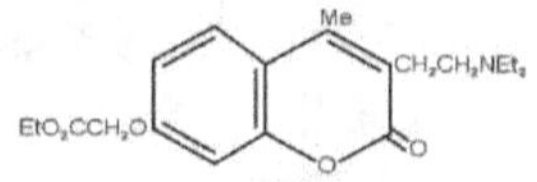

(B)

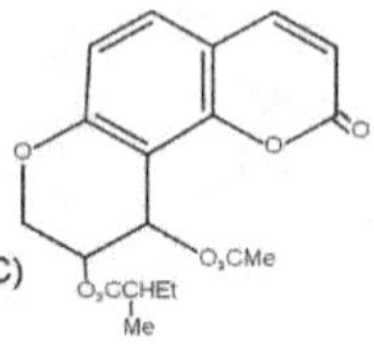

(C)

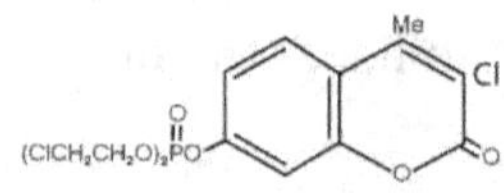

(D)

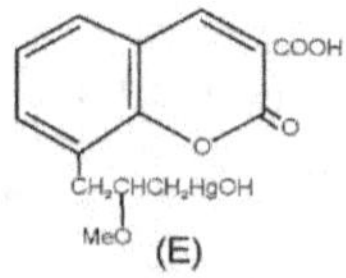

(E)

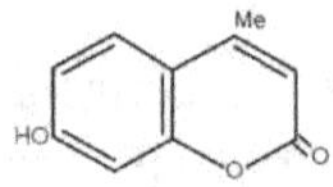

(F)

6. **Intal** (chemically disodium cromoglycate) is a successful drug for prevention of asthmatic attacks.

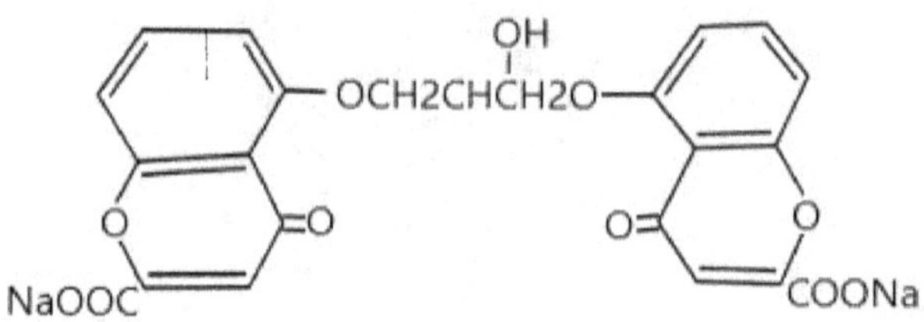

7. Flavone derivative (**G**) of 3-methyl chromone, flavoxate (Urispas) acts as antispasmodic agent to reduce smooth muscle spasm in the urinary tract. Formononetin (**H**) helps as a diuretic while Dimefline (**I**) is a respiratory stimulant.

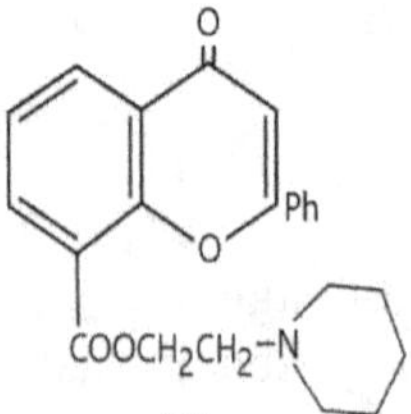

(G)

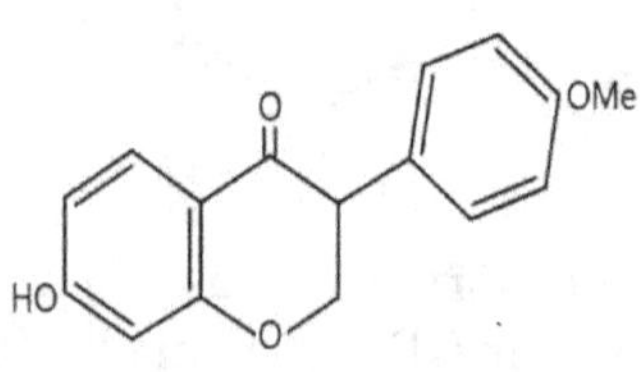

(H)

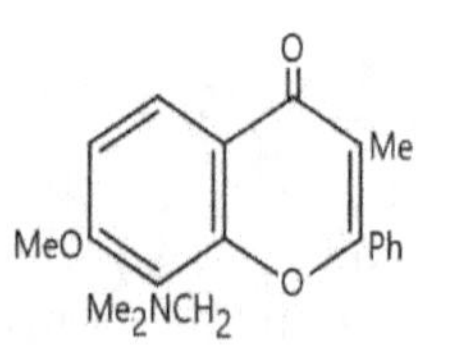

(I)

2.12 THIOPYRANS AND THEIR BENZO-DERIVATIVES

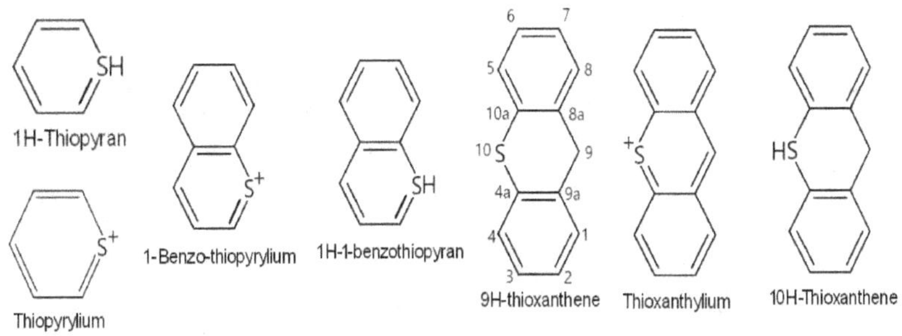

2.12.1 SYNTHESIS OF THIOPYRANS AND THEIR BENZO-DERIVATIVES

Although thiopyrans have been less extensively investigated, yet there are varied methods for their synthesis. **Cyclization** of o-thiosalicyclic esters with acrylonitrile is one such method.

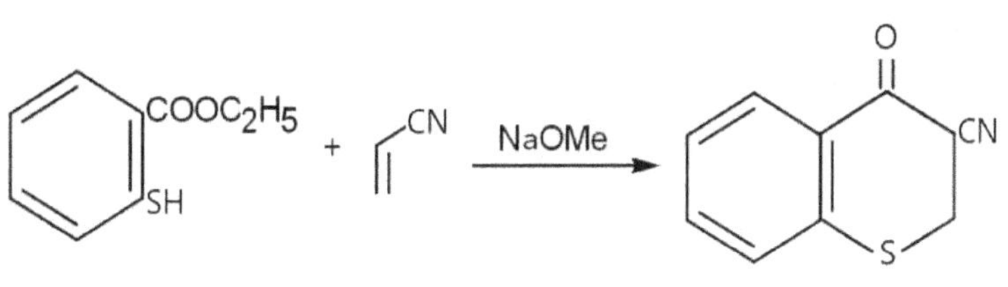

In the same way, enamine and a thio ketone derivative cyclize to form the product.

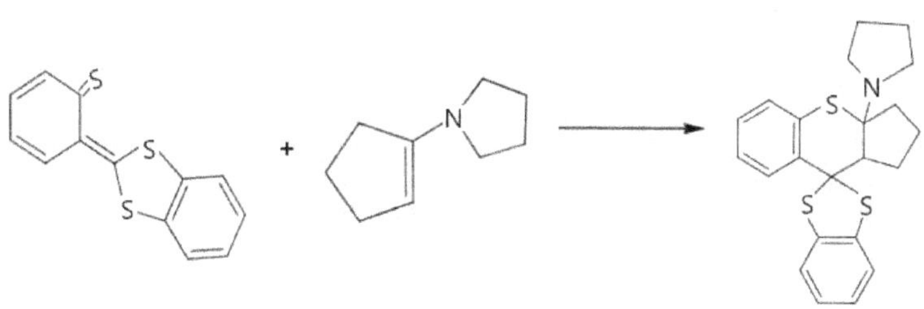

Synthesis of ring by Y-closure is a very convenient method of obtaining thiochromones and thioxanthones. This takes place with the help of intramolecular **Friedel Crafts condensation** of an aromatic ring and carboxylic acid.

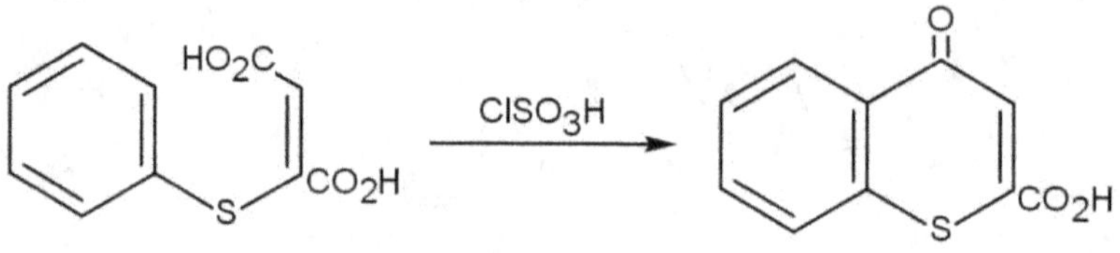

Thiopyran system can also be produced by the elimination of the amino group or through condensation.

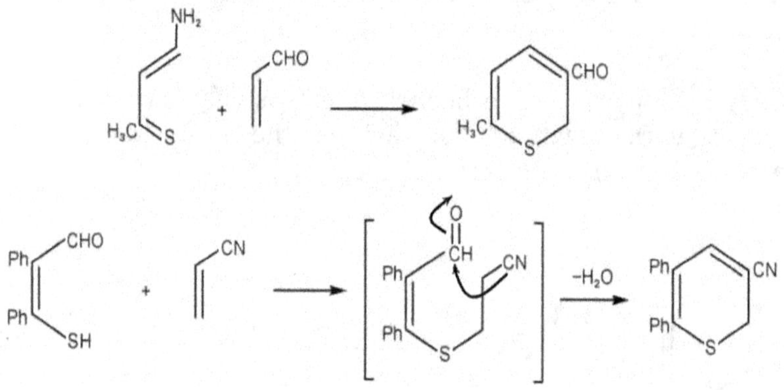

Substituted thiopyrans are prepared by reduction of thiopyrylium salts or by **elimination** reaction.

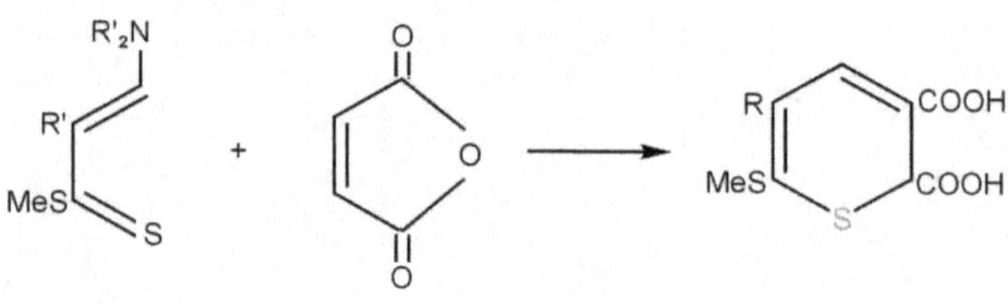

Thiopyrylium derivatives can be obtained by the action of perchlorate salt on thiopyranones.

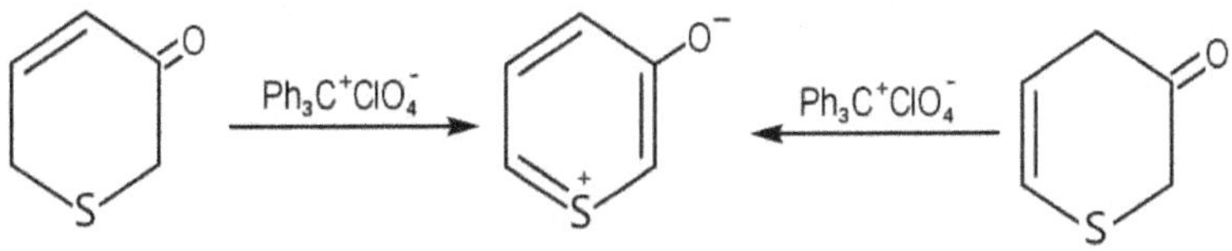

Alkylthio substituents are prepared by making cationic species.

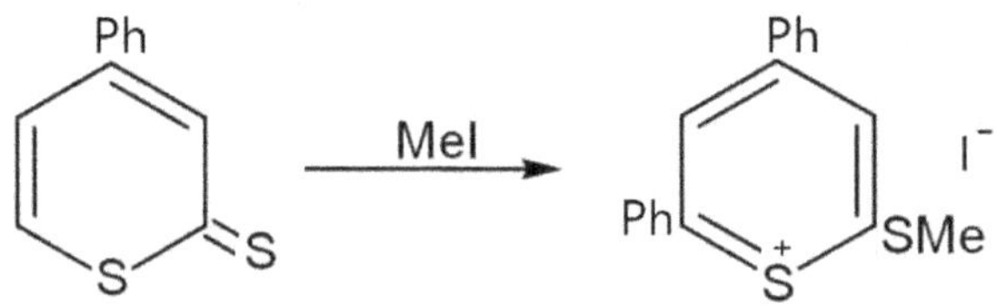

Compounds substituted with halogens are rare. They are prepared by elimination from an intermediate.

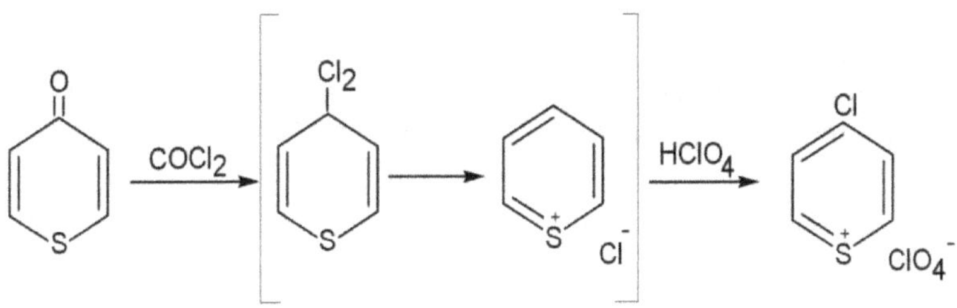

2.12.2 PROPERTIES OF THIOPYRANS AND THEIR BENZO-DERIVATIVES

Sulphur atom dominates the heterocyclic compound when electrophiles approach. The carbon attacks 5,6-dihydro-2H-thiopran and results in a product at sulphur.

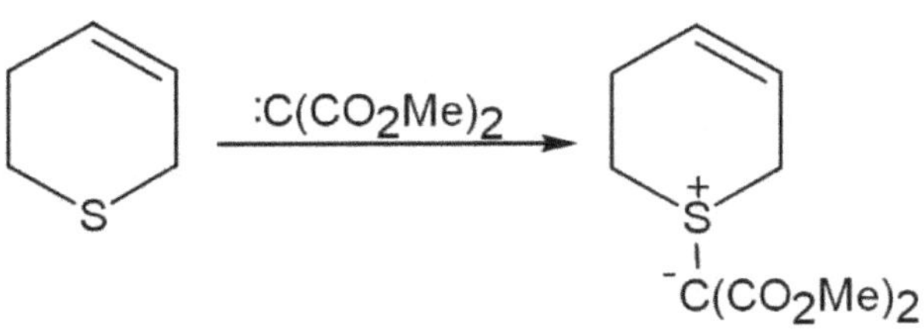

Thiopran-3-one undergoes reactions and forms fused products.

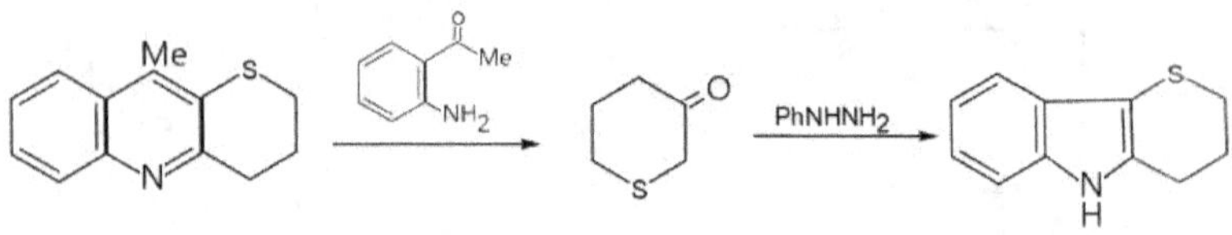

Cyclisation of certain benzo-thio compounds takes place under special conditions.

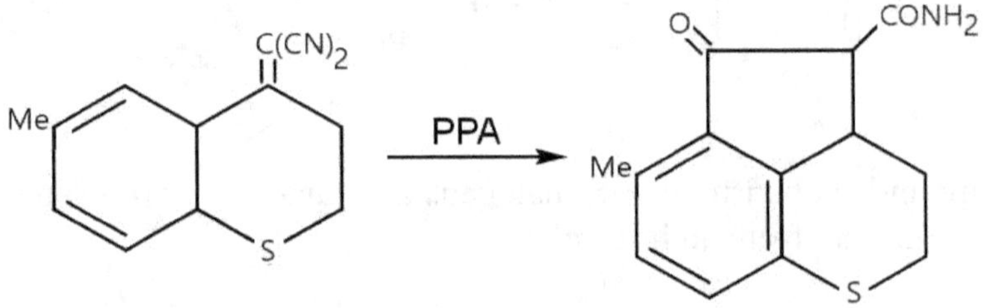

There occurs an interesting inversion in the presence of light resulting in the formation of spirothietanone.

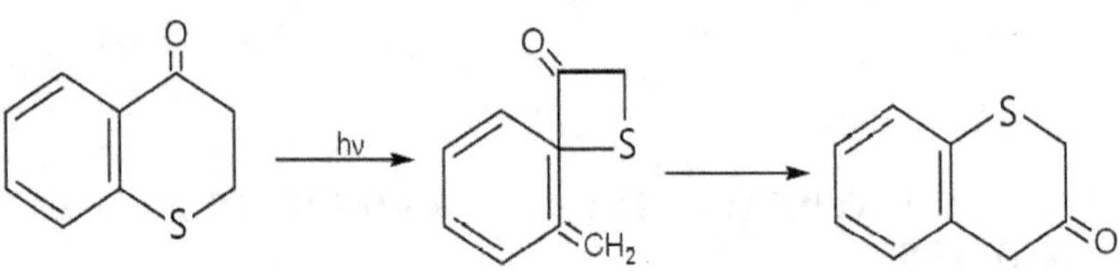

When reaction of 2H-thiopyrans is brought about with acids, it results in a mixture of thiopyrylium salt and tetrahydro thiopyran.

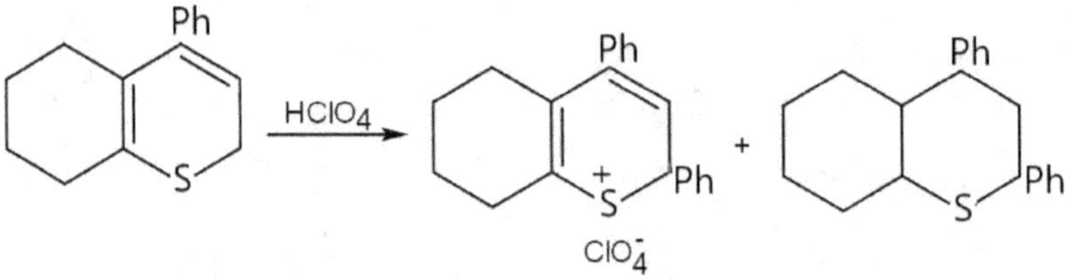

A range of some characteristic reactions of 4H-thiopryan-4-ones has been illustrated below.

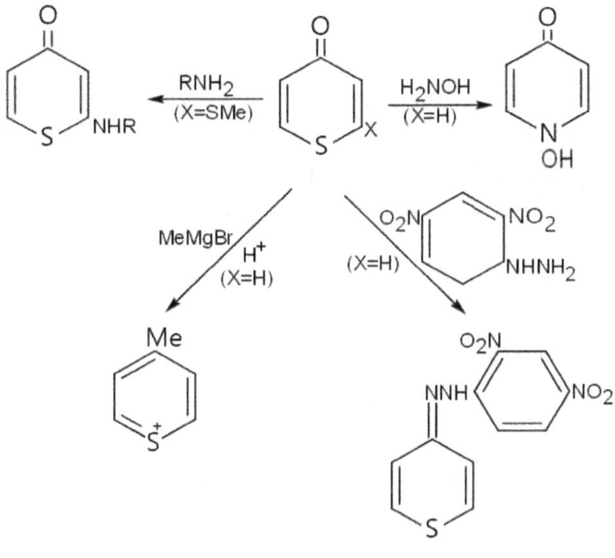

Action of azides on 2,6-diphenyl-4-thiopyran gives thiazepinone.

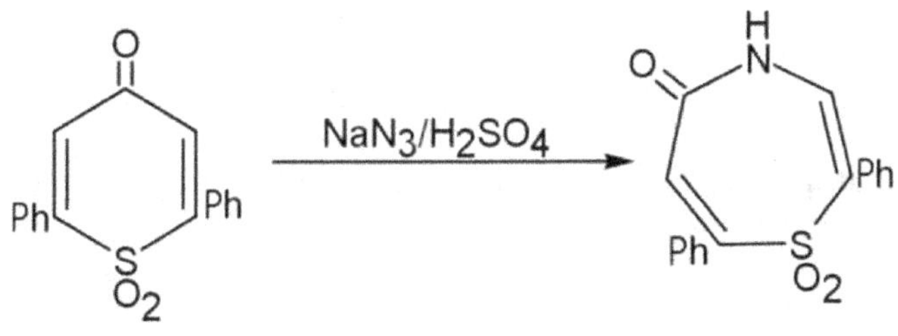

Condensation of compounds containing aldehydic group at position–3 with o-diamines occur easily.

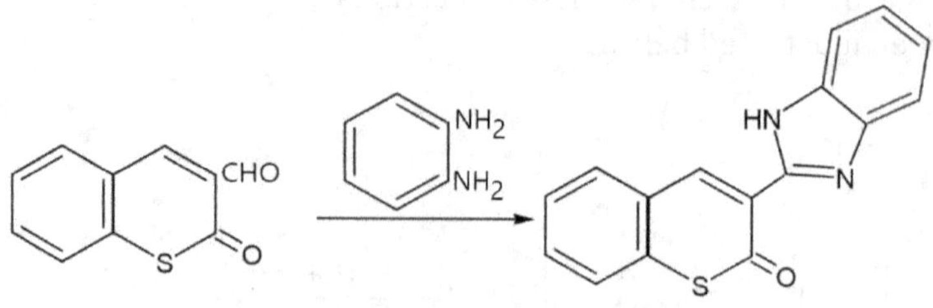

When halogen is substituted at position-4, then it acts as a leaving group and gives 4-amino, oxy or thio derivatives.

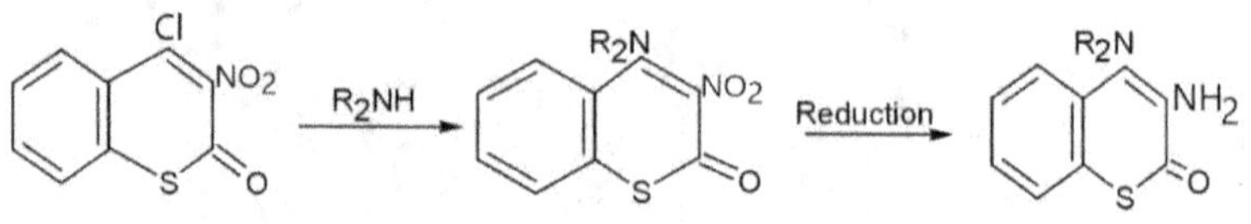

When 4-chlorothiopyrylium salts react with amines, they tend to result in ring opening.

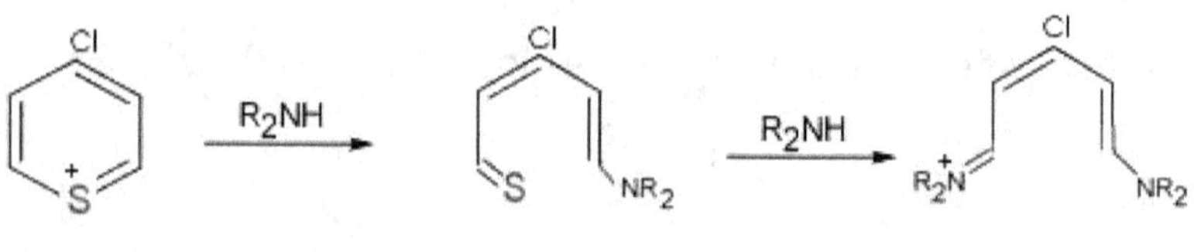

Thiopyrylium compounds, substituted with methyl groups, react with electrophiles, and give conjugated systems.

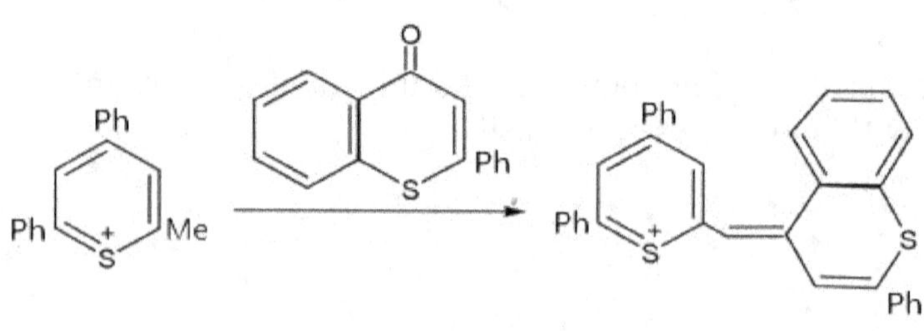

The reaction with organosilicon compound is interestingly summarized.

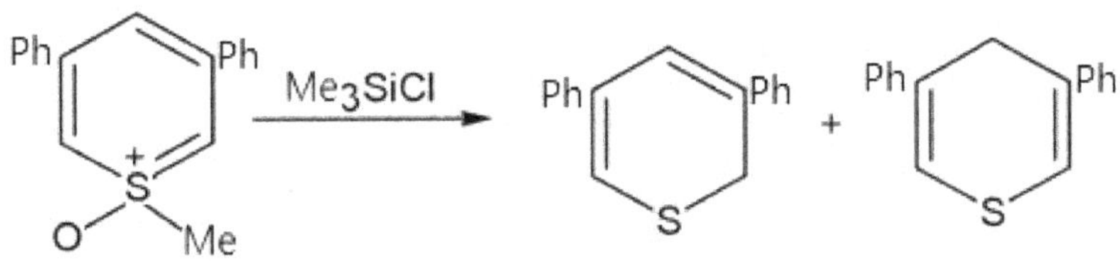

2.12.3 APPLICATIONS OF THIOPYRANS AND THEIR BENZO-DERIVATIVES

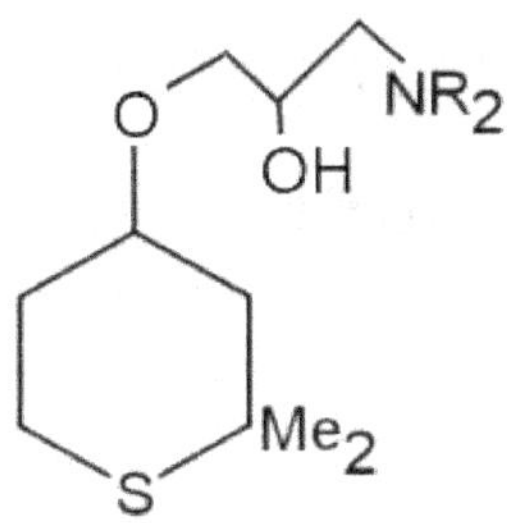

It shows cardiovascular effects.

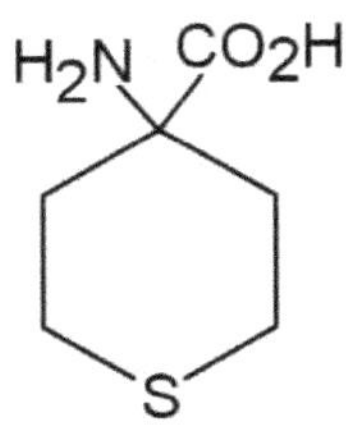

It acts as a good inhibitor of transport system in certain tumours cells.

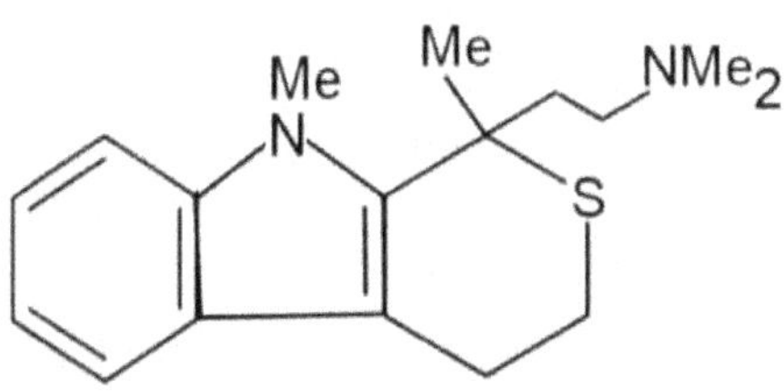

Tandamine; as a potent inhibitor of biogenic amine uptake.

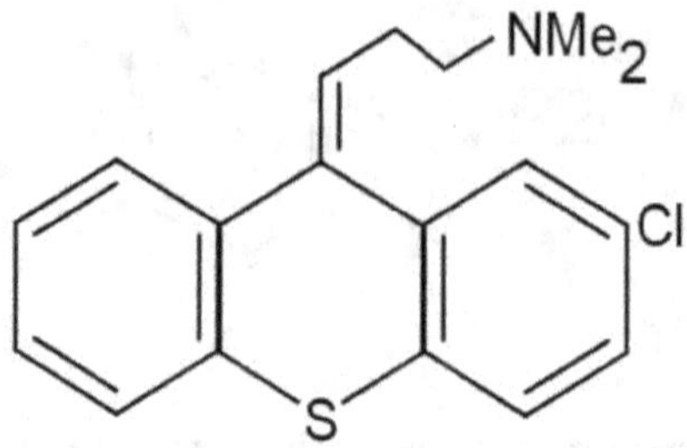

Tranquillizing drug, **Chlorprothixene**.

2.13 OXAZINE, THIAZINE AND THEIR BENZO-DERIVATIVES

The six – membered heterocyclic compounds that have nitrogen as one atom and oxygen or sulphur as the other heteroatom.

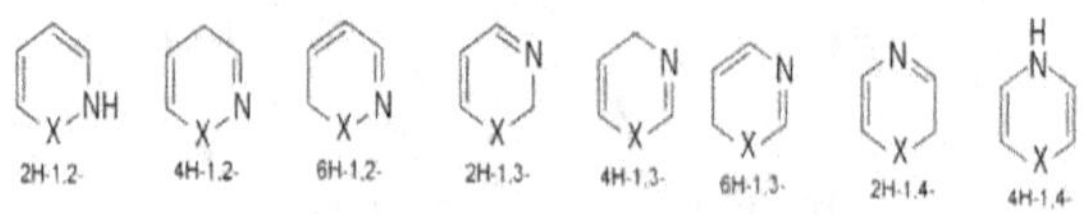

Oxazines (X = O) and Thiazines (X = S)

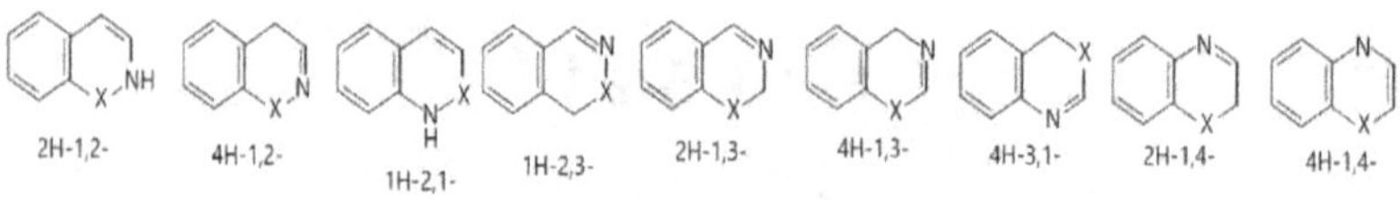

Benzoxazines (X = O) and Benzothiazines (X = S)

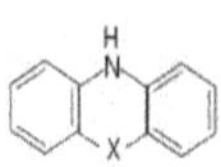

Phenoxazine (X = O) and Phenathiazine (X = S)

2.13.1 SYNTHESIS OF OXAZINE, THIAZINE AND THEIR BENZO-DERIVATIVES

Cyclization of the mono-oximes of α, β-unsaturated-1,4-dicarbonyl compounds yields 1,2- oxazines. Hence, 2,3-benzoxazin-1-one is obtained from oximes of 2-acyl benzoic acid.

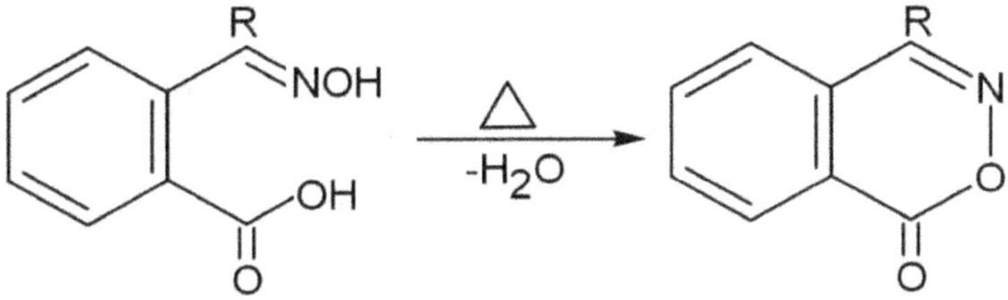

Dihydrobenzothiazinone is obtained because of **ring expansion** of 1,2-benzothiazoles on heating in presence of sodium ethoxide.

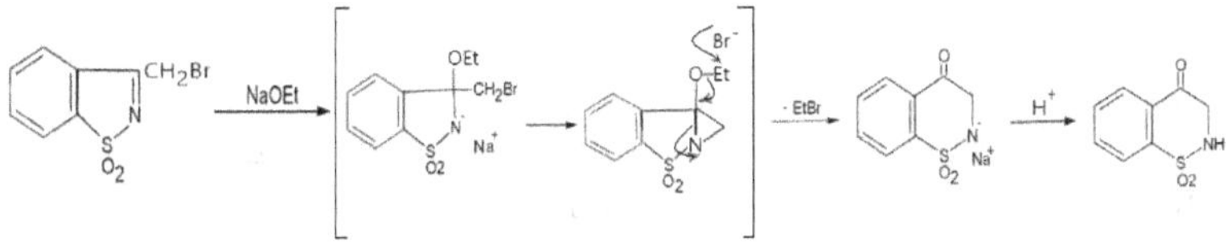

Cyclization of amide or thioamides with acidic reagents gives 4H-1,3-oxazines and 4H-1,3–thiazines.

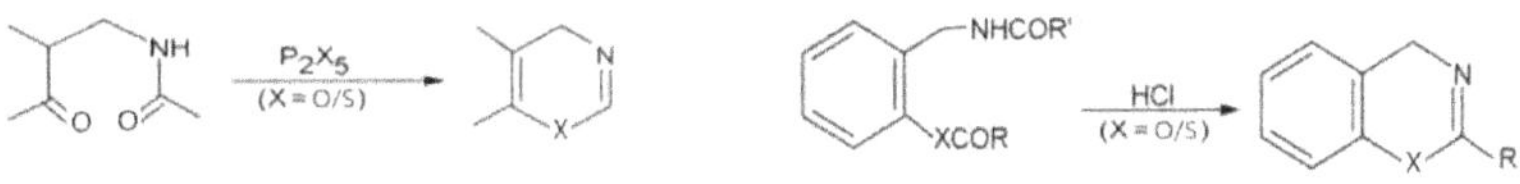

Thiazin-4-ones are obtained from 1,3-oxazinium salt by treatment of hydrogen sulphide and potassium carbonate while oxazin-4-ones can be obtained when **cycloaddition reaction** takes place between isocyanates and ketones or alkynes or between nitriles and acylketenes.

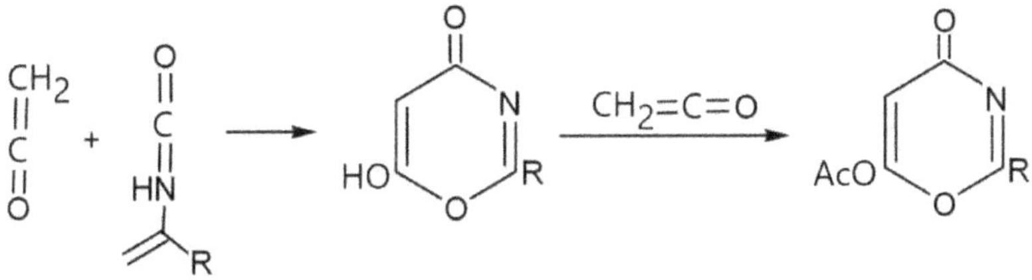

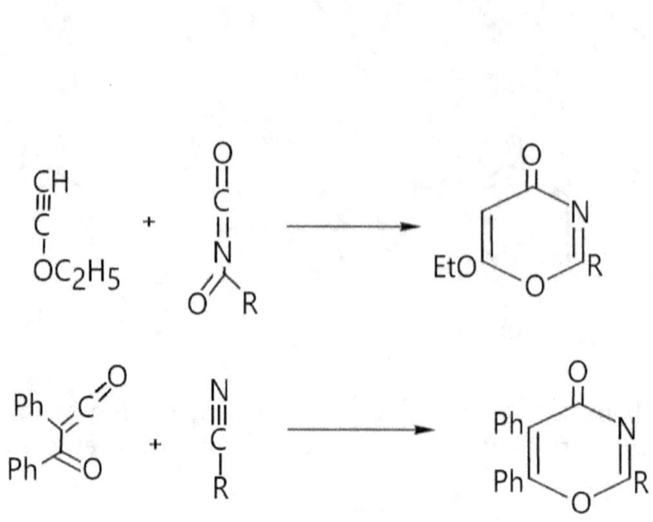

Action of potassium cyanide on isothiazolium chloride leads to ring expansion thereby forming imine derivative.

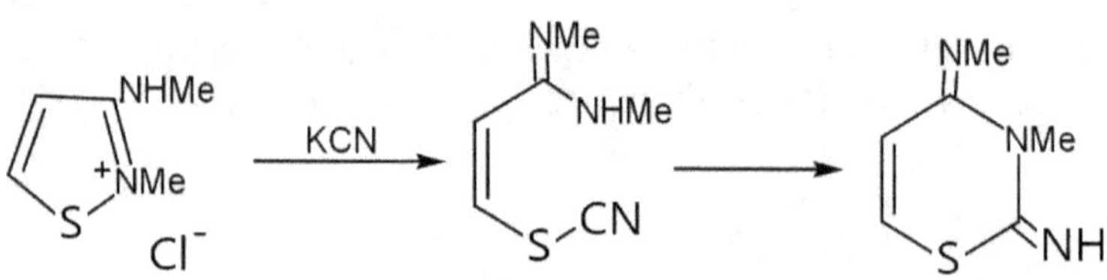

2H-1,4-Benzoxazines are produced when acetals (X = O) cyclize in acidic solutions. In the same way, **reduction** of 2-nitrophenoxyacetophenone, with Zn dust and ammonia, results in the formation of respective N-oxide variety.

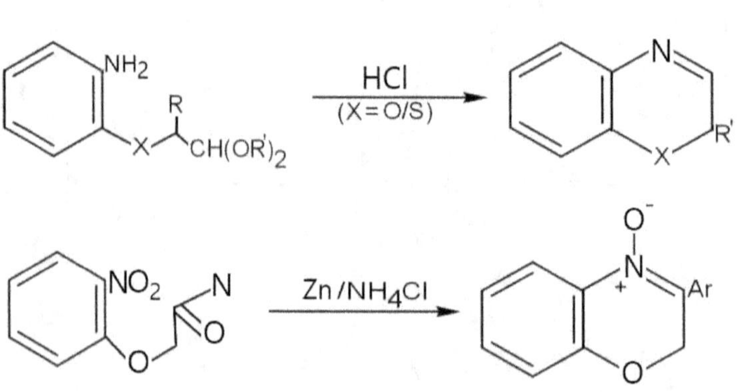

Reductive cyclohydration of 2-nitrophenxyacetic acids or their thioxy counterparts lead to the formation of 2,4-dihydro-1,4-benzoxazin-3-ones and -benzothiazin-3-ones because the reactant heterocyclic compounds have methylene group.

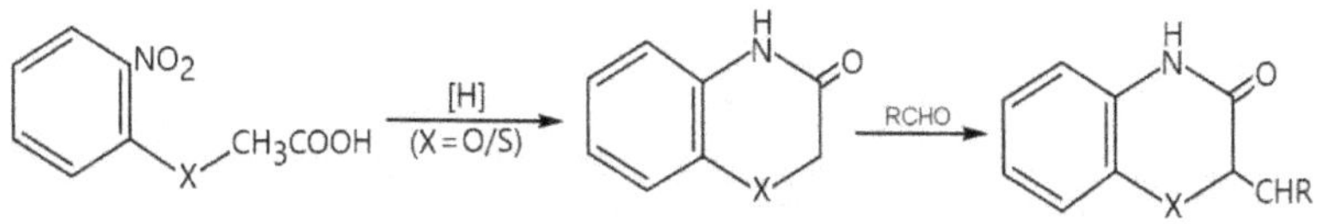

2.13.2 REACTIONS OF OXAZINE, THIAZINE AND THEIR BENZO-DERIVATIVES

Ring opening reaction takes place on exposure to aqueous base. At room temperature, 4H-1,3-benzo thiazine is very stable and do not respond to bases, but at high temperatures heterocyclic ring is destroyed.

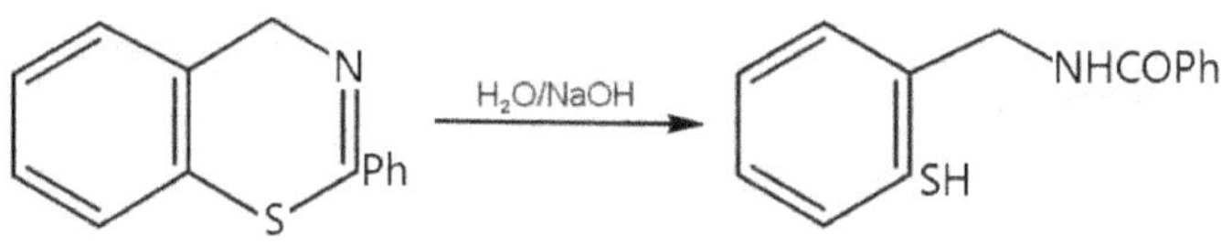

If both carbonyl and thionyl functional groups are part of the ring, then primary amines form imino derivatives as thiocarbonyl sites are selectively attacked, however, reaction at position-2 is observed is dithioxo compound.

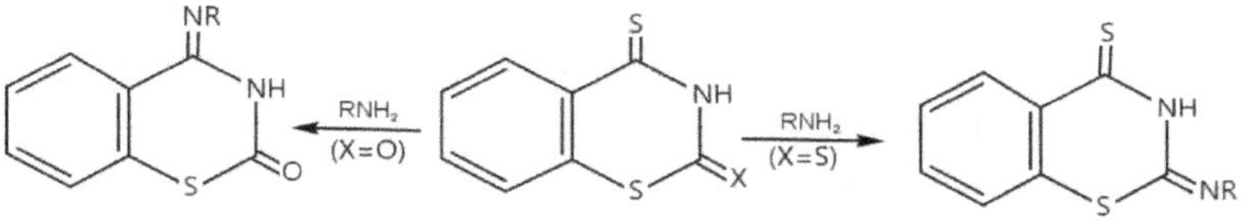

Imine bond of dihydro-1,3-oxazines reduce to tetra hydro derivative in the presence of sodium borohydride. In the same way, dihydro-1,3-benzo thiazinones gets reduced in the presence of lithium aluminium hydride to form dihydrobenzothiazines.

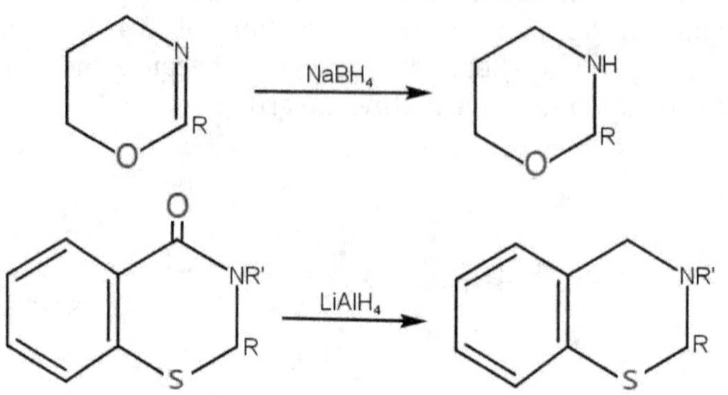

The **addition reaction** gives bicyclic adducts when oxiranes react with the imine bond of 5,6-dihydro-4H-1,3-oxazines, while ketenes give cepham varieties.

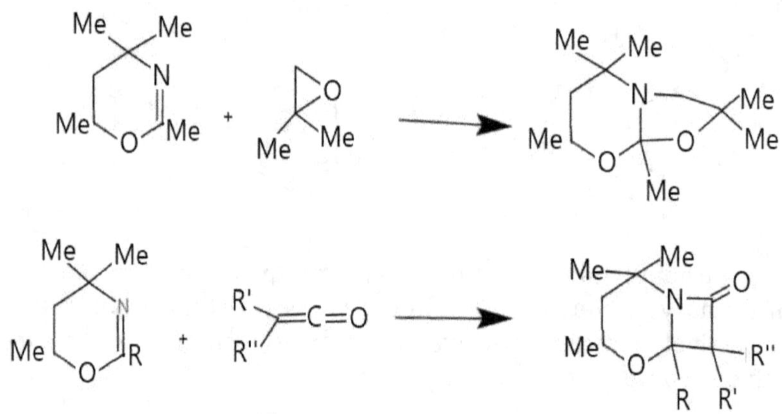

Sodium in liquid ammonia brings about **cleavage** of dihydro-1,4-thiazinones. Dihydro-1,4-oxazinones, on catalytic hydrogenation, gives tetra hydro-1,4-oxazin-2-ones.

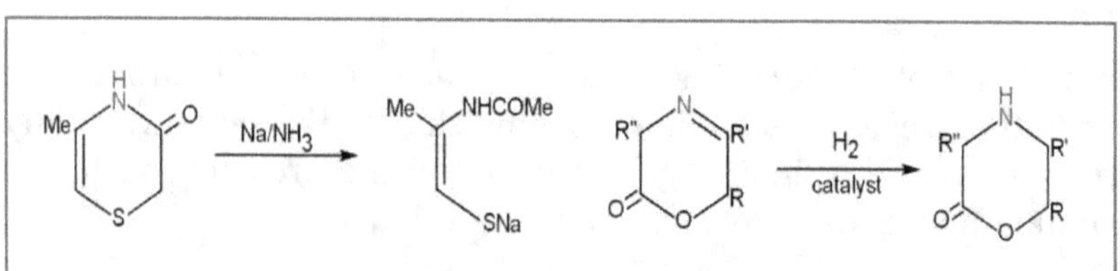

REFERENCES

1. J. A. Moore and E. C. Capaldi (1964); J. Org. Chem., **29**, 2960.
2. O. E. Edwards and T. Singh (1954); Can. J. Chem., **32**, 683.
3. T. Kametani, K. Nyu and T. Yamanaka (1971); Chem. Pharm. Bull., **19**, 1321.
4. R. T. Coutts (1969); J. Chem. Soc. (C), 713.
5. Y. Ahmad and S. A. Shamsi (1996); Bull. Chem. Soc. Jpn., **39**, 195.
6. C. W. Muth, N. Abraham, M. L. Linfield, R. B. Wotring and E. A. Pacovsky (1980); J. Org. Chem., **25**, 736.
7. F. Elay and A. Deryckere (2000); J. Heterocycl. Chem., **7**, 1191.
8. M. W. Gittos, M. R. Robinson, J. P. Verge, R. V. Dories, B. Iddon and H. Suschitzky (1986); J. Chem. Soc., Perkin Trans. 1, **33**, 196.
9. A. Moheen, M. E. Qmar and S. Yamada (1966); Chem. Pharm. Bull., **14**, 842.
10. G. Gast, J. Schmutz and D. Sorg (1977); Helv. Chim. Acta, **60**, 1644.
11. E. Ziegler, H. Sterk and W. Sterger (1991); Monatch. Chem., **101**, 762.
12. H. C. Van der Plas, H. Jongejan, G. Guertsen and M. C. Vollering (1971); Recl. Trav. Chim. Pays–Bas., **90**, 1246.
13. T. Tezuka, O. Seshimoto and T. Mukai (1975); Tetrahedron Lett., **5**, 1067.
14. C. Kaneko, I. Yokoe and M. Ishikawa (1967); Tetrahedron Lett., **7**, 5237.
15. O. R. Rodig (1974); Chem. Heterocycl. Compd., **14**(1), 309.
16. J. A. Zoltewicz and L. W. Dcady (1978); Adv. Heterocycl. Chem., **22**, 71.
17. Y. Okamoto and Y. Shimakawa (1980); J. Org. chem., **35**, 3752.
18. T. Asano and W. J. le Noble (1978); Chem. Rev., **78**, 407.
19. A. R. Katritzky, H. O. Tarhan and S. Tarhan (1970); J. Chem. Soc.(B), 114.
20. R. C. De Selms (1968); J. Org. Chem., **33**, 478.
21. T. J. Kress, I. L. Moore and S. M. Costantino (1976); J. Org. Chem., **41**, 93.
22. K. M. Dyumaev and L. D. Smirnor (1995); Russ. Chem. Rev. (Engl.) Transl., **44**, 823.

23. M. L. Ash and R. G. Pews (1981); J. Heterocycl. Chem., **18**, 939.
24. J. Rokach and Y. Girard (1978); J. Heterocycl. Chem., **15**, 683.
25. A. F. Pozharskii, A. M. Simonov and V. N. Doron'kin (1978); Russ. Chem. Rev. (Engl. Transl.), **47**, 1042.
26. H. J. N. Van den Haak, H. C. Van der Plas and B. van Veldhuizen (1981); J. Heterocycl. Chem., **18**, 1349.
27. R. E. Lyle and E. White, V (1971); J. Org. Chem., **36**, 772.
28. F. Minisci, R. Bermardi, F. Bertini, R. Galli and M. Perchinummo (1971); Tetrahedron, **27**, 3575.
29. L. D. Smirnov, N. A. Andronova , V. P. Lerina and K. M. Dyumaev (1972); lzr. Akad. Nauk SSSR, Ser. Khim., 457 (Chem. Abstr., **77**, 34, 278).
30. R. K. Smalley (1977); Chem. Heterocycl. Compd., **32(1)**, 526.
31. E. Lebenstedt and W. Schunack (1975); Arch. Pham. (Weinheim, Ger.), **308**, 413.
32. J. D. V. Hanotier and M. Hanotier–Bridoux (1974); U. S. Pat. 3. **829**. 432(Chem. Abstr., 1975, 82, 43, 197)
33. F. W. Vierhapper and E. L. Eliel (1975); J. Org. Chem., **40**, 2729.
34. M. Honel and F. W. Vierhapper (1980); J. Chem. Soc., Perkin Trans. **1**., 1933.
35. E. Ager, B. Iddon and H. Suschitzky (1970); J. Chem. Soc. (C), 1530.
36. G. Odham and E. Stenhagen (1971); Acc. Chem. Res., **4**, 12.
37. A. Kleeman (1977); Chem. -Ztg., **101**, 389.
38. B. Kaminsky and R. I. Meltzer (1976); J. Med. Chem., **11**, 160.
39. E. H. Wiseman, Y. H. Chang and J. G. Lombardino (1976); Arzniem. Forsch., **26**, 1300.
40. I. Antonini, F. Claudi, G. Cristalli, P. Franchetti, M. Grifantini and S. Martelli (1981); J. Med. Chem., **24**, 1181.
41. F. Bossert, H. Meyer and E. Wehibger (1981); Angew. Chem. Int. Ed. Engl, **20**, 762.
42. H. Junch, A. Hermetter and H. Fisher-Colbrie (1976); chem. Ber., **109**, 1787.
43. L. Distefano and R. N. Castel (1968); J. Hetrocycl. chem., **5**, 111.
44. G. A. Reynolds, J. A. Van Allan and J. F. Tinker (1969); J. Org. Chem., **24**, 1205.
45. B. Stanovnik, M. Tisler, J. Bradac, B. Budic, B. Koren and B. Mozetic–Rescic (1979); Heterocycles, **12**, 457.

46. S. Gelin (1979); J. Org. Chem., **44**, 3053.
47. W. L. F. Armarego (1961); J. Chem. Soc., **2**, 677.
48. H. B. Redereeh, F. Effenberger and H. J. Treiber (1963); Chem. Ber., **96**, 1505.
49. K. Folkers and T. B. Johnson (1933); J. Am. Chem. Soc., **55**, 3361.
50. V. D. Alam and R. C. Anderson (1974); Synthesis, **9**, 286.
51. D. J. Brown (1970); J. Soc. Chem. Ind., **69**, 353.
52. L. Doub and V. Krolls (1970); J. Heteroycl. Chem., **7**, 527.
53. V. I. Gunar, L. F. Ovechkina, l. A. Mikhailopulo and S. I. Zav'yalov (1968); Izv. Akad. Nauk SSSR, Ser. Khim., **1**, 918.
54. H. G. Peer and A. Van der Heijden (1969); Recl. Trav. Chim. Pays-Bas, **88**, 1335.
55. Y. Ohtsuka (1999); J. Org. Chem., **44**, 827.
56. O. S. Wolfbeis (1977); Synthesis, **7**, 136.
57. E. J. J. Grabowshi, E. W. Tristram, R. Tull and P.I. Pollak (1968); Tetrahedron Lett., **8**, 5931.
58. C. E. Mixan and R. C. Pews (1977); J. Org. Chem., **42**, 1869.
59. H. M. Liebich, D. R. Donglas, A. Zlatkis, F. Mueggler-charan and A. Donzel (1972); J. Agric. Food Chem., **20**, 96.
60. K. Watanabe and Y. Sato (1971); J. Agric. Food Chem., **19**, 1017.
61. J. P. Walradt, A. O. Pittet, T. E. Linkin, R. Muralidhara and A. Sanderson (2001); J. Agric. Food Chem., **19**, 972.
62. A. Ohsawa, H. Ortrai, H. Ohrishi and H. Igeta (1981); J. Chem. Soc. Chem. Commun., **1**, 1174.
63. E. Banberger and A. von Gold Berger (1898); Ber., **31**, 2636.
64. H. Neurnhoffer and P. F. Wiley (1978); Chem. Heterocycl. Compd., **33**, 91.
65. U. Golik and W. Taub (1975); J. Heterocycl. Chem, **12**, 1155.
66. B. Stanovaik and M. Tisler (1971); J. Heterocycl. Chem., **8**, 785.
67. G. Ege, P. Arnold, E. Beisiegel and I. Lehrer (1976); Liebigs Ann. Chem, **7**, 946.
68. D. H. Hey, C. W. Ress and A. R. Todd (1962); Chem. Ind. (London), **2**, 1332.
69. A. J. Barker, T. McC. Patersen, R. K. Smalley and H. Suschitz-ky (1979); J. chem. Soc., Perkin Trans **1**, 2203.
70. H. Neurloffer and P. F. Wiley (1978); Chem. Heterocycl. Compd., **33**, 58.

71. A. Mustafa, A. K. Mansour and H. A. A. Zaher (1971); J. Prakt. Chem., **313**, 699.
72. A. C. Lovelitte (1979); J. Heterocycl. Chem., **16**, 555.
73. V. V. Dovlatyan and A. V. Dovlatyan (1980); Khim. Geterotsikl. Soedin., 411(Chem. Abstr., **93**, 94 549).
74. G. A. Laughran, G. F. L. Ehlers and J. L. Burkett (1966); J. Heterocycl. Chem., **3**, 137.
75. H. H. Takimats and G. C. Denautt (1996); Tetrahedron Lett., **6**, 5369.
76. C. Grundamann and A. Kreutzberger (1957); J. Am. Chem. Soc., **79**, 2839.
77. I. Ya. Postovskii and V. A. Ershav (1971); Khim. Getrotsikl. Soedin., **7**, 708.
78. P. R. Wiley (2000); Chem. Heterocycl. Compd., **33**, 1094.
79. P. F. Wiley (1978); Chem. Heterocycl. Compd., **33**, 1152.
80. D. D. Libman and R. Slack (1956); J. Chem. Soc., **5**, 2253.
81. E. E. Glover and G. Jones (1959); J. Chem. Soc., **9**, 1686.
82. J. M. Vierfond, Y. Mettey, R. Joubin and M. Miocque (1979); J. Heterocycl. Chem., **16**, 753.
83. C. K. Bradsher and D. A. Hunt (); J. Org. Chem., **45**, 4248.
84. O. Westphal (1967); Swiss Pat. 417 601(Chem. Abstr., **67**, 60 695).
85. D. D. Chapman (1975); J. Chem. Soc. Chem. Commun., **19**, 489.
86. G. M. Sanders, M. Van Dijk and H. C. Van der Plas (2008); Heterocyc., **15**, 213.
87. A. Fozard, L. S. Davies and C. K. Bradsher (1971); J. Chem. Soc.(C), **97**, 3650.
88. T. L. Houge and G. Jones (1986); J. Chem. Soc.(C), **68**, 1088.
89. D. Farquhar, T. T. Gogue and D. Leaver (1976); J. Chem. Soc., Perkin Trans **1**, 341.
90. A. Forzard and G. Jones (1964); J. Chem. Soc., **4**, 2760.
91. E. M. Hawes, D. K. J Gorecki and D. D. Johnson (1973); J. Med. Chem., **16**, 849.
92. E. M. Hawes, D. K. J Gorecki and D. D. Johnson (1974); J. Heterocycl. Chem., **11**, 151.
93. A. Decormeille, G. Queguiner and P. Pastour (1975); C. R. Hebd. Seances Acad. Sci. Ser. C, **280**, 381.
94. V. G. Granik, E. O. Sochnera and N. P. Solov'era (1980); Khim. Geterotsikl. Seodin., 416.

95. V. G. Granik, E. O Sochnera and N. P. Solov'era, E. F. Kuleshora and O. S. Anisimora (2009); Khim. Geterotsikl. Soedin., **80**, 1120.
96. Y. G. Gravik, A. M. 2hidkova and R. A. Dubinshii (1982); Khim. Geterotsikl. Soedin., **2**, 518.
97. R. Tan and A. Taurins (1966); Tetrahedron Lett., 1233.
98. F. Alhaique, F. M. Riccieri and E. Santucci (1975); Gazz. Chem. Ital., **105**, 1001.
99. Y. Sato, T. Iwashige and T. Miyadera (1962); Jpn. Pat. 62 4149 (Chem. Abstr.; **59**, 2823).
100. Y. Tamura, L. C. Chen, M. Fujita and Y. Kita (1982); Chem. Pharm. Bull., **30**, 1257.
101. W. Roszkiewicz and M. Wozniak (1976); Synthesis, 691.
102. R. Herbert and D. G. Wibberley (1969); J. Chem. Soc.(C), **16**, 1505.
103. F. Krohnke, K. Ellegast and E. Bertram (1956); Liebigs Ann. Chem., **600**, 198.
104. B. Stanovnik and M. Tisler (1978); Vestn. Slov. Kem. Drus., **25**, 159 (Chem. Abstr., 89, 129, 364).
105. W. Sliwa (1978); Pr. Nauk. Inst. Chem. Org. Fiz. Politech. Wroclaw, **13**, 1.
106. A. I. Mikhalev and M. E. Konshin (1977); Khim. Geterotsikl. Soedin., **7**, 1241.
107. S. Von Niementowski and E. Sucharda (2006); J. Prakt. Chem., **94**, 193.
108. E. V. Brown (1965); J. Org. Chem., **30**, 1607.
109. H. C. Van der Plas and M. Wozniak (1976); J. Heterocycl. Chem., **13**, 961.
110. W. Czuba and M. Wozniak (1993); Rocz. Chem., **47**, 2361.
111. E. V. Brown and S. R. Mitchell (1975); J. Org. Chem., **40**, 660.
112. E. Eichler, C. S. Rooney and H. W. R. Williams (1986); J. Heterocycl. Chem., **13**, 43.
113. W. Czuba and T. Kowalska (1979); Zesz. Nauk. Uniw. Jagiellon. Pr. Chem., **24**, 13.
114. D. R. Buckle, B. C. C. Cantello, H. Smith and B. A. Spicer (1975); J. Med. Chem., **18**, 726.
115. H. C. Van der plas, M. Wozniak and H. J. W. van der Haak (1983); Adv. Heterocycl. Chem., **33**, 95.
116. W. W. Paudler and T. J. Kress (1990); Adv. Heterocycl. Chem., **11**, 123.
117. R. A. Bowie (1970); Chem. Commun., **19**, 565.

118. R. A. Bowie, M. J. C. Mullan and J. F. Unsworth (1972); J. Chem. Soc., Perkin Trans. **1**, 1106.
119. R. A. Bowie and B. Wright (1975); J. Chem. Soc., Perkin Trans 1, **5**, 1109.
120. Y. Hamada, M. Sugiura and M. Hirota (1978); Yakugaku Zasshi, **98**, 1361.
121. Y. Hamada and K. Shigemura (1979); Yakugaku Zasshi, **99**, 1225.
122. Y. Hamada, I. Takeuchi and M. Hirota (1971); Chem. Pharm. Bull., **19**, 1751.
123. J. W. Bunting (1874); J. Chem. Soc., Perkin Trans. **1**, 1833.
124. J. W. Bunting and N. G. Meathrd (1972); Can. J. Chem., **50**, 917.
125. W. Czuba and J. A. Bajgrowicz (1989); Bull. Acad. Pol. Sci. Ser. Sci. Chem., **27**, 571.
126. S. Carboni, A. Da Settimo and I. Tonetti (1970); J. Heterocycl. Chem., **7**, 875.
127. X. Y. Zhung, Y. Xia, F. H. Gao, H. Z. Guo and C. Chen (1979); Yao Hsueh. Hsuch. Pao, 1979, **14**, 736.
128. A. Shiozawa, Y. Ishikawa, M. Ichikawa, H. Miyazaki and H. Yamakana (1982); Fr. Pat., **2**, 492, 825.
129. R. M. Tithrova, A. S. Elina, E. N. Padeiskaya and L. M. Polukhina (1975); Khim. Form. Zh., **9**, 10.
130. B. Stanovnik, M. Tister, V. Golob, I. Hvala and O.Nikoli (1980); J. Heterocycl. Chem., **17**, 733.
131. K. E. Schulte, V. von. Weissenborn and G. L. Tittel (1970); Chem. Ber., **103**, 1250.
132. J. Bailey and J. A. Elvidge (1973); J. Chem. Soc., Perkin Trans. **1**, 823.
133. J. A. Bristol and R. G. Lovey (1980); J. Org. Chem., **45**, 1918.
134. V. A. Chuiguk and N. N. Vlasova (1977); Chem. Geterotsikl. Soedin., **77**, 1484.
135. H. Yuki and Y. Naka (1976); (Yoshitomi Pharm. Ind. Ltd.), Jpn. Kokai **76**, 139 633.
136. A. Srinivasan, P. E. Fagerness and A. D. Broom (1978); J. Org. Chem., **43**, 828.
137. E. M. Levine, C. K. Chu and T. J. Bordos (1997); J. Heterocycl. Chem., **14**, 611.
138. G. Evans and P. Caluwe (1979); Macromolecules, **12**, 803. (Chem. Abstr., 1979, 91, 158 125)

139. K. Noda, A. Nakagawa, Y. Nakajima and H. lde (1977); (Hisamitsu Pharm. Co. lnc.), Jpn. Kokai. **77** 85 194.
140. I. Maeba and R. N. Castle (1989); J. Heterocycl. Chem., **16**, 249.
141. M. Kramberger, B. Stanovnik and M. Tisler (1972); Croat. Chem. Acta, **44**, 419.
142. M. G. DcAmezua, M. Lora-Tamayo, J. L. Soto and E. O. Toro (1990); An. Quim., 66, 561.
143. D. Schelz (2008); Helv. Chim. Acta., **61**, 2452.
144. E. C. Taylor, Jr (1990); in' Chemistry and Biology of Ptredine', ed. k. Iwai, M. Akino, M. Goto and Y. Iwanami; International Academic Printing Co., Tokyo, **79**.
145. J. Weinstock, R. Y. Dunoff, B. Sutton, B. Trost, J. Kirkpatrick, F. Farina and A. S. Straub (1968); J. Med. Chem., **11**, 549.
146. T. S. Osdene, in' Pteridine Chemistry' ed. W. Pfliederer and G. C. Taylor (1964); Pergamon, Oxford, **65**.
147. J. Pachter in' Pteridine Chemistry' ed. W. Pfliederer and G. C. Taylor (1974); Pergamon, Oxford, **47**.
148. A. Stuart, D. W. West and H. C. S. Wood (2004); J. Chem. Soc., **4**, 4769.
149. C. Schopf and K. H. Ganshirt (1962); Angew. Chem. Ed. Engl., **1**, 115.
150. M. Bohme (1981); Ph.D. Thesis, University of Konstane.
151. W. Pfleider, E. Liedek and M. Rukwied, Chem. Ber., **95**, 803.
152. G. Konrad and W. Pfleiderer (1970); Chem. Ber., **103**, 735.
153. A. F. Kluge and C. P. Lillya (1971); J. Org. Chem., **36**, 1988.
154. H. Henry, J. M. Jablonski and C. T. McCarty (1972); J. Chem. Soc., Perkin Trans. **1**, 2903.
155. J. A. Panetta and H. Rapoprt (1982); J. Org. Chem., **47**, 946.
156. M. A. Butt and J. A. Elvidge (1963); J. Chem. Soc., **3**, 4483.
157. R. A. Finnegan and K. E. Merkel (1972); J. Org. Chem., **37**, 2986.
158. R. Graham and J. R. Lewis (1978); J. Chem. Soc., Perkin Trans. **1**, 876.
159. L. C. King and F. J. Ozog (1956); J. Org. Chem., **20**, 448.
160. R. W. H. O'Donnell, F. P. Reed and A. Robertson (1936); J. Chem. Soc., **6**, 419.
161. J. Bandrenghien, J. Jadot and R. Huls (1953); Bull. Cl. Sci. Acad. Rog. Belg., **39**, 105.
162. C. L. McIntosh and O. L. Chapman (1973); J. Am. Chem. Soc., **95**, 247.

163. E. V. Kuznetsov, D. V. Pruchkin, A. I. Pyshchev and G. N. Dorafunko (1978); Chem. Heterocycl. Compd. (Engl. Transl.), **14**, 1067.
164. D. L. Dare, I. D. Entwistle and R. A. W. Johnstone (1973); J. Chem. Soc., Perkin. Trans. **1**, 1130.
165. W. C. Still, Jr. and D. J. Goldsmith (1970); J. Org. Chem., **35**, 2282.
166. I. Sigg, G. Hoas and T. Winkler (1982); Helv. Chim. Acta, **65**, 275.
167. G. Barker, G. P. Ellis and D. A. Wilson (1971); J. Chem. Soc.(C), **1**, 2079.
168. Y. Duteustre, J. Lemaire and R. Vessiere (1977); Bull. Soc. Chem. Fr., **7**, 911.
169. R. Paul, O. Riobe and M. Maumy (1976); Org. Synth., **55**, 62.
170. A. Ninagawa, R. Nomura and H. Matsuda (1979); Bull. Chem. Soc. Jpn., **52**, 1169.
171. P. Margaretha (1973); Tetrahedron, **27**, 4498.
172. A. P. Monsenko (1973); Zh. Org. Khim., **9**, 775(Chem. Abstr., 1973, 79, 31, 717)
173. F. Ishii, R. Okazaki and N. Inamoto (1976); Tetrahedron. Lett., **9**, 4283.
174. J. P. Pradere, Y. T. N'Guessan, H. Quiniou and F. Tonnard (1975); Tertahedron, **31**, 3059.
175. M. Weissenfels and S. Illing (1993); Z. Chem., **13**, 130.
176. R. Kalish, A. E. Smith and E. J. Smeetry (1971); Tetrahedron Lett., **97**, 2099.
177. S. Baklien, P. Groth and K. Undhein (1975); J. Chem. Soc., Perkin Trans. **1**, 2241.
178. A. Sultan Afridi, A. R. Katritzky and C. A. Ramsden (1976); J. Chem. Soc., Chem. Commun., **6**, 899.
179. S. Yoneda, T. Jugimoto, O. Tanaka, Y. Moriya and Z. Yoshida (1975); Tetrahedron, **31**, 2669.
180. W. Ando, H. Higuchi, and T. Migita (1977); J. Org. Chem., **42**, 3365.
181. A. Ceoisy, A. Ricci, M. Jancevska, P. Jacquiguon and D. Balucani (1976); Chem. Lett., **5**.
182. S. W. Schneller, D. R. Moore and M. A. Smith (1976); J. Hetrocycl. Chem., **13**, 123.
183. S. K. Klimenko and V. G. Kharehenko (1971); Khim. Geterotsikl. Seodin., sb **3**, 85.
184. W. Ried and H. Bopp (1978); Synthesis, **8**, 211.

185. R. Neidlein and I. Korber (1978); Arch. Pharm. (Weinheim, Ger.), **311**, 236.
186. A. G. Hortmann, R. L. Harris and J. A. Miles (1974); J. Chem. Soc., **96**, 6119.
187. A. O. Tosunyan, M. R. Bagdasaryan, S. A. Vartanyan, O. M. Avakyan and O. S. Noravyan; Khim. Farm. Zh., **12**, 56.
188. N. J. Lewis, R. L. Inloes, J. Hes, R. H. Matthews and G. Milo (1978); J. Med. Chem, **21**, 1070.
189. W. Lippman and T. A. Pugsley (1976); Biochem. Phamacol., **25**, 1179.
190. R. A. Abramovitch, K. M. More, I. Shinkai and P. C. Srinivasan; J. Chem. Soc., Chem. Commun., **6**, 771.
191. Z. Eckstein and T. Urbanski (1978); Adv. Heterocycl. Chem., **23**, 1.
192. M. Sainsbury (1978); in 'The chemistry of Carbon Compounds', ed. S. Coffey; Elsevier, Ameterdam, 2nd ed., vol. IV, p. 427.
193. Y. Yamamoto, Y. Azuma and S. Ohnishi (1981); Heterocycles, **15**, 851.
194. J. Rokach, P. Hamel, Y. Girard and G. Reader (1979); Tetrahedron Lett., **7**, 1281.
195. H. Bartsch, W. Kropp and M. Pailer (1979); Monatsh. Chem., **110**, 257.
196. P. Battistoni, P. Bruni and G. Fava (1999); Tetrahedron, **35**, 1771.
197. F. Eiden and F. Meinel (1979); Arch. Pham., **312**, 302.
198. J. Szabo (1989); Khim. Geterotsikl. Soedin., 291 (Chem. Abstr., 1979, 91, 39, 354).
199. J. Szabo, L. Fodor, I. Varga, E. Vinkler and P. Sohar (1977); Acta Chim. Acad. Sci. Hung., 92, 317.
200. F. Moll and H. J. Wieland (1975); Arch. Pharm., **308**, 481.
201. S. Hoff, A. P. Block and E. Zwanenburg (1972); Tetrohedron Lett., **2**, 5199.
202. V. Caplor, A. Lisini, F. Kajfez, D. Kolbah and V. Sunjic (1978); J. Org. Chem., **43**, 1355.

CHAPTER 3

SEVEN MEMBERED HETEROCYCLIC COMPOUNDS

3.1. INTRODUCTION

The seven membered heterocyclic ring compounds are large molecules containing mostly one or more nitrogen, oxygen, or sulphur as hetero atom. The heterocyclic ring system with one nitrogen is called **AZEPINE**, the ring containing oxygen as hetero atom is termed as **OXEPINE** and ring system with one sulphur is called **THIEPIN**.

3.2 AZEPINE

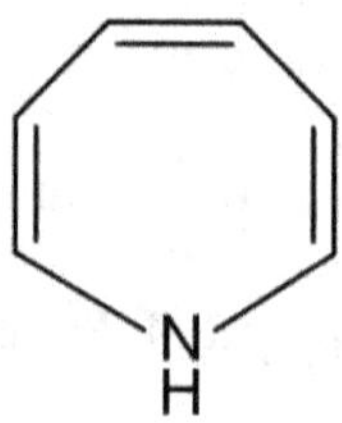

3.2.1 SYNTHESIS OF AZEPINE

At temperature of 155-160°C and N_2 pressure, **ring expansion** occurs on intermolecular insertion of nitrene in the presence of benzene and substituted azepine is obtained.

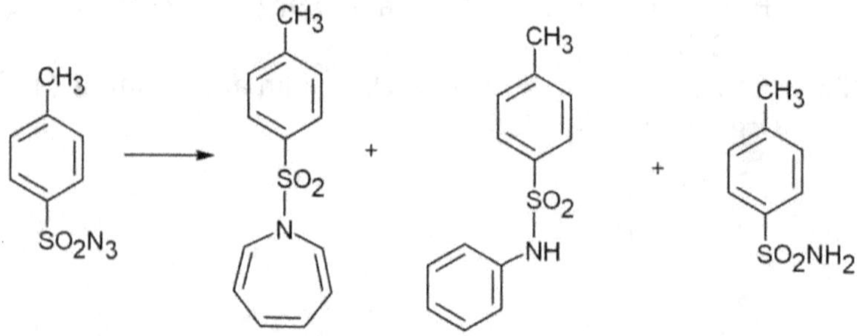

Azepine can also be synthesized in the presence of polyethylene glycol at 60°C and best result are obtained with methyl and ethyl as alkyl groups.

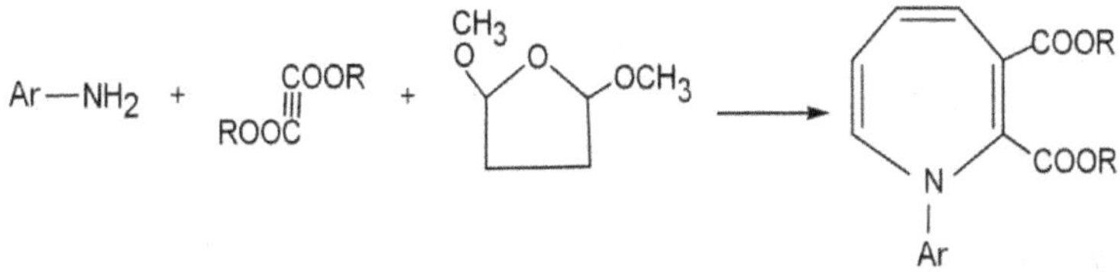

Cycloaddition of 2,5-dimethyl(3,4-diphenyl) cyclopentadienone and 2-phenylazirine in the presence of benzene yields substituted azepine. The reaction proceeds via an intermediate stage whose product is shown in the box.

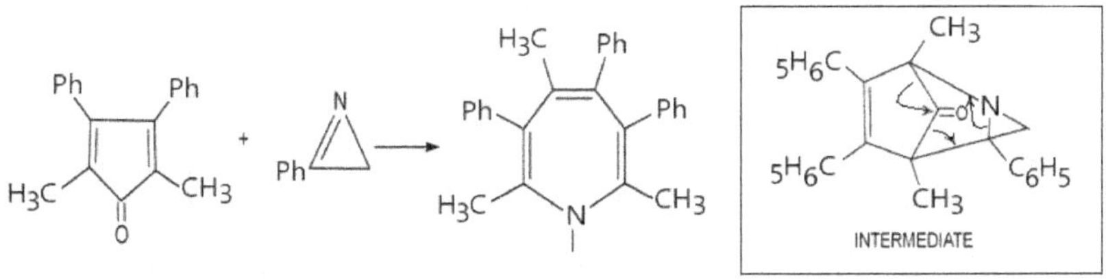

Another cycloaddition reaction is observed between cyclopropenes and 1,2,4–triazines to yield 4-H azepine.

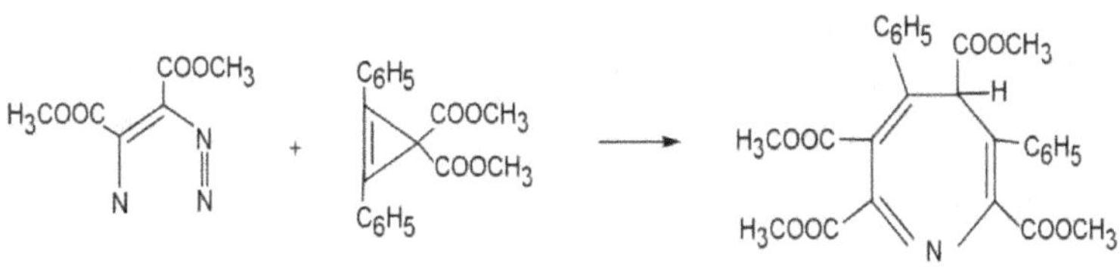

3.2.2 PROPERTIES OF AZEPINE

Azepine (C_6H_7N) undergoes photo chemical **ring contraction** reaction to yield 2-azabicyclo [3.2.0] hepta-3,6-diene. However, the reaction can be thermally reversed via ring opening mechanism.

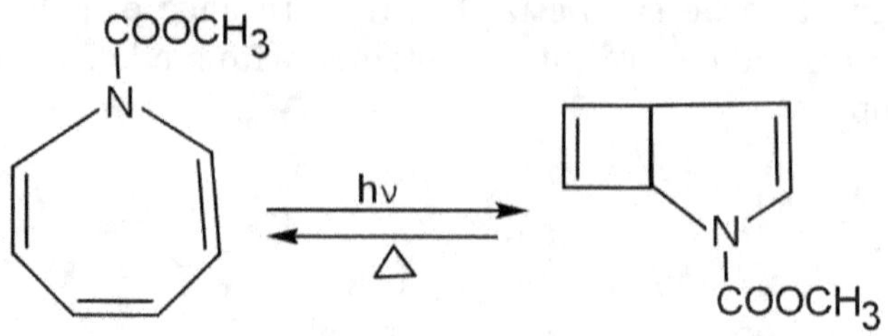

N-Acylation and **mesylation** can be observed in azepine. Methane sulphonyl chloride (MsCl) in the presence of triethyl amine (TEA) and chloroform at 20°C to yield azepine sulphonate.

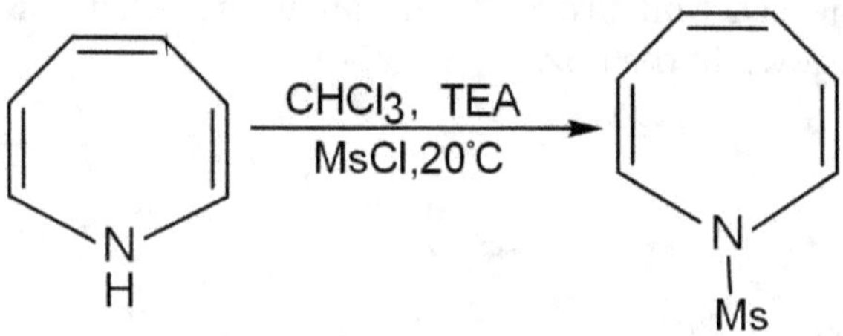

Oxidation of methyl-2,7-dimethyl azepine-1-carboxylate, in the presence of selenium oxide and 1,4-dioxane, gives a dialdehyde.

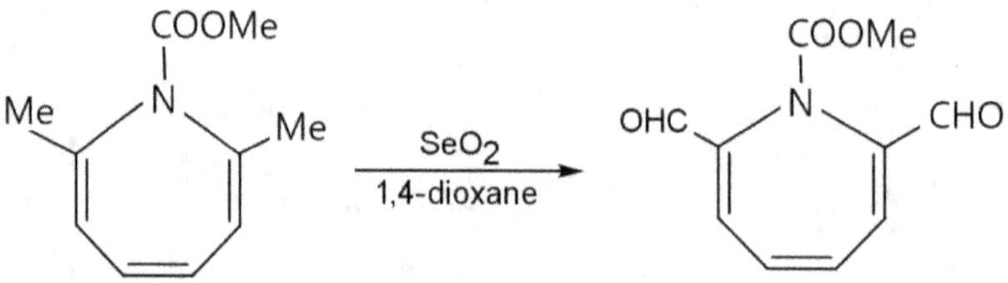

3.2.3 APPLICATIONS OF AZEPINE

1. **Imipramine** is a tricyclic antidepressant and is used in the treatment of depression, anxiety, and panic disorders. It also finds usage in curbing bedwetting.

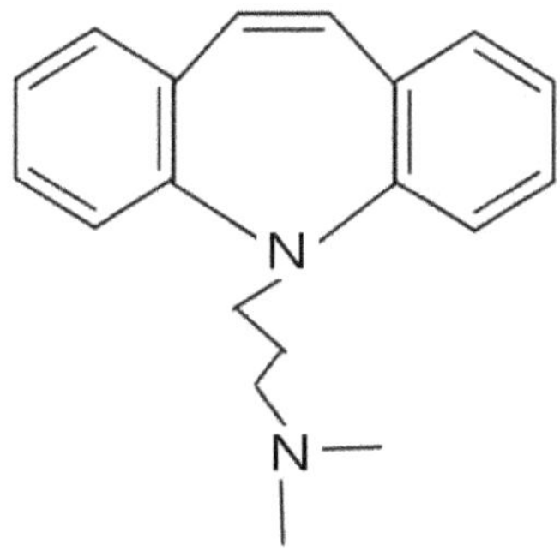

2. **Carbamazepine** is used to prevent and control seizures. It is an anti-convulsant or an anti-epileptic drug.

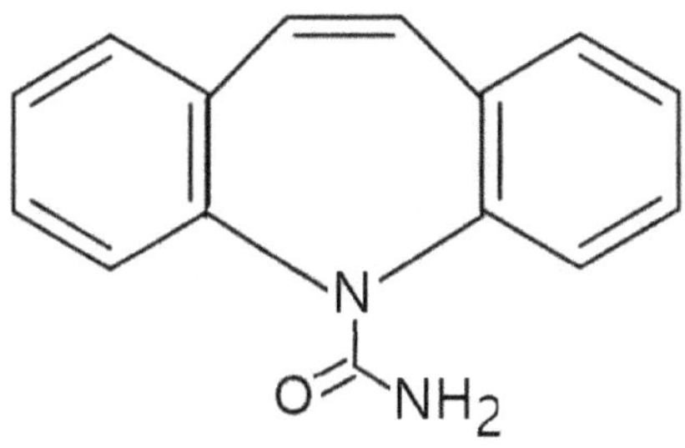

3.3 OXEPINE

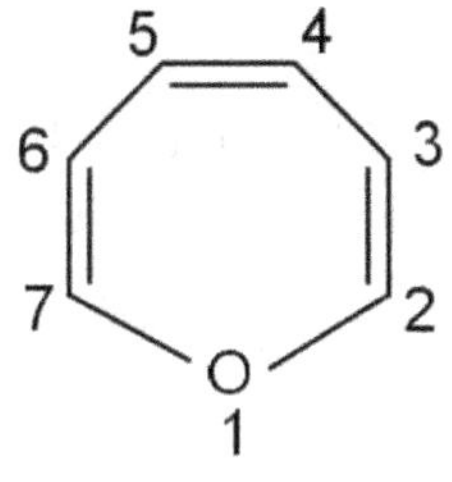

3.3.1 SYNTHESIS OF OXEPINE

The synthetic methods of Oxepines have been reported as old as in 1964 by Vogel et al in which unsubstituted or substituted cyclohexanes yielded unsubstituted or substituted Oxepine.

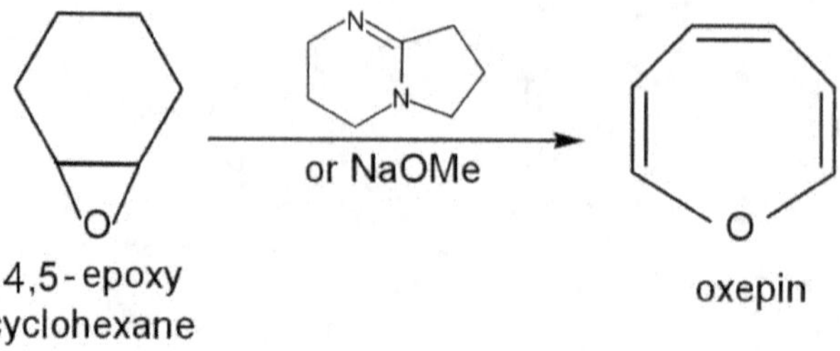

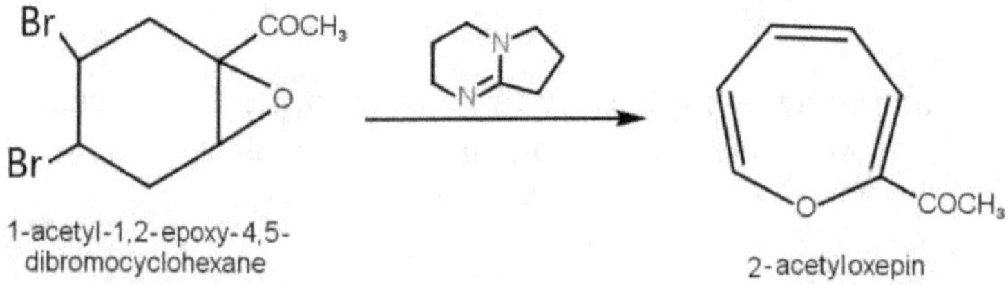

These methods have been modified to get other varieties of substituted Oxepines. Oxepines substituted by one or two methyl groups on bromination form dibromo products which change into 2-methyl Oxepine with sodium methoxide.

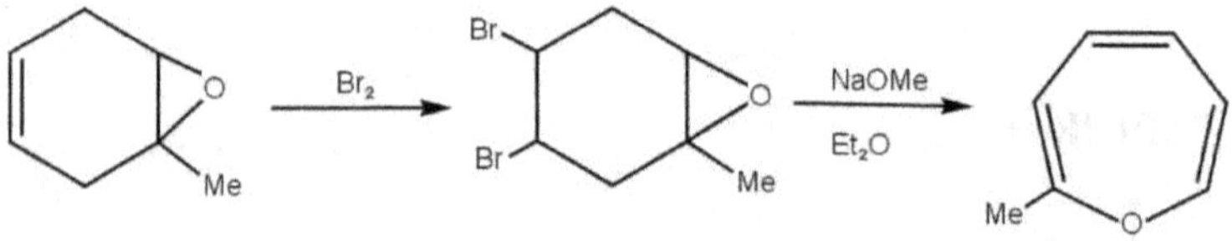

There are many substituted varieties which give same or similar product with other agents as N-bromo succinimide and potassium tert-butoxide.

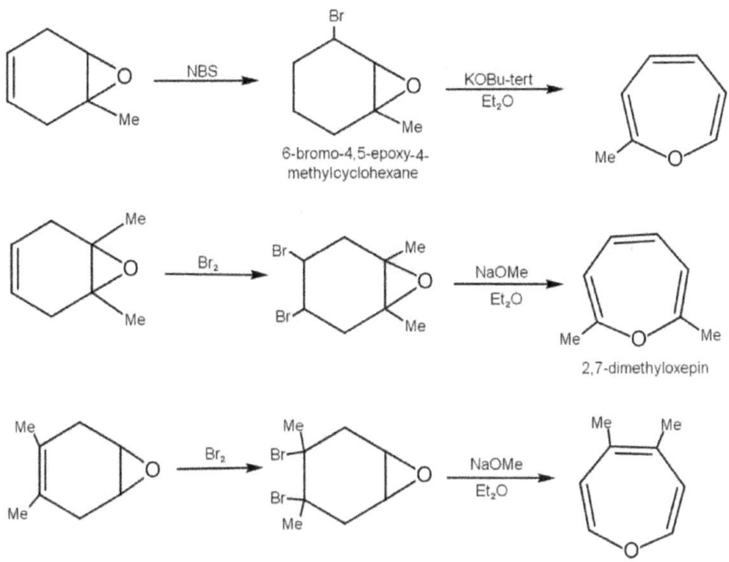

The best reported method till date is still one devised by Van Tame Lon and Carty which involves **epoxidation** of a bicyclic compound in the presence of m-chloroperbenzoic acid and then heating it.

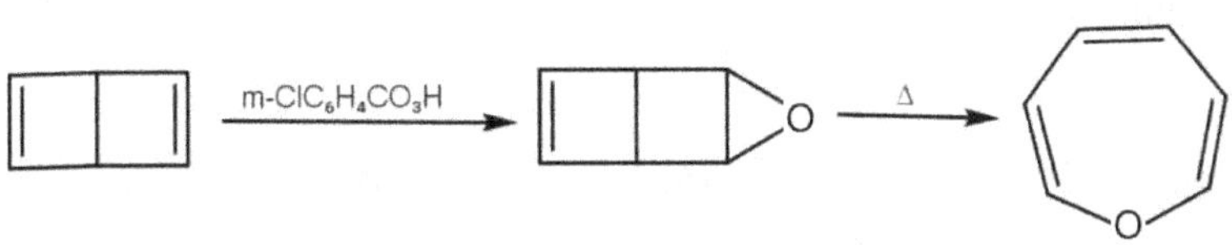

3.3.2 PROPERTIES OF OXEPINE

There have been quite a few chemical reactions reported of oxepine among which catalytic reduction is the most significant.

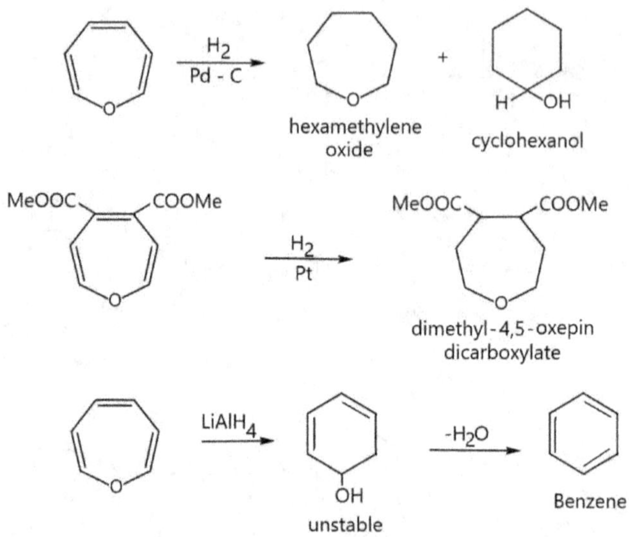

The **catalytic reduction** of Oxepine to hexamethylene oxide and catalytic reduction of dimethyl-4,5-Oxepine dicarboxylate to hexahydro derivative is such a facile rearrangement that even under mild hydrogenating conditions cyclic alcohol is produced as side product. Benzene is obtained as the final product on reduction with lithium aluminium hydride after an unstable intermediate loses water.

The **photochemical rearrangement** leads to the formation of bicyclic product whether the Oxepine is substituted or unsubstituted.

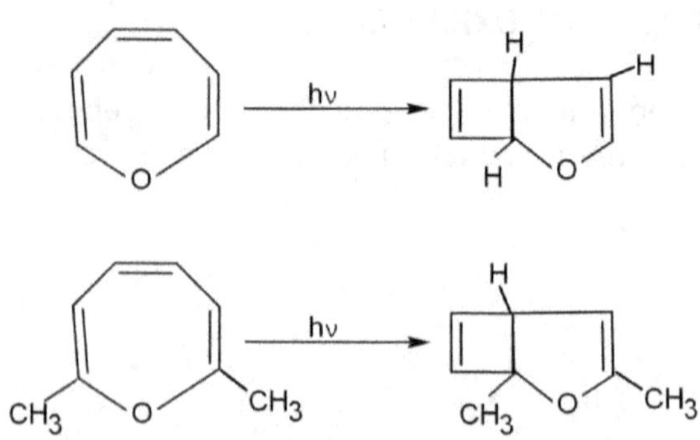

3.3.3 TYPES OF OXEPINE

3.3.3.1 REDUCED OXEPINE

3.3.3.1.1 DIHYDRO OXEPINE

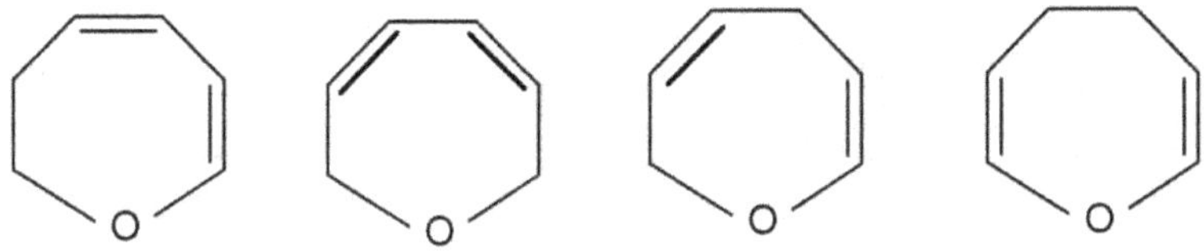

2,3-dihydrooxepin 2,7-dihydrooxepin 2,5-dihydrooxepin 4,5-dihydrooxepin

Of these theoretically available four tautomer, 2,7-dihydroOxepine is the least stable because it has no conjugation between oxygen atom and diene system hence it converts to 2,3-dihydroOxepine.

SYNTHESIS

2,3,6,7-TetrahydroOxepine is converted by **epoxidation** into 4,5-epoxy-2,3,4,5,6,7-hexahydro-oxepine and finally pyrolysis leads to the formation of 2,3-dihydro-oxepine.

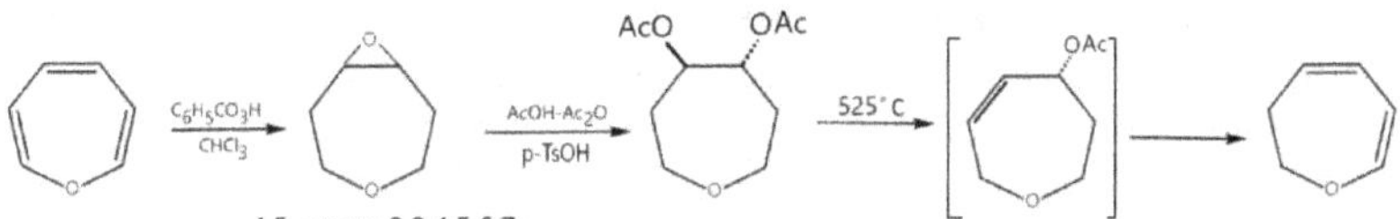

4,5-epoxy-2,3,4,5,6,7-hexahydrooxepin diacetate

4,5-Dihydro Oxepine is prepared by **pyrolysis** of 2-acetoxy-2,3,4,5-tetrahydroOxepine at 470°C.

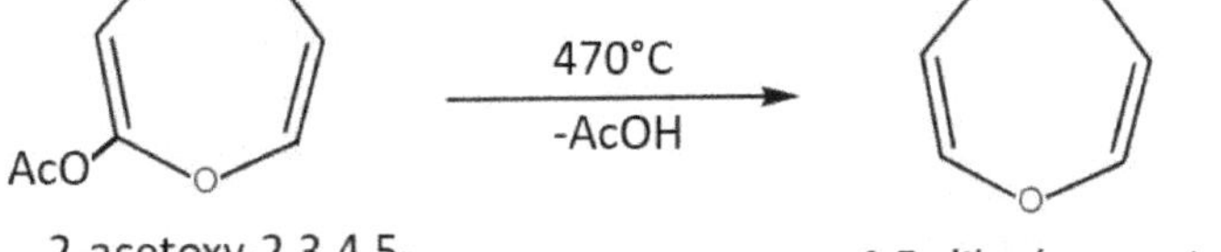

2-acetoxy-2,3,4,5-tetrahydrooxepin 4,5-dihydrooxepin

Condensation of two molecules of 2-acetoxy-4-bromo-2-butene followed by direct hydrolysis of the product by acid leads to the formation of 2,7-dimethyl-4,5-dihydro-oxepine. 2,5-Dihydro-oxepine is prepared by the thermal rearrangement of cis-2-vinylcyclopropane carboxaldehyde which is prepared from cis-2-vinylcyclopropanecarbonyl chloride.

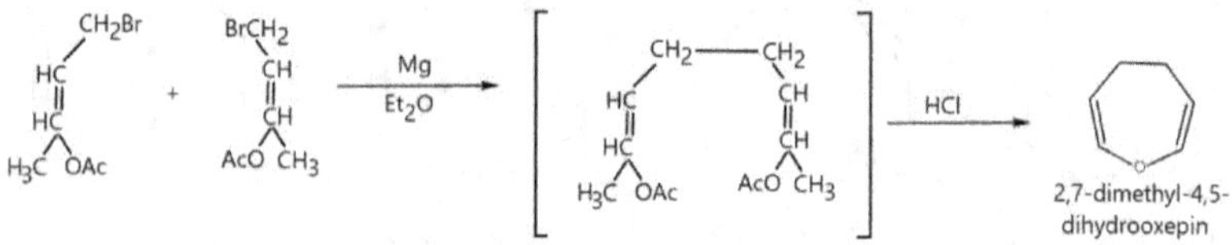

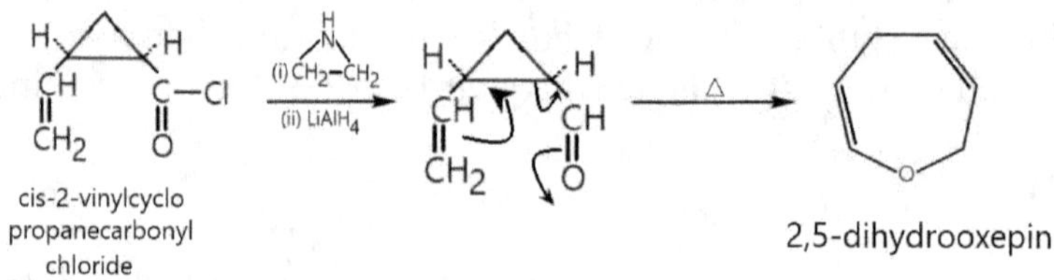

REACTIONS

Reaction of 2,3-dihydro Oxepine with hydrogen in the presence of catalyst forms hexamethylene oxide; however, if the Oxepine is substituted with chlorine then ring cleavage occurs.

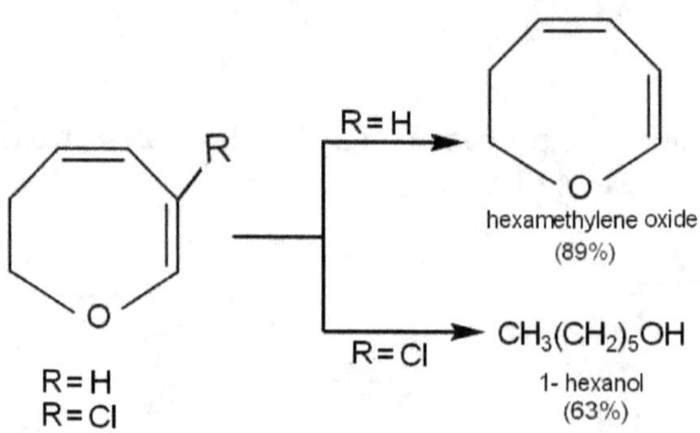

Photoisomerization of 2,3-dihydro-oxepine yielded bicyclic product after rearrangement.

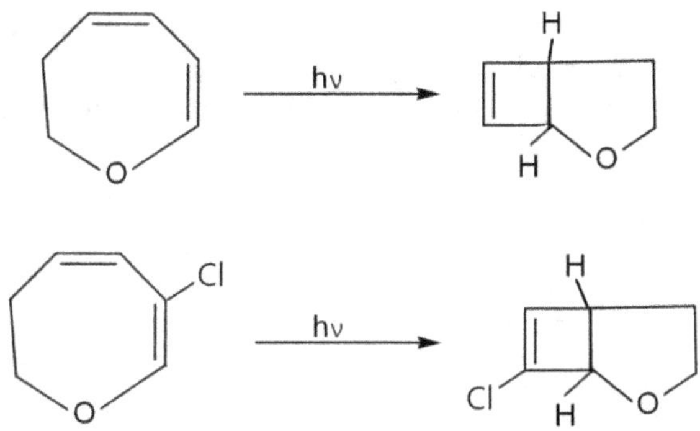

Reduction of 4,5–dihydro oxepine leads to high yield of hexahydro product in the presence of platinum oxide. On treating with acetic acid (warm and aqueous) for 10 min., cyclopentene-1-carboxaldehyde was formed.

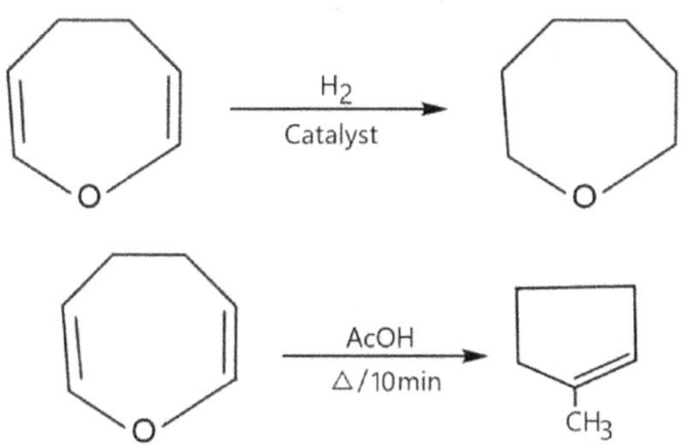

3.3.3.1.2 TETRAHYDROOXEPINE

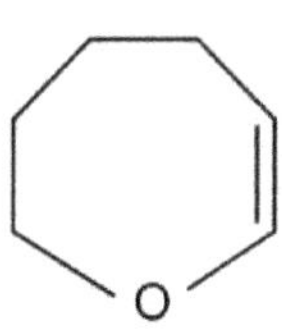

2,3,4,5- tetra hydrooxepin

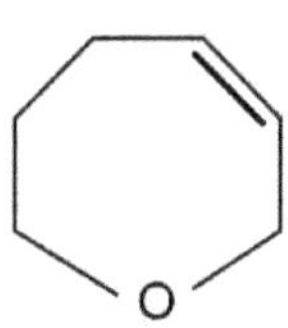

2,3,4,7- tetrahydrooxepin

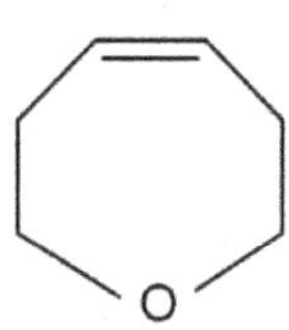

2,3,6,7- tetrahydro-oxepin

SYNTHESIS

Synthesis of 2-acyloxy–2,3,4,5–tetrahydro Oxepine involves air **oxidation** of cyclohexene in the presence of an acyl anhydride and an appropriate radical initiator (such as azobisisobutyronitrile). Thus, with different values of R, we get 2-acetoxy–2,3,4,5–tetrahydro Oxepine (R = Me) and 2,3,4,5–tetrahydro–2–propionyl oxy-oxepine (R = Et).

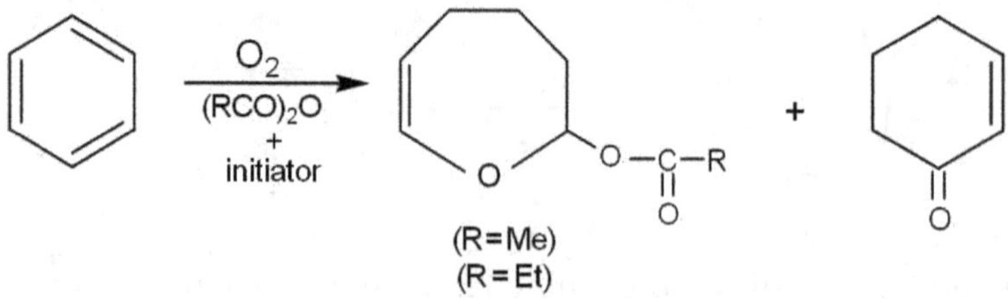

REACTIONS

Catalytic reduction leads to the formation of hexamethylene oxide and per benzoic acid oxidation forms 4,5–epoxy–2,3,4,5,6,7–hexahydro-oxepine.

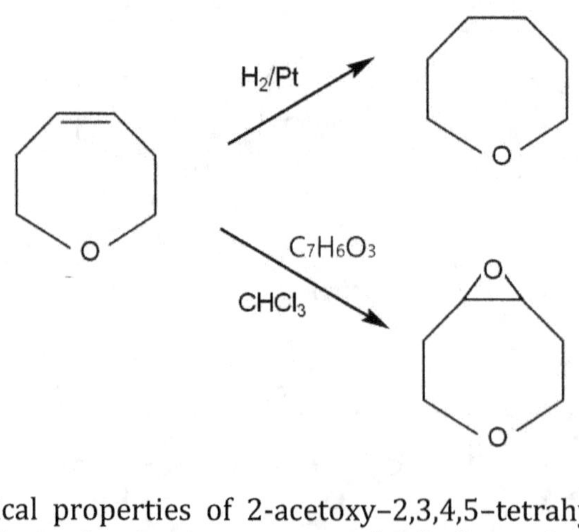

The chemical properties of 2-acetoxy–2,3,4,5–tetrahydro oxepine (**A**) show that reduction in presence of palladium–charcoal gives 2-acetoxy–2,3,4,5,6,7-hexahydroOxepine (**B**). When platinum was used as catalyst, (**A**) was reduced to 1,6-hexanediol (**C**) and acetic acid. Oxidation with neutral potassium permanganate yielded glutaric acid. Similarly, (**A**) was converted into adipaldehyde dioxime (**E**) and cyclopentene–1-carboxaldehyde oxime (**F**) on treatment with hydroxylamine hydrochloride.

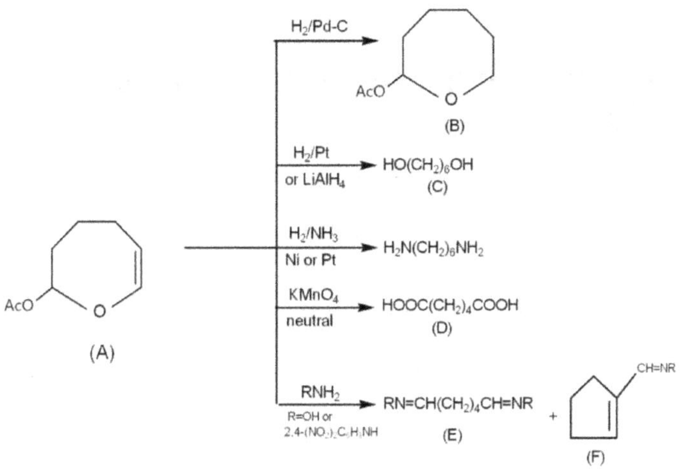

3.3.3.1.3 HEXAHYDROOXEPINE (OXEPAN)

SYNTHESIS

Dehydration of 1,6–hexanediol forms hexamethylene oxide, 2-methyltetrahydro pyran and 2-ethyltetrahydrofuran.

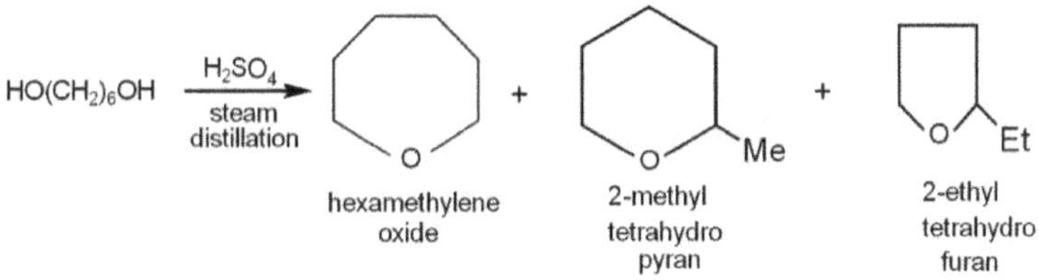

Substituted hexamethylene oxide derivatives have been prepared by **intramolecular cyclization** of 2,7-dimethyl–2,7-octanediol with warm dilute sulfonic acid resulting in the formation of 2,2,7,7–tetramethyl oxepan.

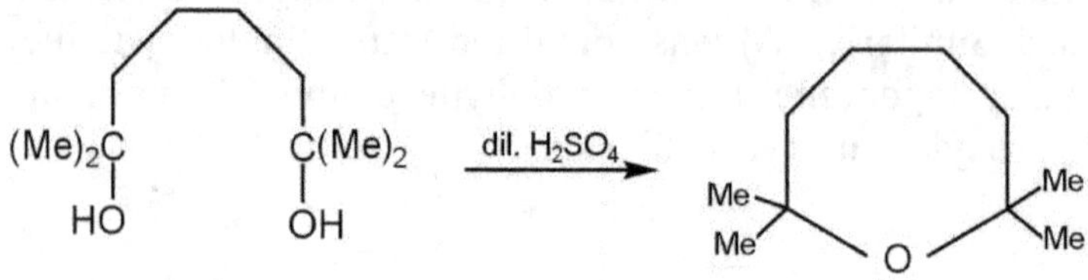

3.3.3.2 OXEPINE RING SYSTEMS CONTAINING TWO RINGS (Fused Ring Systems)

3.3.3.2.1 2-OXABICYCLIC OCTANE

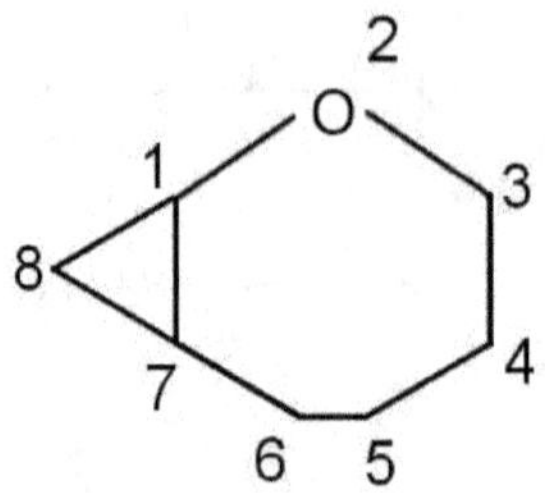

Addition of dichlorocarbene to 2,3,4,5–tetrahydro oxepine in the presence of ethylene oxide gives substituted 2-oxabicyclo octane.

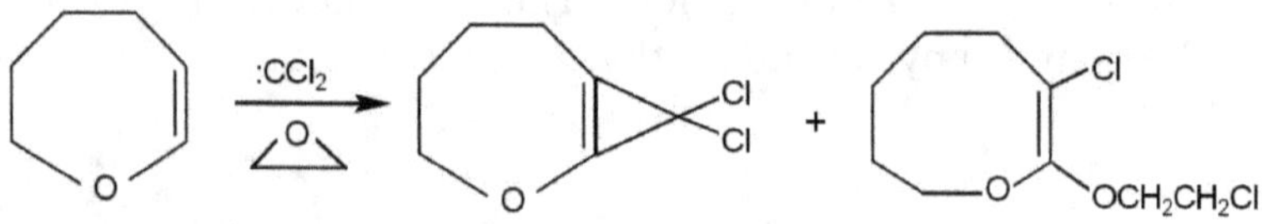

3.3.3.2.2 2-OXABICYCLIC NONANE

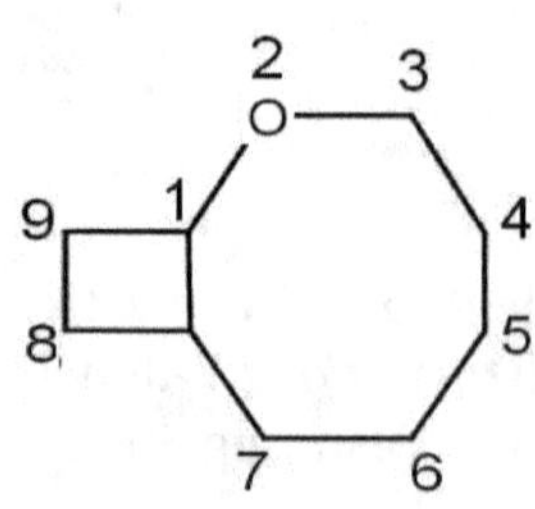

Photo-condensation of 3-methyl-2-cyclohexanone and isobutylene leads to the formation of 6,8,8-tri acetylbicyclo octan-1-one which on oxidation with peracetic acid gave 7,9,9-trimethyl-2-oxabicyclononan-3-one (**A**). Chromic acid oxidation of (**A**) gives cyclobutanone structure.

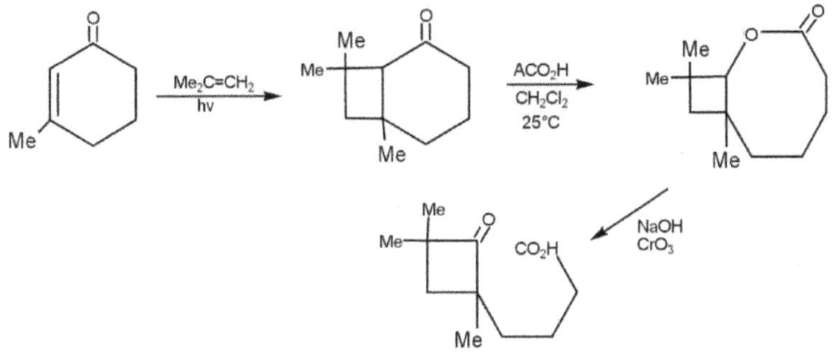

3.3.3.2.3 2H-CYCLOPENT [b] OXEPINE

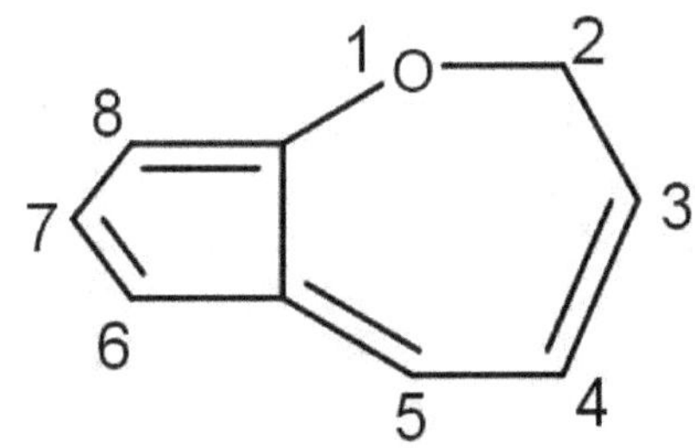

Treatment of ethyl 4-(2-hydroxycyclopentyl) butyrate with alkali caused ring closure and formed lactone 3,4,5,5a,6,7,8,8a–octahydro–2H–cyclopent[b]oxepine–2-one. This compound on reduction with lithium aluminium hydride afforded 4-(2-hydrocyclopentyl) butyric aldehyde while on heating with phosphoric acid showed ring contraction.

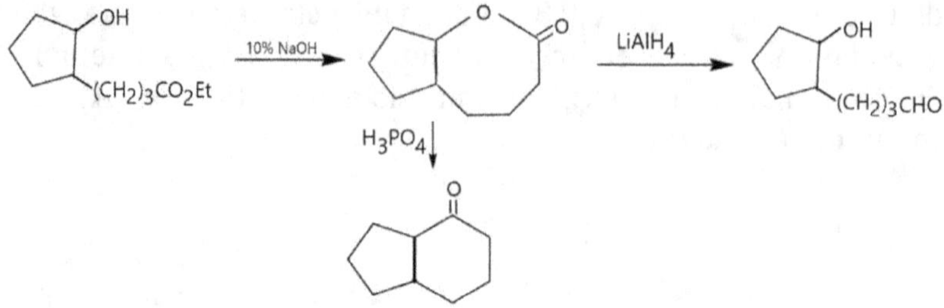

3.3.3.2.4 FURO OXEPINE

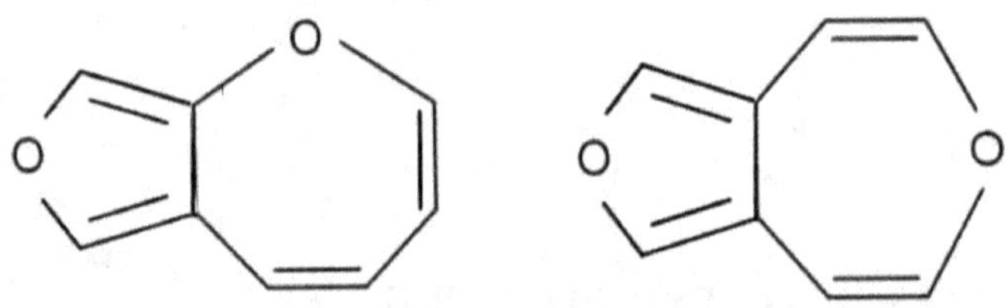

Furo[3,4-b]oxepin Furo[3,4-d]oxepin

6-Benzyuloxy-3-oxohexanoate(**A**) reacts with 2-acetoxypropionyl chloride gave ester(**B**) which transformed to (**C**) on alkaline hydrolysis followed by lactonization with sulphuric acid. Catalytic hydrogenation of (**C**) in presence of palladium-charcoal with HCl, gave carolic acid (**D**) and without HCl led to the formation of deoxy derivative (**E**).

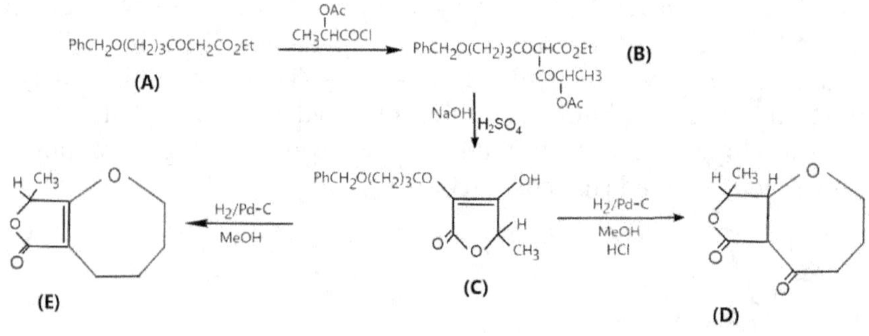

3.3.3.2.5 THIENO[3,4-d] OXEPINE

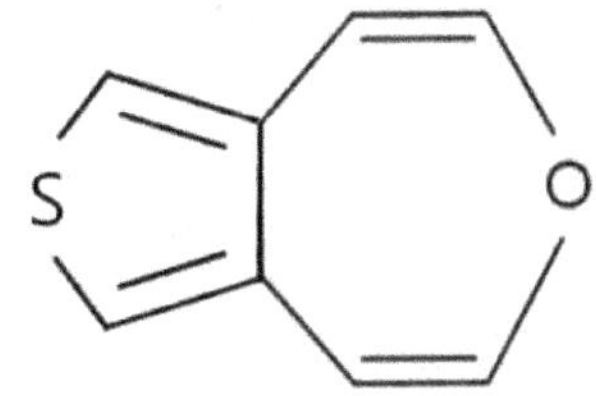

Bis-chloromethylation of 2,5–dimethylthiophene followed by reaction with sodium methoxide and oxidation with dinitrogen tetraoxide led to 2,5–dimethylthiophene–3,4-di aldehyde whose condensation with bis(2–bromoethyl)ether, tri-phenylphospine and sodium methoxide gave 1,3-dimethylthieno[3,4-d]Oxepine.

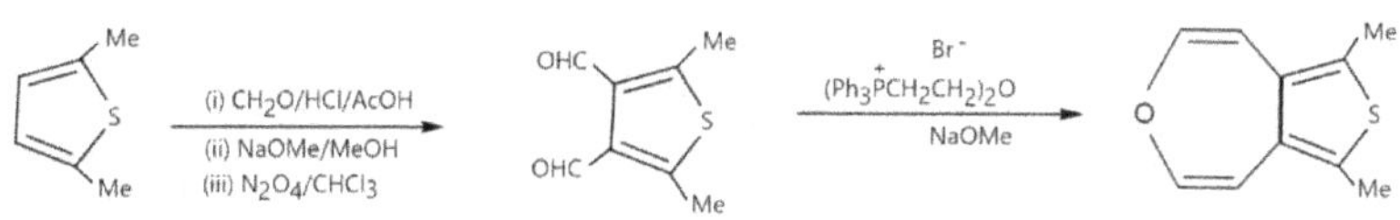

3.3.3.2.6 1-BENZOXEPINE

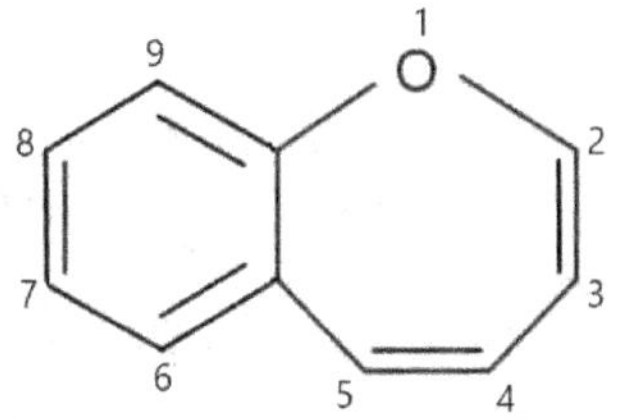

Levulinic acid, on **condensation** with salicylic acid in presence of dry hydrogen chloride, gives derivates of 2,3-dihydro-1-benzOxepinee.

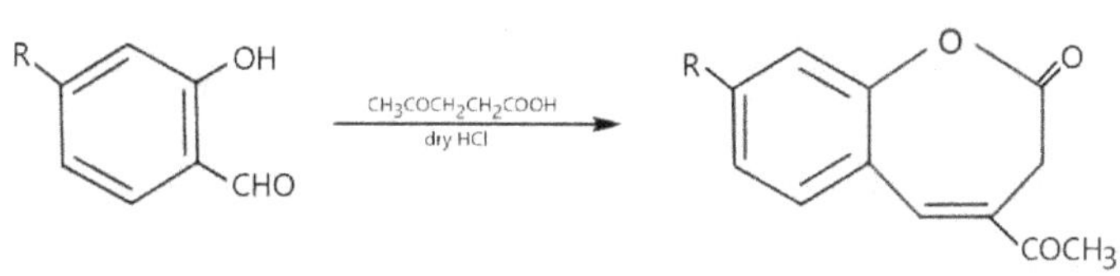

R = H [4-acetyl-2,3-dihydro-1-benzoxepin-2-one]
R = OH [4-acetyl-2,3-dihydro-8-hydroxy-1-benzoxepin-2-one]

4,5–dihydro-1-benzOxepine can be prepared from tetralin by **photo–oxidation** followed by heating of resultant 2-acetony-2,3,4,5–tetrahydro-1-benzOxepine to eliminate acetic acid.

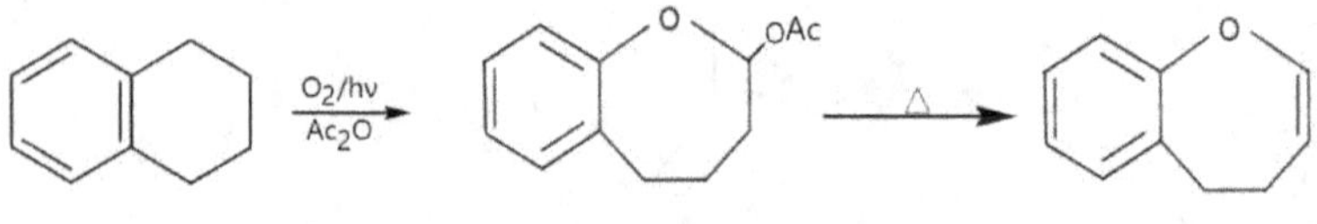

In the similar fashion, 4-(3',4'–dimethyl phenoxy) butyric acid can easily be transformed into 2,3,4,5–tetrahydro–7,8–dimethyl-1-benzoxepine–5–one by xylene in polyphosphoric acid.

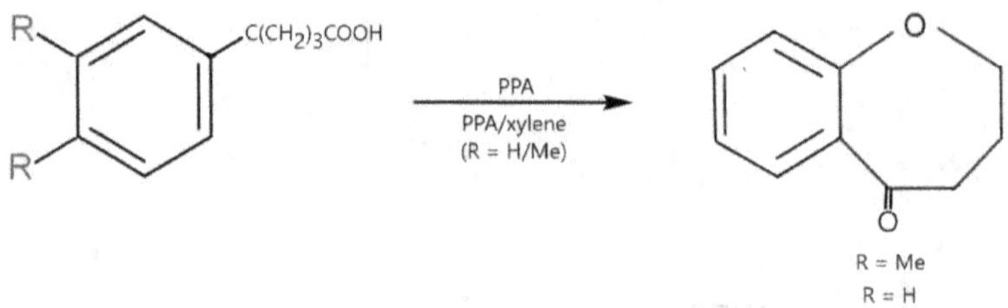

REACTIONS

Reaction of 2,3,4,5-tetrahydro–7,8-dimethyl-1-benzOxepine-5-one (**A**) have been most widely studied and reported. Reduction of (A) with aluminium isopropioxide gives 2,3,4,5- tetrahydro-5–hydroxy–7,8-dimethyl-1-benzOxepine (**B**). Addition of methyl magnesium bromide leads to 2,3,4,5–tetrahydro-5-hydroxy-5,7,8-trimethyl-1-benzOxepine (**C**).

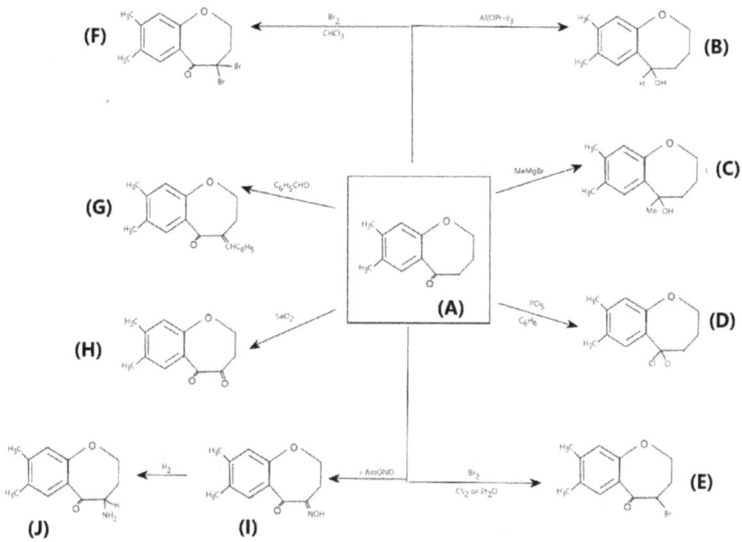

Treatment of (**A**) with phosphorous pentachloride in benzene gives 5,5–dichloro–2,3,4,5–tetrahydro-7,8–dimethyl-1-benzOxepine (**D**). Bromination in different solvents affords different products; In carbon disulphide, produces 4-bromo-2,3,4,5-tetrahydro-7,8-dimethyl-1-benzOxepine-5-one (**E**) and in chloroform yields 4,4-dibromo-2,3,4,5-tetrahydro-7,8-dimethyl-1-benzOxepine-5-one (**F**). Reaction of (**A**) further include condensation with benzaldehyde, oxidation with selenium oxide and oximation with isoamyl nitrite. The products obtained are 4-benzylidene-2,3,4,5-tetrahydro-7,8-dimethyl-1-benzoxepine-5-one (**G**), 2,3,4,5-tetrahydro-7,8-dimethyl-1-benzoxepine-4,5-dione (**H**) and 2,3,4,5-tetrahydro-7,8-dimethyl-4-oximino-1-benzoxepine-5-one (**I**), respectively. (I), on catalytic reduction, leads to the formation of 4-amino-2,3,4,5-tetrahydro-7,8-dimethyl-1-benzOxepine-5-one (**J**).

3.3.3.2.7 2-BENZOXEPINE

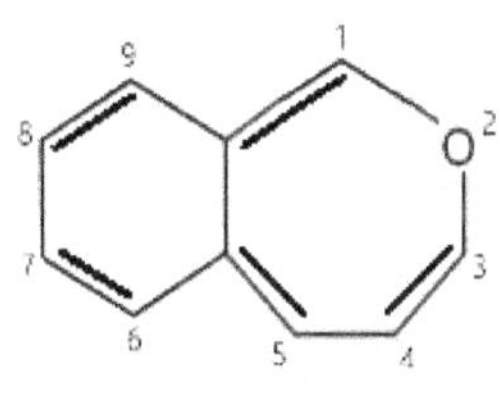

Chloromethylation of 3-phenyl-1-propanol with formaldehyde and concentrated hydrochloric acid gives chloromethyl-3-phenylpropyle ether which on intra-molecular Friedel-Crafts reaction under the influence of aluminium chloride in carbon disulphide results in 4,5-dihydro-1H,3H-2benzOxepine (R=H) and 4,5-dihydro-1H,3H-2-benzOxepine (R=Me).

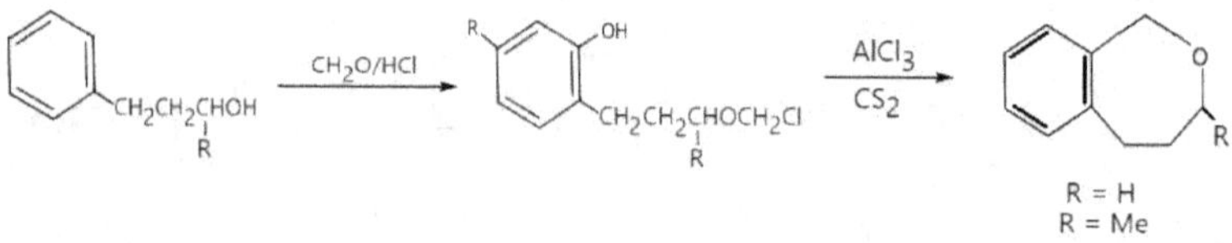

Oxidation of 4,5-dihydro-1H,3H-2-benzOxepine with alkaline potassium permanganate gives phthalic acid; on heating with zinc chloride in acetic acid it forms a diacetate and a dibromide with fuming hydrobromic acid. **Photocatalyzed oxidation** at 100°C gives 4,5-dihydro–1H,3H-2-benzoxepine-1-one.

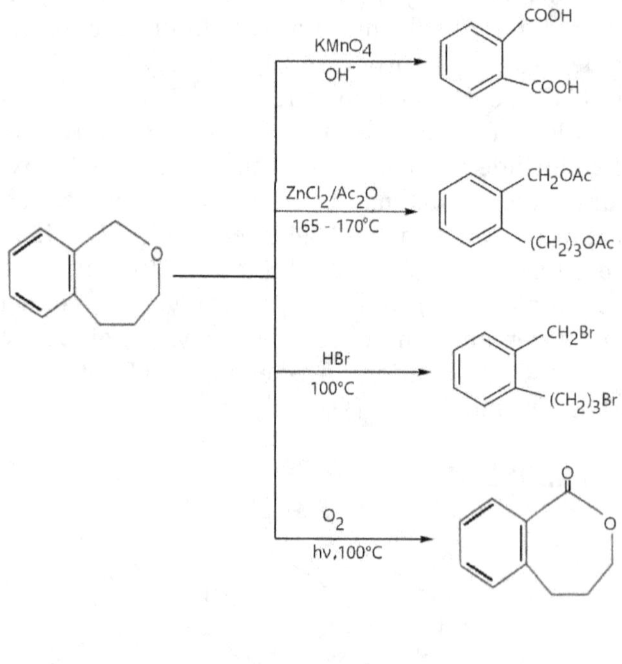

Further, oxidation of 4,5-dihydro-1H,3H-2-benzOxepine with chromic acid yielded 3-(o-carboxyphenyl)-1-propane and oxidation with selenium dioxide in boiling xylene gave a mixture of 3-(o-hydroxymethyl) phenylpropion-aldehyde and 3(o-formyl phenyl) propionaldehyde (R=CHO).

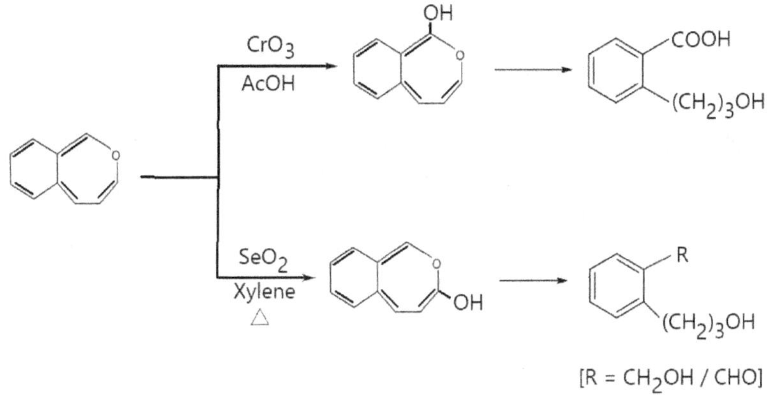

Oxidation of 4,5-dihydro-1H,3H-2-benzOxepine with nitric acid to 3-(o-carboxyphenyl) propionic acid and catalytic hydrogenation on the presence of Raney nickel gave 3-(o-tolyl)-1-propanol.

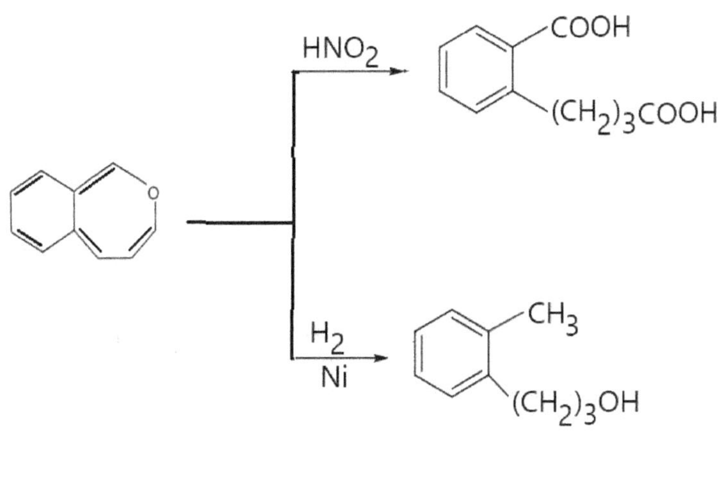

3.3.3.2.8 3-BENZOXEPINE

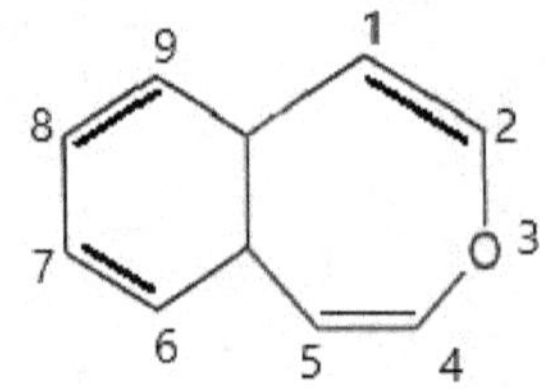

Dimethyl 3-benzOxepine-2,4-dicarboxylate can be obtained by the reaction of phthalaldehyde and dimethyl-2,2'-oxydiacetate in presence of sodium methoxide or potassium tert-butoxide as condensing agent. Hydrolysis of this product gave 3-benzoxepine-2,4-dicarboxylic acid.

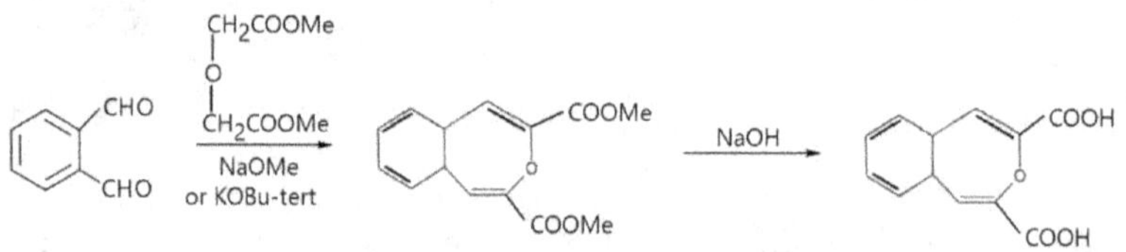

A dibromide is obtained from dimethyl-3-benzOxepine-2,4-dicarboxylate (R=Me) or 3-benzoxepine - 2,4, - dicarboxylate (R=H) with zinc metal. **Pyrolysis** at 300°C under reduced pressure gave 2-hydroxyl–3-naphthoic acid while distillation with copper powder produced naphthalene-2,3-dicarbocylic acid.

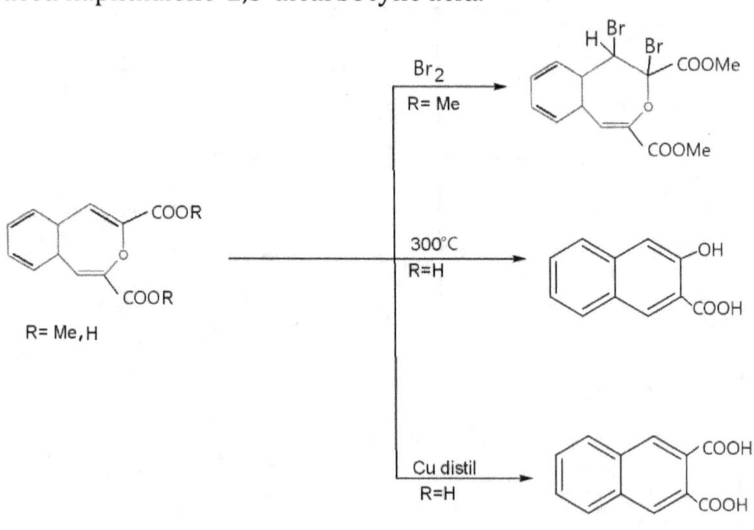

In the same way, there are reports by Jorgenson of other set of chemical properties as well. Reduction of diester(R=Me) with hydrogen over palladium carbon afforded dimethyl-1,2-dihydro-3-benzoxepine-2,4-dicarboxylate while catalytic reduction of diacid(R=H) led to uptake of two molecules of hydrogen producing 1,2,4,5–tetrahydro-3-benzOxepine-2,4-dicarboxylic acid. When the diester is reduced with lithium aluminium hydride or its mixture with aluminium chloride, it gave 3-benzOxepine-2,4-dimethanol. The dimethanol on oxidation with manganese dioxide gave monoaldehyde and on treatment with acetic anhydride in pyridine gave diacetate.

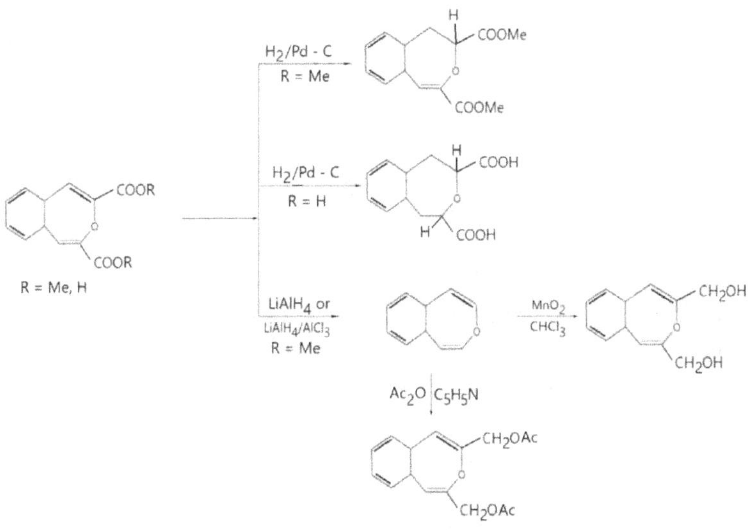

3.3.3.3 OXEPINE RING SYSTEMS CONTAINING THREE RINGS

3.3.3.3.1 DIBENZ[b, d]OXEPINE

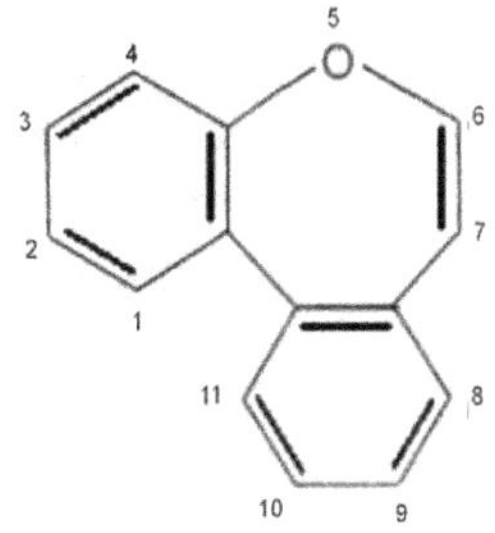

6,7-Dihydrodibenz[b, d]Oxepine was prepared by condensation of potassium salt of o-nitrophenol with phenethyl bromide.

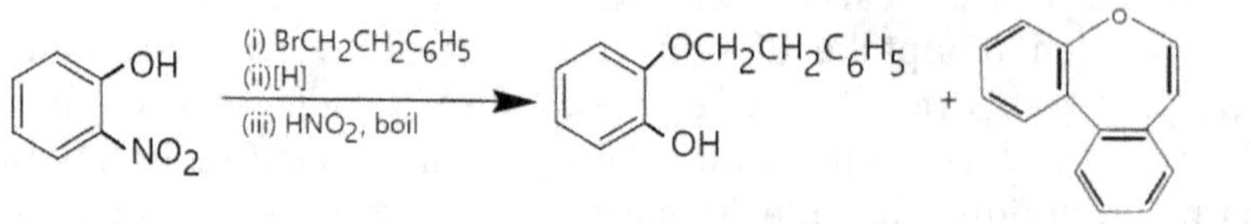

6,7-dihydro
dibenz[b,d]oxepin

3.3.3.3.2 DIBENZ [b, e]OXEPINE

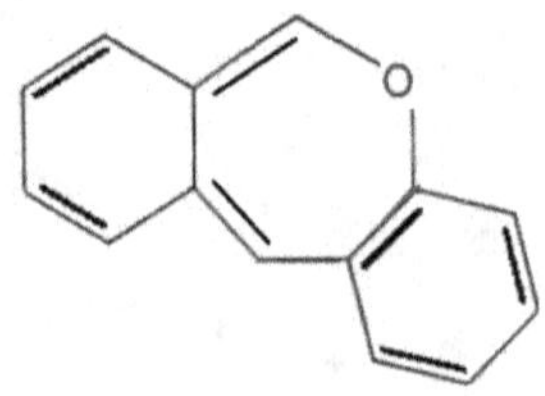

Pyrolysis of 2-chloro-3-(2,4-dihydroxyphenyl)-2-methyl-3-phenyl-1-indanone gave 3-hydroxy-11-α-methyl–benzylidenedibenz[b, e] Oxepine-6-(11H)-one. This product, with hot acetic anhydride in the presence of traces of sodium acetate, gave acetoxy lactone.

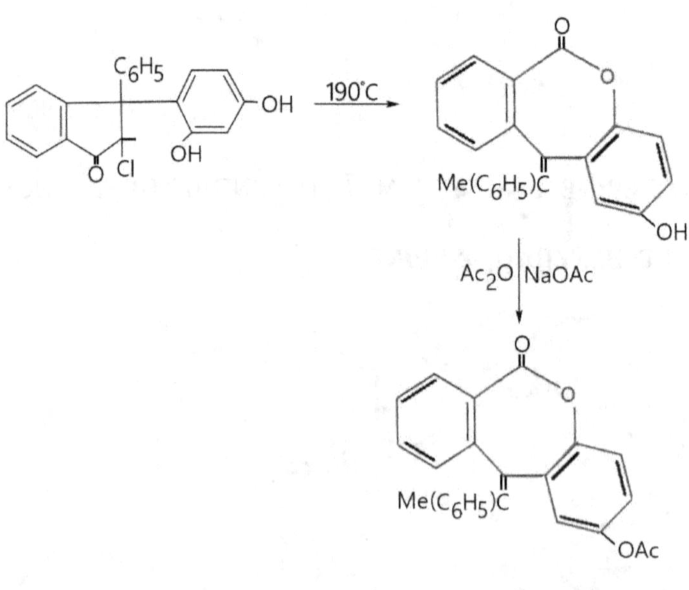

3-Methoxydibenz [b, e] Oxepine-6,11-dione, on being subjected to various reagents, leads to cleavage of seven-membered lactone ring. Thus, on exposure to acetic acid, ethanol, ethylamine, and diethyl amine gave hydroxy acid, its ethyl ester, its N-ethyl amide and its N, N-diethyl amide respectively. Treatment with ammonia gave iso-indole and condensation with hydrazine resulted in phthalazine.

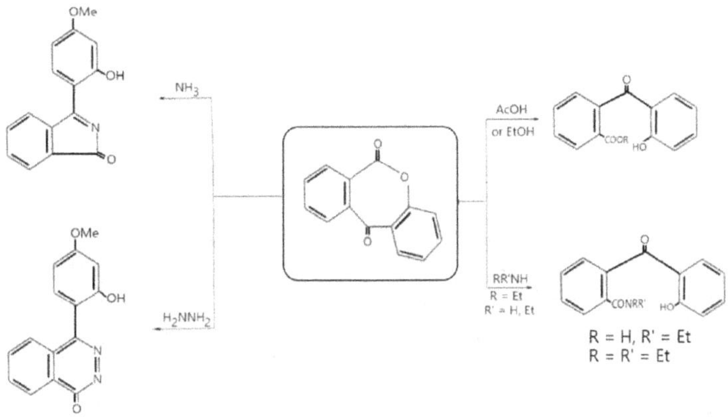

3.3.3.3.3 DIBENZ [b, f]OXEPINE

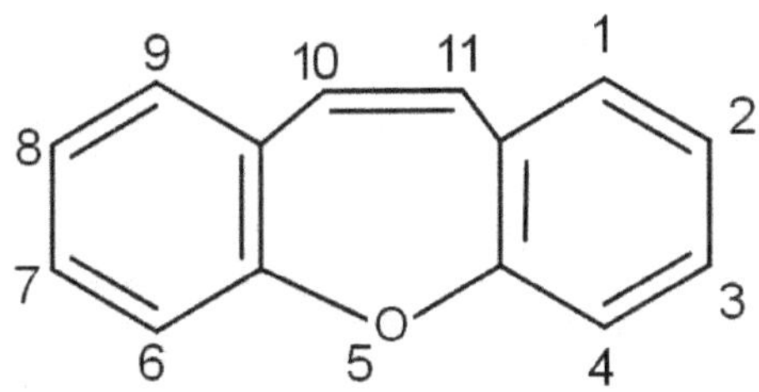

Reduction of xanthene-10-carboxylic acid with lithium aluminium hydride resulted in alcohol which in refluxing xylem in presence of phosphorus pentoxide generated seven membered ring system. Methyl xanthene-10-carboxylate was treated with sodium amide in liquid ammonia as base and methyl iodide as alkylating agent resulted in ester with was reduced to respective alcohol which got converted to tosylate ester. This product, on reaction with formic acid, gave 10-methyldibenz[b, f] Oxepine (R'=Me).

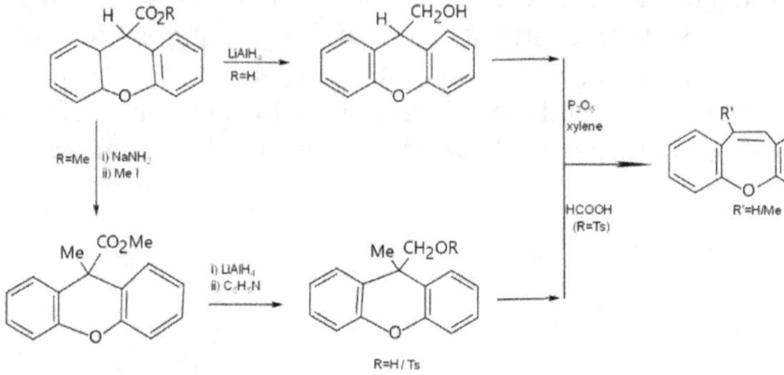

In another method, **bromination** of 4,4-dibromo-2,2'-dimethyl diphenyl ether with N- bromo-succinimide in benzoyl peroxide gave tetrabromide as intermediate which in phenyl lithium resulted in 2,8-dibromo-10,11-dihydrodibenz [b, f] Oxepine along with a dimer product which was formed when the starting material is a non-bromo compound.

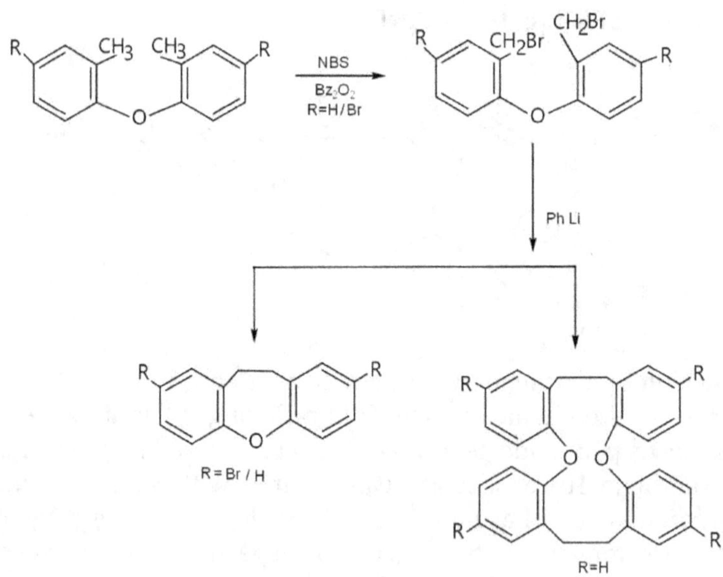

Dibenz [b, f] oxepine-10(11H)–one (**A**) with sodium amalgam or palladium charcoal in hydrochloric acid leads to 10-amino-10,11-

dihydrodibenz [b, f] oxepine (**B**). If (**A**) is reduced catalytically N-acetyl derivative (**C**) is obtained. When (**A**) oxidised with selenium dioxide in aqueous dioxane, a diketone, dinner [b, f] oxepine-10,11-dione (**D**) is produced which can be reduced to 10,11-diacetoxydibenz [b, f] oxepine (**E**). A facile ring contraction of the benzilic rearrangement was in the presence of a base resulting in α-hydroxy or α-methoxy acid (**G**) depending on the alcoholic or aqueous base. Ring contraction was reported with sodium xanthene-10-sulfonate (**F**).

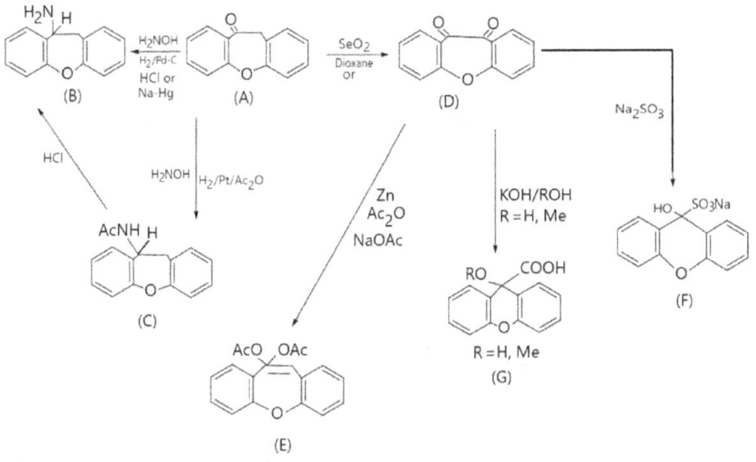

In the same way, reaction of substituted dibenz[b, f]Oxepine-10(11H)-one have also been studied and reported extensively. Reaction of 2,3,6-trimethoxydibenz [b, f] Oxepine-10(11H)-one(**Z**) with hydroxyl amine gave a mixture of 2,3,6-trimethoxy-10[11H]–oximinodibenz[b, f]Oxepine (**X**) and 2,3,6-trimethoxy-10,11-bis–(oximino) dibenz [b, f] Oxepine (**Y**). When **X** was reduced catalytically, it produces 10-amino-10,11-dihydro–2,3,6-trimethoxydibenz [b, f] Oxepine whereas on reduction of **Y** with lithium aluminium hydride 10,11-diamino-10,11- dihydro-2,3,6-trimethoxydibenz [b, f] Oxepine was obtained. Direct catalytic reduction of **Z** produced 10,11-dihydro 2,3,6-trimethoxydibenz [b, f] Oxepine-10-ol.

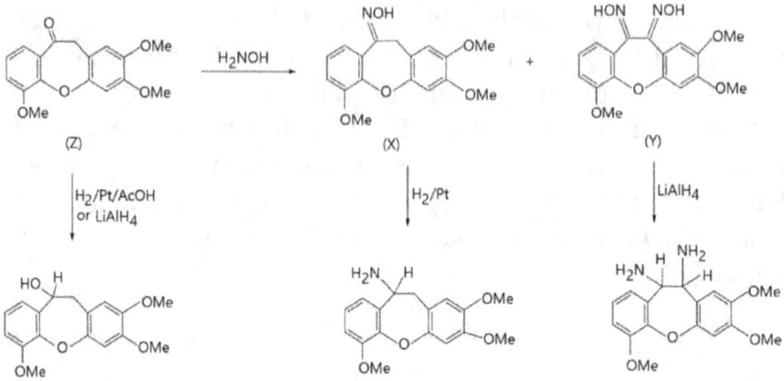

3.3.3.3.4 DIBENZ [c, e]OXEPINE

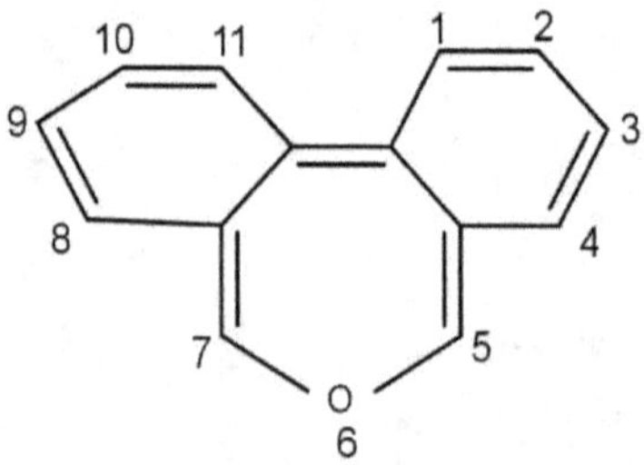

Phosphorus pentachloride or phosphorus oxychloride with diphenic acid afforded a mixture of dibenz[c, e] Oxepine-5,7-dione and fluorenone–4–carboxylic acid. In the same way, **catalytic oxidation** of phenanthrene at 300°C over vanadium pentoxide also resulted in dibenz [c, e] oxepine-5,7-dione.

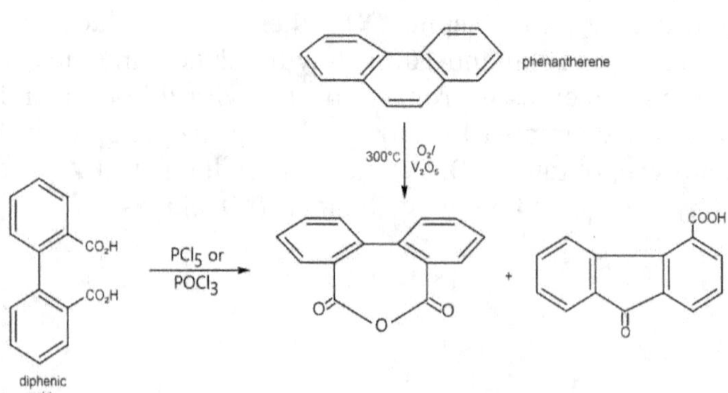

Dibenz [c, e] oxepine-5,7-dione (or diphenic anhydride) undergoes thermally induced decarboxylation reaction forming fluorenone. Under mild conditions of treatment with alcohols, monoesters are produced (depicting cleavage of seven membered ring). Monophenyl ester was obtained when heating was brought about with phenol at 125°C for 5 hours. However, in the presence of stannic chloride, a mixture of diphenyl ester and another compound (Fries Rearrangement product). In aluminium chloride, diketone is obtained.

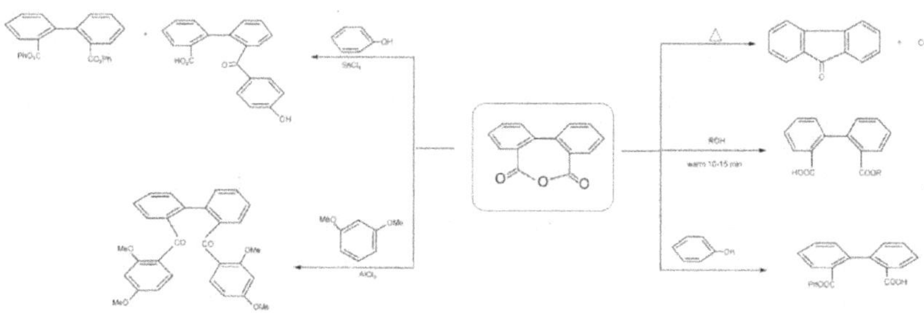

3.3.3.3.5 NAPHTHO-OXEPINE

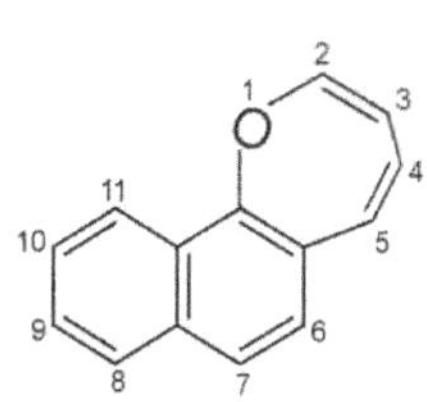

Naphth(1,2-b) oxepin

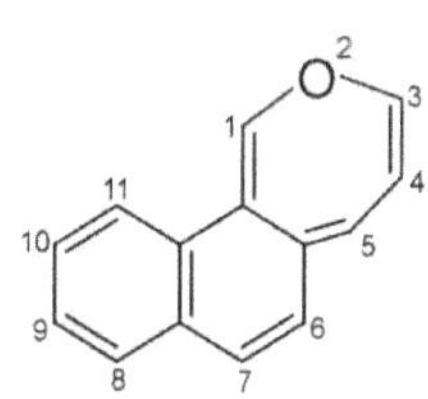

Naphth [1,2-c] oxepin

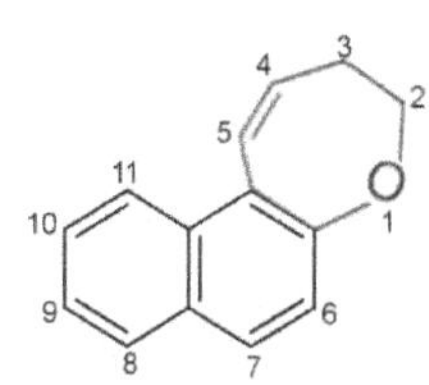

Naphth(2,1-b) oxepin

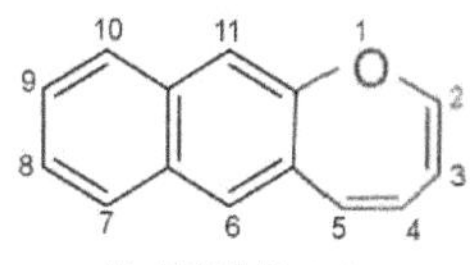

Naphth(2,3-b)oxepin

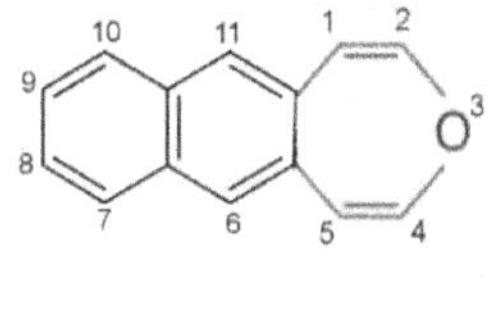

Naphth(2,3-d) oxepin

When sodium salt of 1-naphthol is condensed with ethyl–4–bromo butyrate, an ester is obtained which on saponification gave 4-(1-naphthoxy) butyric acid (**A**). The treatment of this product (**A**) with hot polyphosphoric acid afforded 2,3-dihydro-4H-naphth[1,2-b] Oxepine-5-one(**B**), on reduction, the yielded product was 2,3,4,5-tetrahydronaphth[1,2-b] Oxepine (**C**).

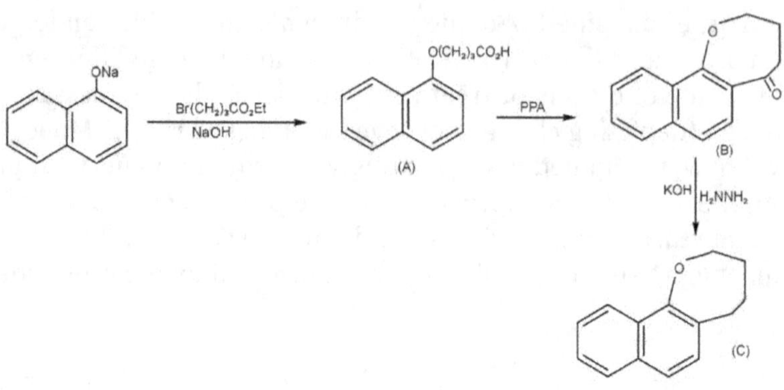

Sodium–2-naphthoxide, on condensation with ethyl–4-bromobutyrate and followed by saponification, gave 4-(2-naphthoxy) butyric acid (R=OH); this, on treatment with hot poly-phosphoric acid, gave a mixture of 3,4-dihydro-2H-naphth [2,3-b] Oxepine-5-one (**E**) and 3,4-dihydro—2H-naphth[2,3-b] Oxepine-5-one (**G**). However, (R=Cl) chloride of 4-(2- naphthoxy) butyric acid, on exposure to stannic chloride in benzene at 0°C gave (**E**) which gave 2,3,4,5-tetrahydronaphth [2,1-b] Oxepine (**F**) on reduction. In the same way, reduction of a mixture of (**E**) and (**G**) gave 2,3,4,5-tetrahydronaphth [2,3-b] Oxepine (**H**).

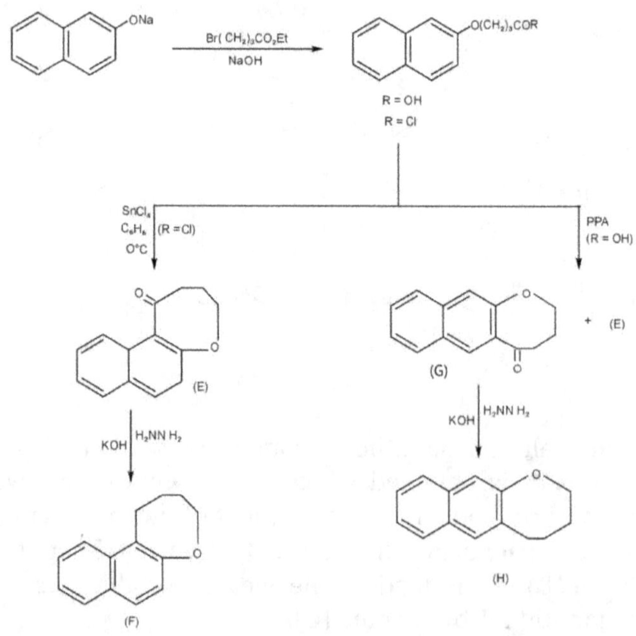

3.3.3.4 OXEPINE RING SYSTEMS CONTAINING MORE THAN THREE RINGS

3.3.3.4.1 TETRACYCLIC SYSTEMS

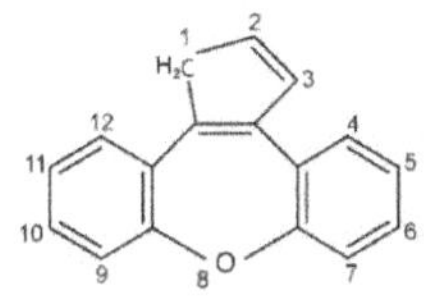

1 H - Dibenzo [b,f] cyclopentoxepin

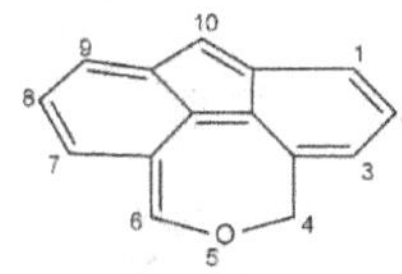

4 H- Fluorenzo [4,5- cde] oxepin

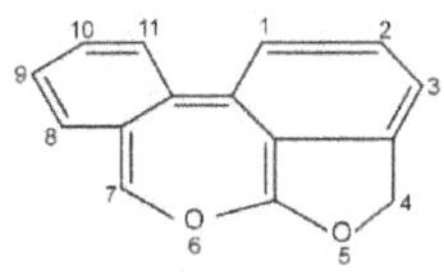

4 H - Isobenzofuro [1,7- cd] [2] benzoxepin

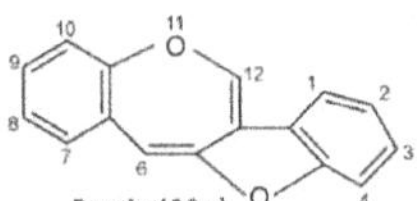

Benzofuro[3,2-c] [1] benzoxepin

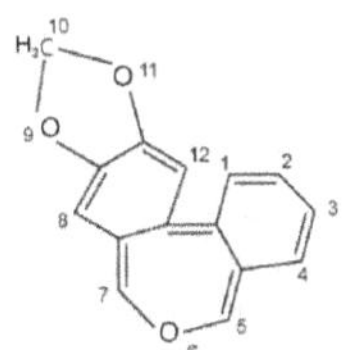

Benzo[d] dioxolo [4,5-b] [2] benzoxepin

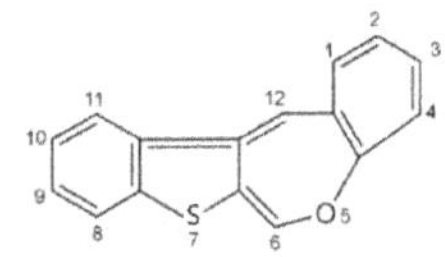

[1] Benzothieno [2,3-c] [1] benzoxepin

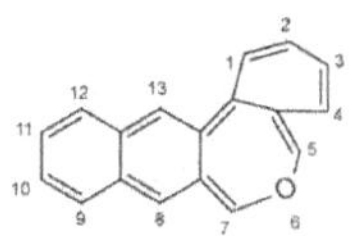

Benzo [c] naphth [2,3-c] oxepin

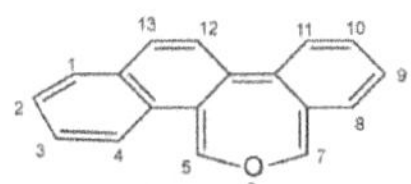

Benzo[c] naphth [2,1-c] oxepin

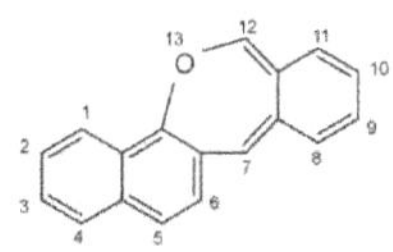

Benzo [e] naphth [1,2-b] oxepin

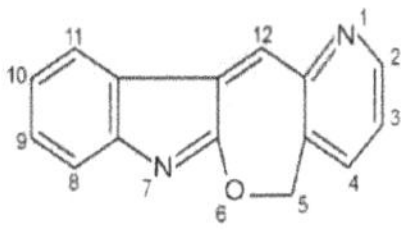

5H- Pyrido [2', 3' . 5,6] oxepino [2,3-b] indole

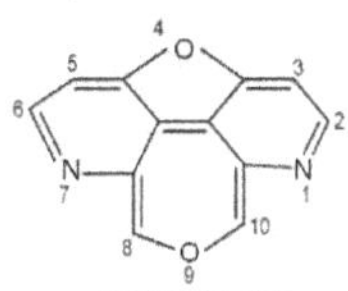

4,9- Dioxa - 1,7- diaza cyclohepta [def] fluorene

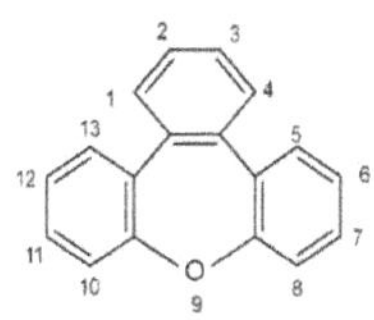

Tribenz [b,d,f] oxepin

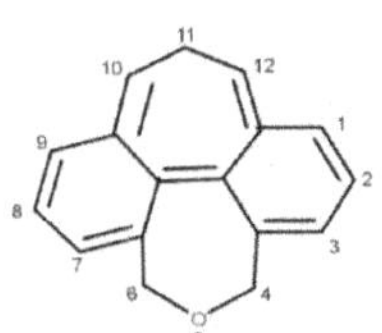

4H,6H-[2] Benzoxepino [6,5,4-def] [2] benzoxepin

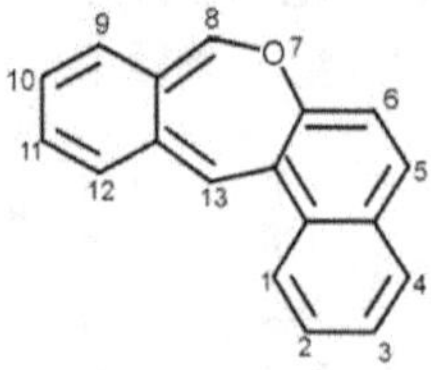

Benzo[e]naphth[2,1-b]
Oxepin

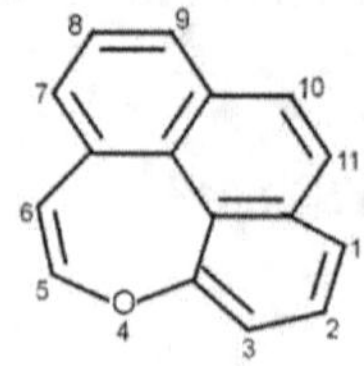

Phenthro[4,6-bcd]
Oxepin

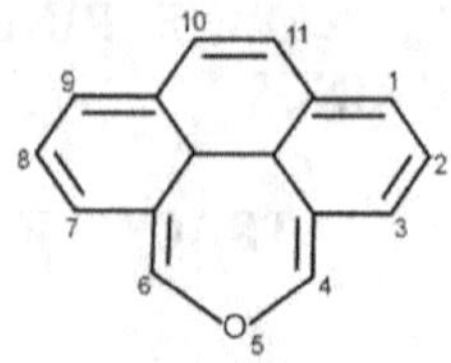

Phenanthro
[4,5-cde]
Oxepin

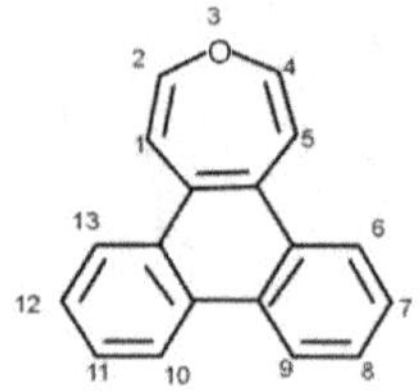

Phenanthro[9,0-d]
Oxepin

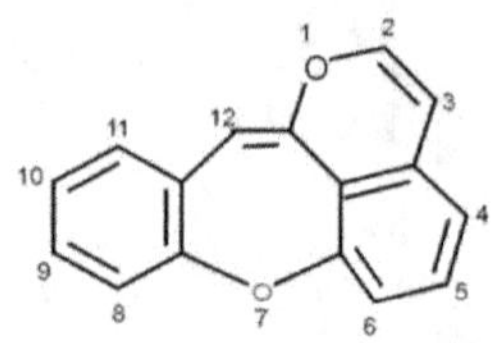

[2]Benzopyrano
[8,1-be][1]benzoxepin

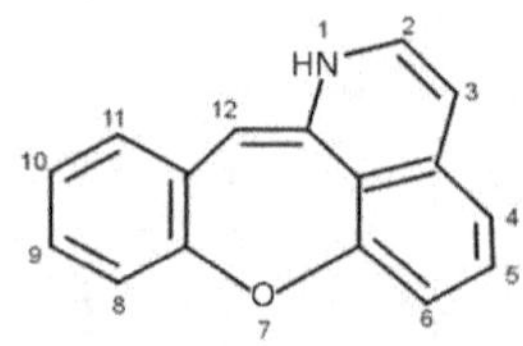

1H-Benz[6,7]Oxepino
[2,3,4-ij] isoquinoline

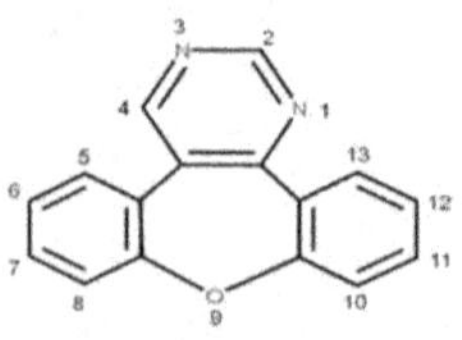

Dibenz[2,3,6,7]Oxepino
[4,5-d] pyrimidines

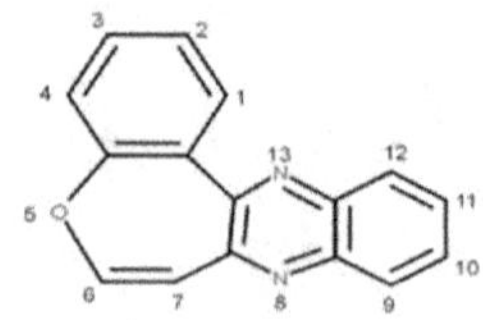

[1] Benzoxepino
[4,5-b]quinoxaline

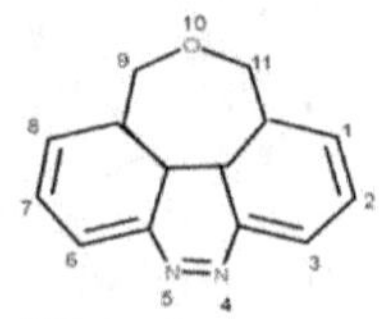

[2]Benzoxpino
[6,5,4-cde]Cinnolines

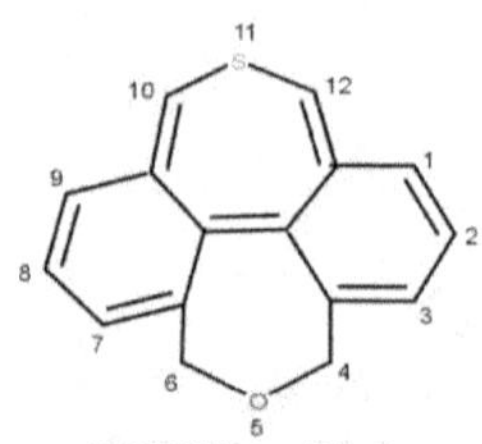

4H,6H-[2]Benzothiepino
[6,5,4-def][2]benzoxepin

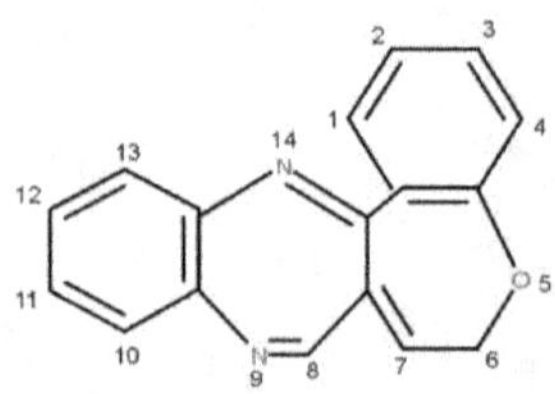

6H-[1]Benzoxepino
[5,4-b][1,5]
benzodiazepine

The formation of 4H-fluoreno[4,5-cde] oxepine is brought about by **thermal cyclization** of fluorenone-4,5-dicarboxylic acid.

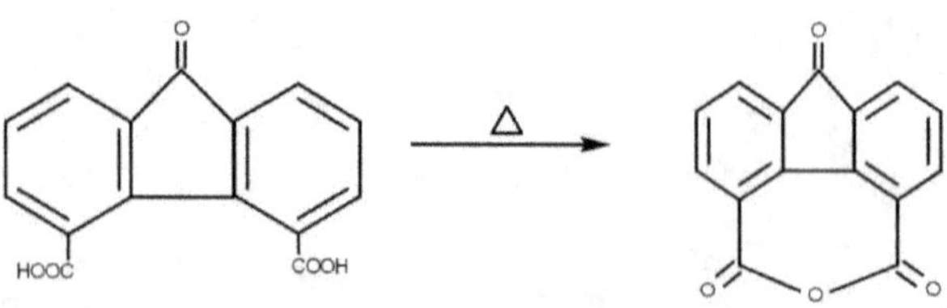

6H-Benzofuro[3,2-c][1] benzoxepine-12-one has been prepared by **isomerization** of 3-(o-hydroxy phenacyl) isocoumaranone with 2-(o-hydroxybenzyl) benzofuran-3-carboxylic acid as intermediate.

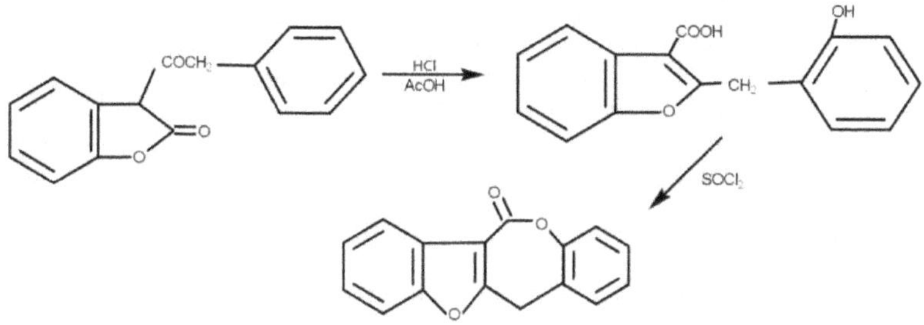

On heating dicarboxylic acid with acetic anhydride, (**B**) is obtained which regenerates the diacid on hydrolysis. A monoester is formed with methanol, monoamide with dilute ammonia and benzo fluorenone on heating.

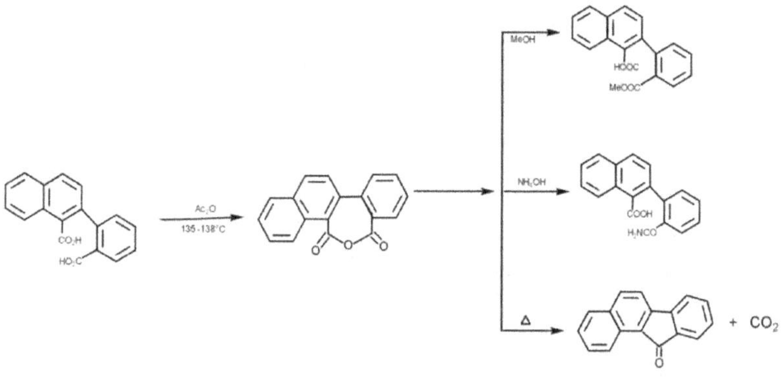

Dehydration of phenanthrene-4,5-dimethanol led to the formation of 4,6-dihydro phenanthro[4,5-cde] Oxepine.

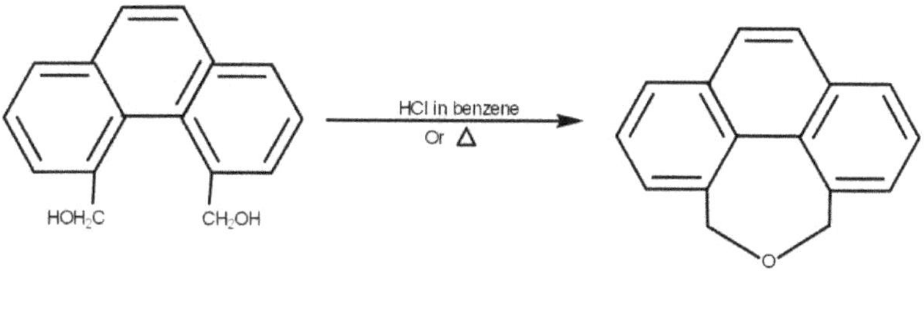

In the same way ring closure of diacid under the influence of polyphosphoric acid gives lactone.

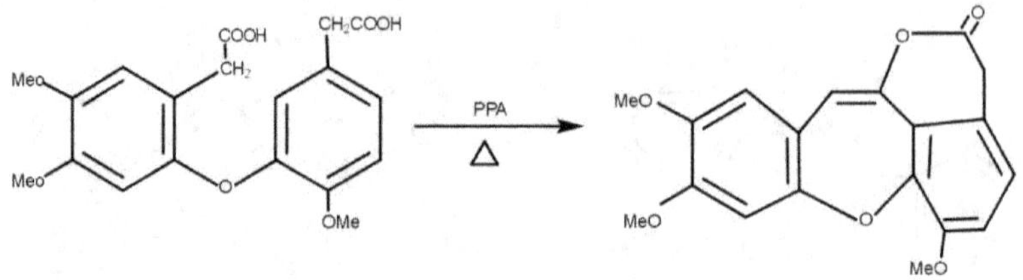

When 2,3-dihydro-7,8-dimethyl-1-benzOxepine-4,5-dione is condensed with o-phenylenediamine, the resulting compound is 6,7-dihydro-2,3-dimethyl[1] benzoxepino[4,5-b]quinoxaline.

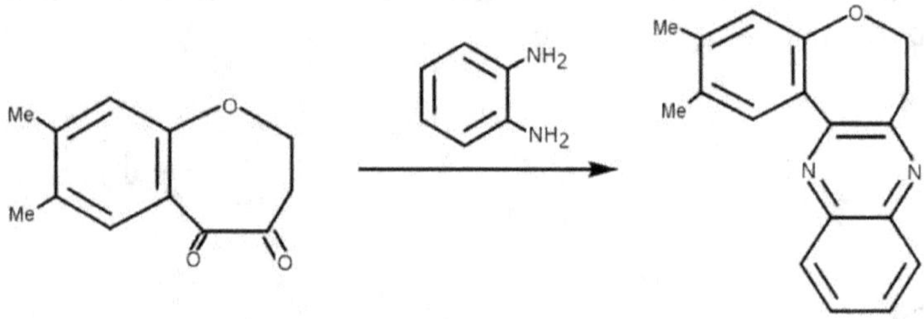

Condensation of o-phenylenediamine and 5-chloro-2,3-dihydro-benzOxepine-4-carboxaldehyde.

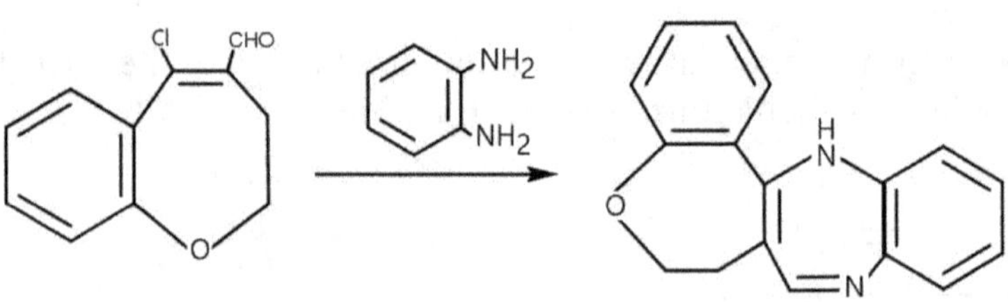

3.3.3.4.2 PENTACYCLIC SYSTEMS

14H-Dibenz[2,3:6,7] Oxepino[4,5-b]indole

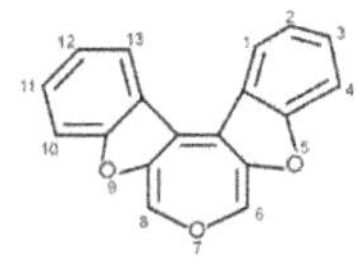

Oxepino[3,4-b:6,5-b] bisbenzofuran

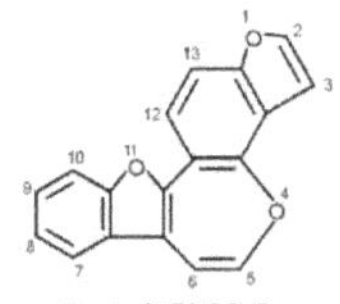

Oxepino[4,5-b:2,3-e'] bisbenzofuran

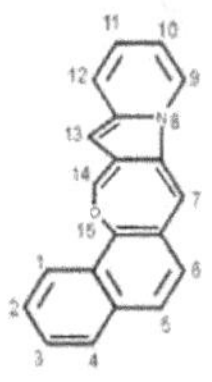

Naphth[2',1':6,7] Oxepino[3,4-b] indolizine

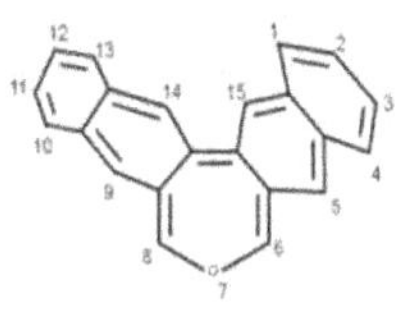

Dinaphth[2,3-c:2',3',-6] Oxepin

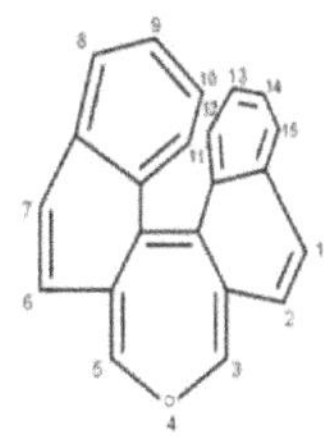

Dinaphth[2,1-C:1',2'-e] Oxepin

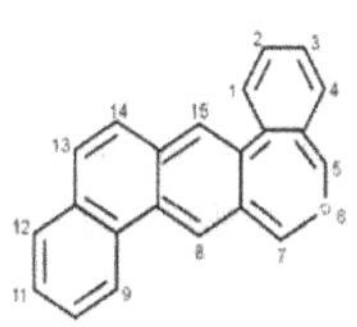

Benzo[c]Phenanthro[2,3-e] Oxepin

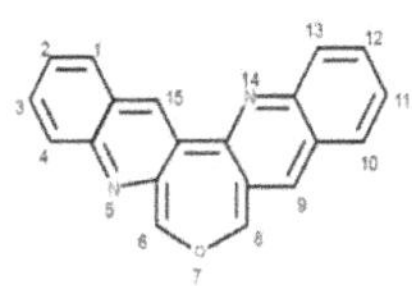

Oxepino[3,4-b:5,6-b'] diquinoline

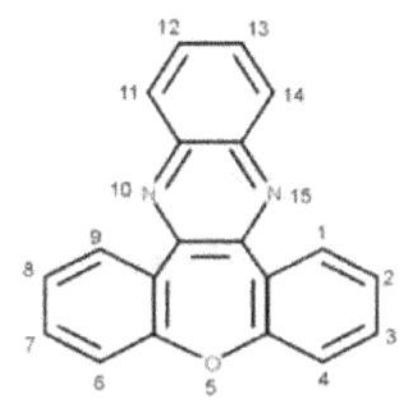

Dibenz[2,3:6,7]Oxepino [4,5-b]quinoxaline

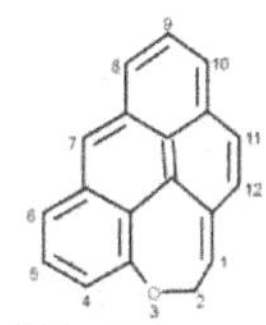

2H-Benzo[8,9] Phenanthro[4,5-bed] Oxepin

The reaction of phenyl hydrazine with 10,11-dihydro-2,3,6-trimethoxydibenz [b, f] Oxepine-10–one via Fisher Indole synthesis results in substituted 14-H-dibenz[2,3:6,7] oxepino[4,5,-b]indole.

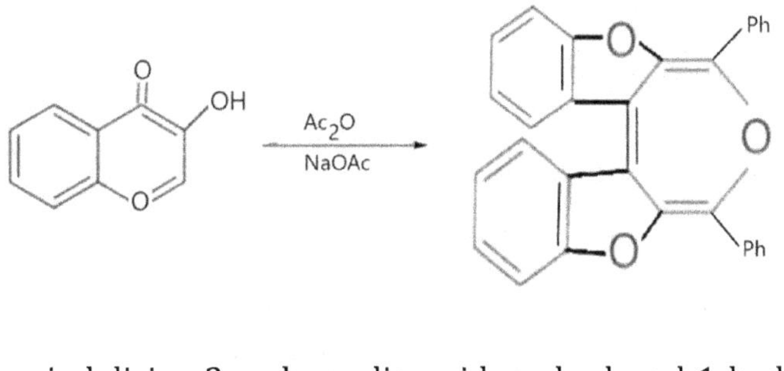

Likewise, indolizine-2-carbonxylic acid and phenyl-1-hydroxy-2-naphthoate at 120-180°C under reduced pressure gives seven membered lactone rings along with release of phenol.

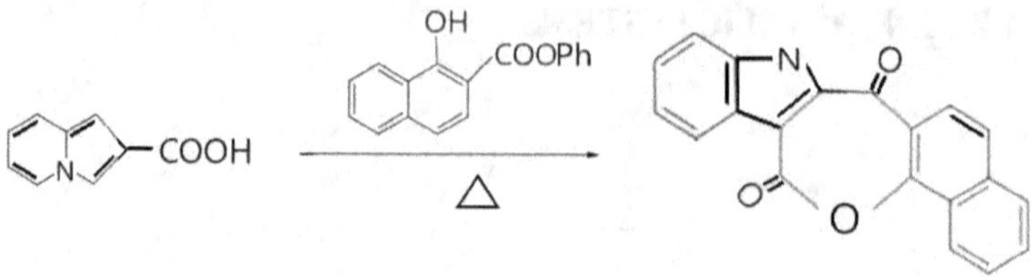

Ring closure can be achieved by passing dry hydrogen chloride gas through a solution of diol in diphenyl either at 100°C.

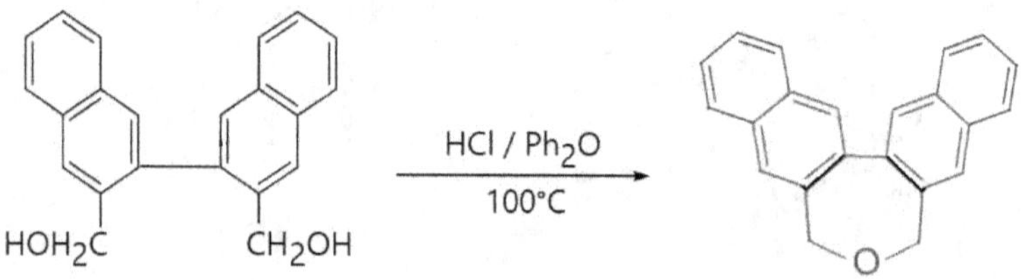

Condensation of dibenz [b, f] oxepine-10,11-dione with o-phenylenediamine leads to the formation of yet another pentacyclic compound.

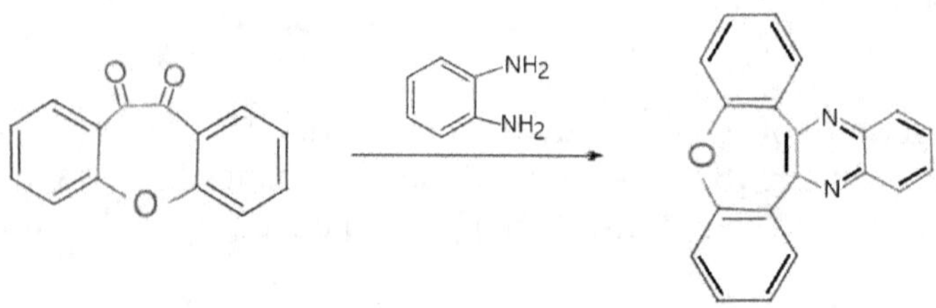

3.3.3.3.4.3 HEXACYCLIC SYSTEMS

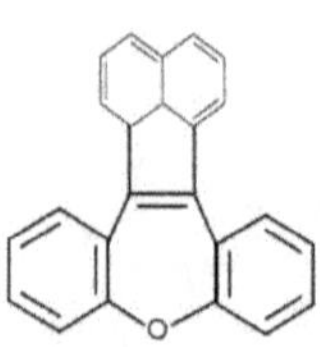

Acenaphtho [1,2-d] dibenz [b,f] oxepin

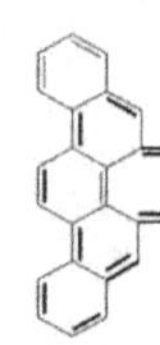

Piceno [6,7-cde] oxepin

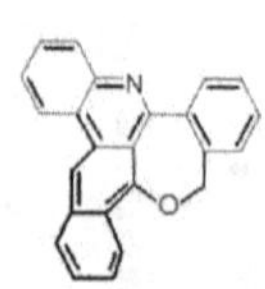

16H-Dibenzo [b,h][2] benzoxepino [3,4,5-de][1,6] naphthyridine

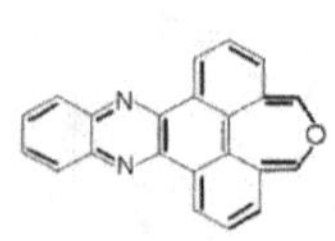

Oxepino [3',4',5',6':4,5] phenanthro [9,10-b] quinoxaline

On heating hydroxy acid with diphenyl ether for 10 minutes, lactone formation is reported.

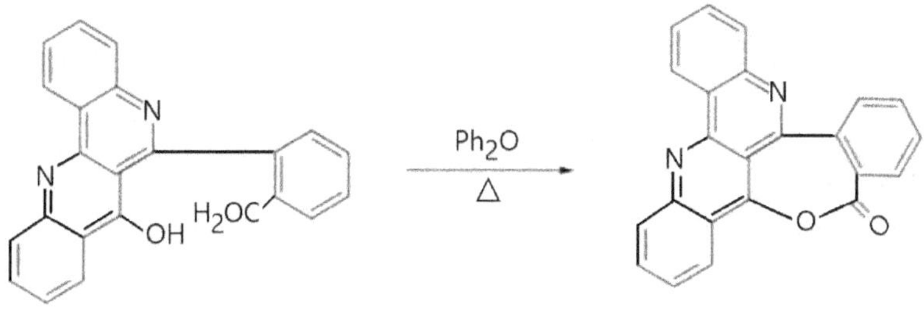

9,10-Phenanthrenequinone-4,5-dicarboxylic acid when condensed with o-phenylenediamine gives a diacid as intermediate which soon crystallizes to hexacyclic derivative.

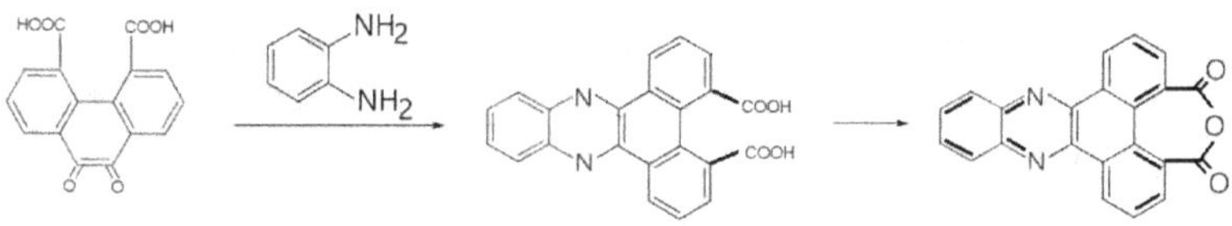

3.3.3.4.4 HEPTACYCLIC SYSTEMS

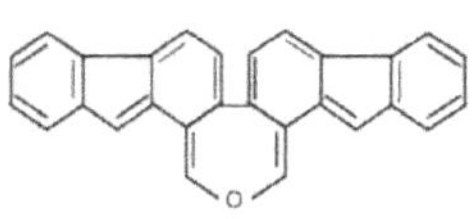
Difluoreno [1,2-c : 2',1'-e] oxepin

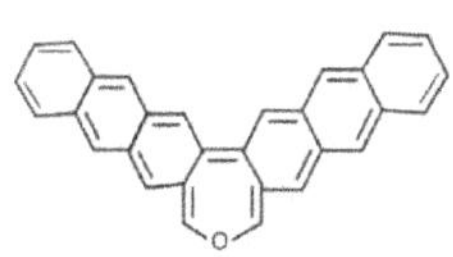
Dianthra [1,2-c : 2',1'-e] oxepin

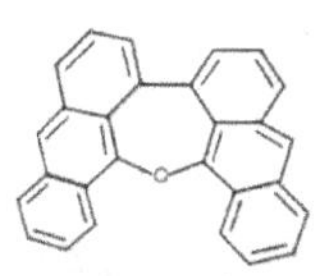
Dianthra [9,1-bc : 1',9'-e,f] oxepin

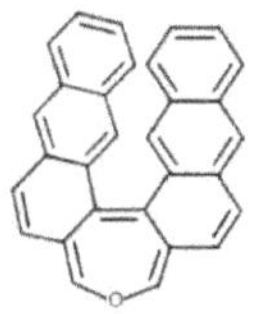
Dianthra [2,1-c : 1',2'-e] oxepin

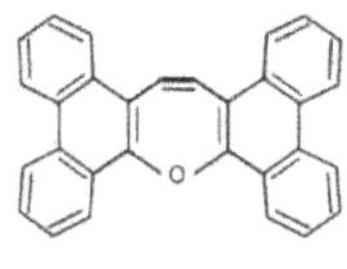
Diphenanthro [9,10-b : 9',10'-f] oxepin

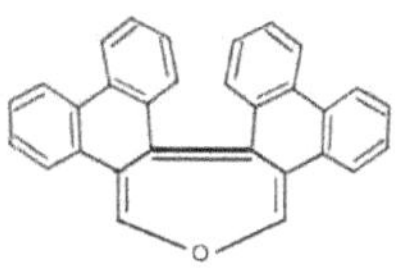
Diphenanthro [9,10-c : 9',10'-e] oxepin

Benzo [6,7] perylo [1,12-cde] oxepin

When 2,2'-bis (fluoren-9-one-1,1'-dicarboxylic acid) is treated with acetic anhydride gives derivative of heptacyclic system.

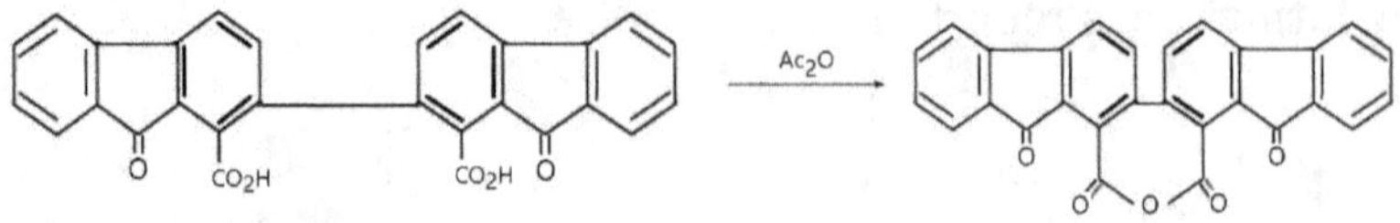

In the same way, acetic anhydride treatment with dicarboxylic axid leads to ring closure.

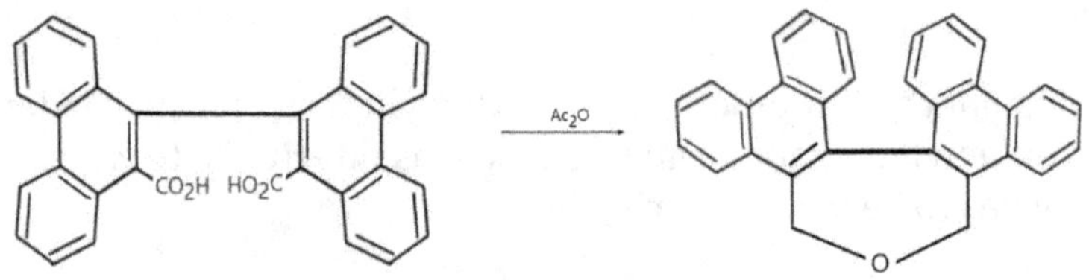

Reaction of 1,2-quinine with alkaline hydrogen peroxide gave a di-acid which on cyclization with nitrobenzene on heating resulted in a highly condensed system.

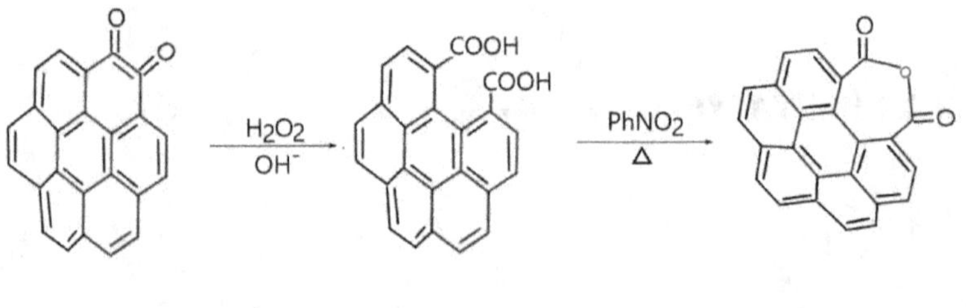

3.3.3.4.5 SYSTEMS CONTAINING MORE THAN SEVEN RINGS

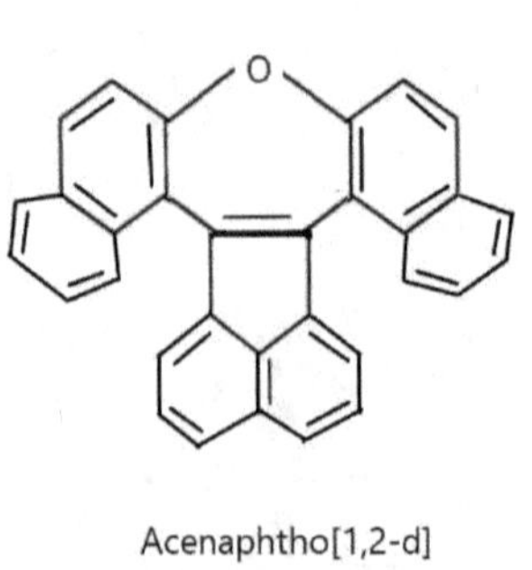

Acenaphtho[1,2-d] dinaphtho[2,1-b:1',2'-f] oxepin

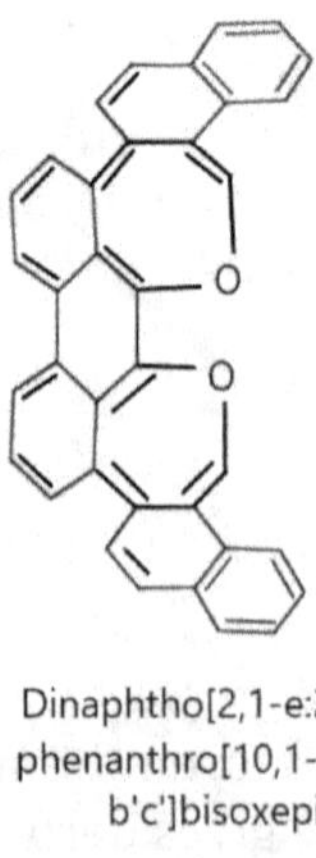

Dinaphtho[2,1-e:2',1'-e] phenanthro[10,1-bc:9,8-b'c']bisoxepin

3.3.3.5 DIOXEPINE RING SYSTEMS

3.3.3.5.1 MONOCYCLIC DIOXEPINE RING SYSTEMS

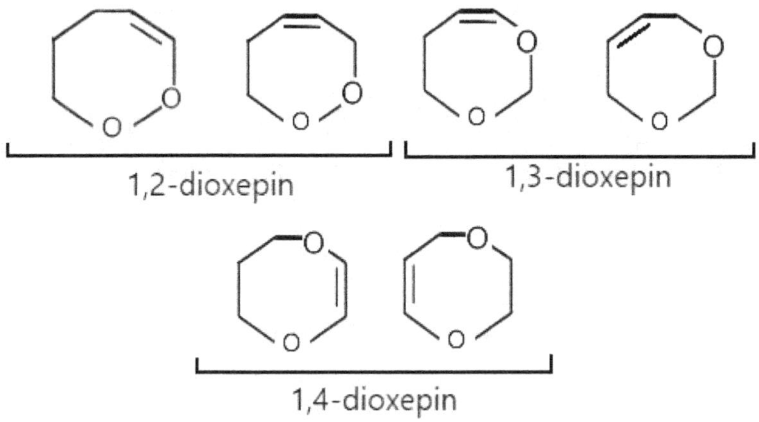

Out of these three varieties 1,3-dioxepine have been most extensively studied.

SYNTHESIS

Reaction of cis-2-butene-1,4-diol with aldehyde, **catalysed** by strong acid (p-toluene sulphonic acid or concentrated sulphuric acid) results in the formation of 1,3-diOxepine derivative at 180-200°C.

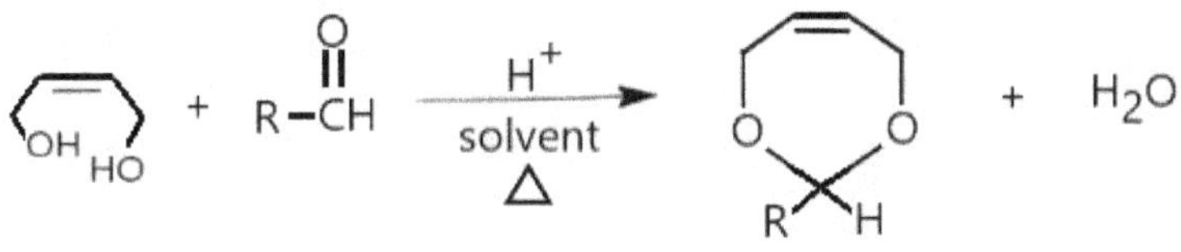

Cyclohexanone reacts with cis-2-butene-1,4-diol gave Spiro derivative of 1,3-dioxepine (7,12-dioxaspiro [5,6] dodec-9-ene) in presence of benzene and p-toluene sulphuric acid.

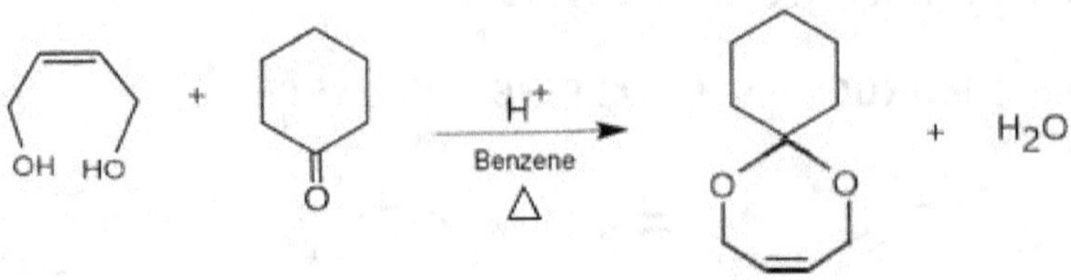

In the same way, cis-2-butene-1,4-diol on reaction with methyl vinyl ether at 140-180°C leads to 2-methyl-4,7-dihydro-1,3-dioxepine.

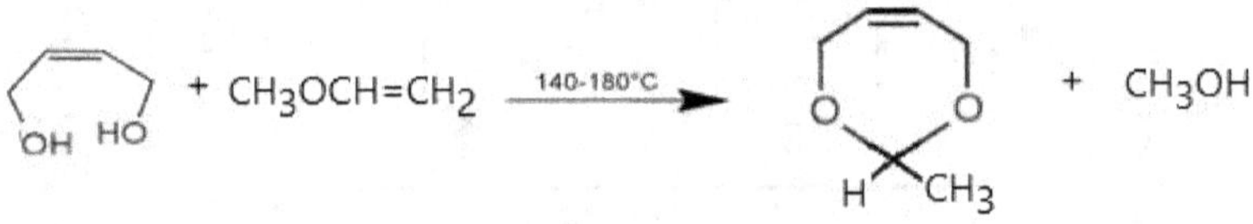

REACTIONS

When 1,3-dioxepine is treated with halogen, it undergoes addition of halogen across the double bond and di-halogen product is obtained.

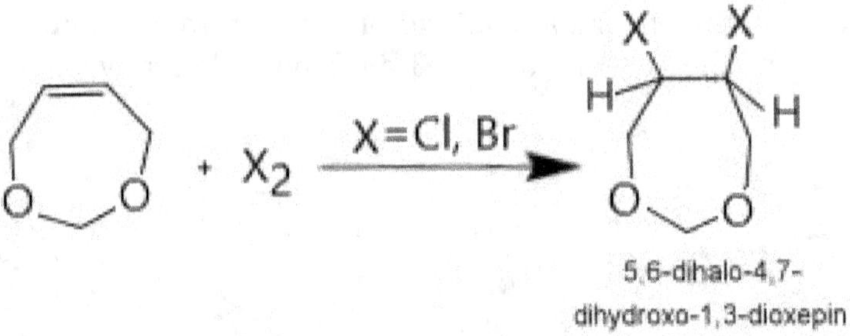

Reduction in presence of Raney Nickel at 100° results in the formation of dihydro product.

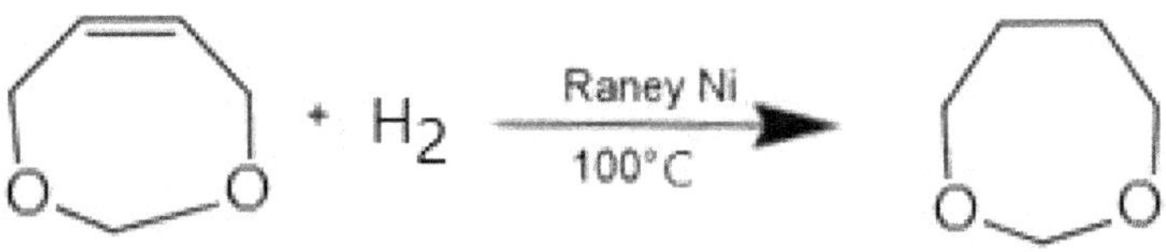

When 1,3- dioxepine reacts with sulphuric acid to give dihydrofuran whereas with more dilute acid leads to ring opening.

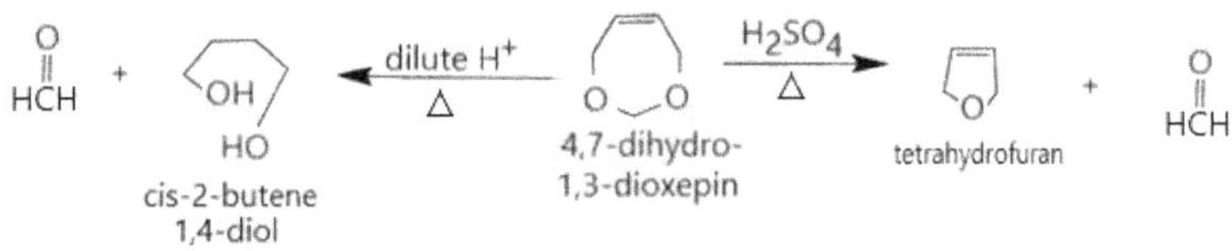

3.3.3.5.2 CONDENSED DIOXEPINE RING SYSTEMS

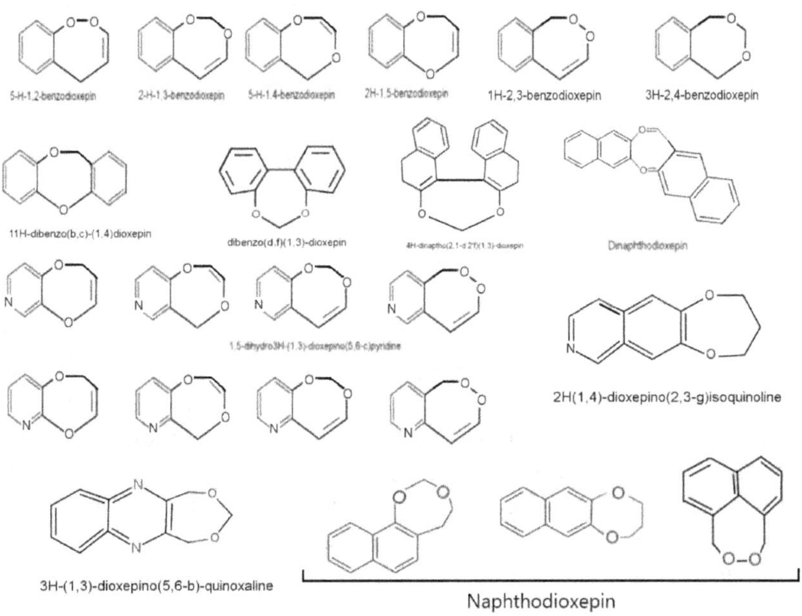

SYNTHESIS

Preparation of 2,3- dihydro-5H-1,4-benzo dioxepine is brought about by the reaction of 2- hydroxy methylphenyl and 1,2-dibromoethane in basic conditions.

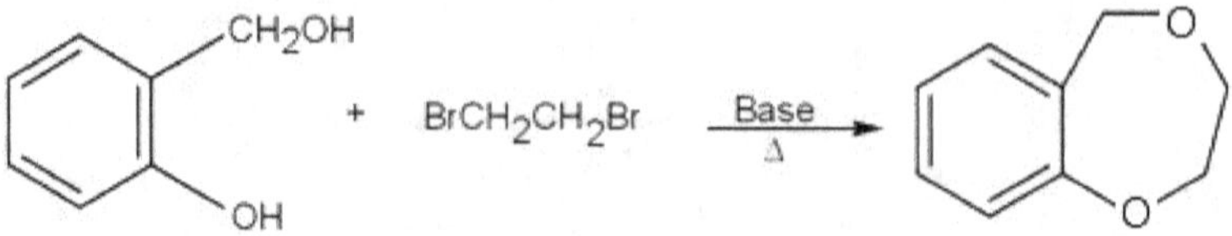

Condensation of 1,3-dibromopropane with 1,2-dihydroxybenzene in the presence of sodium methoxide in boiling methanol.

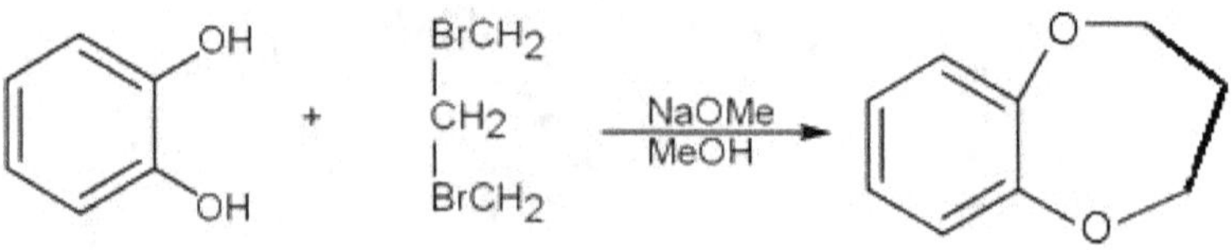

1,3-Dichloro-2-methylene propane and 1,2-dihydroxybenzene on being **refluxed** in ethanol for 24 hours resulted in 3-methylene-2,4-dihydro-3H-1,5-benzodioxepine.

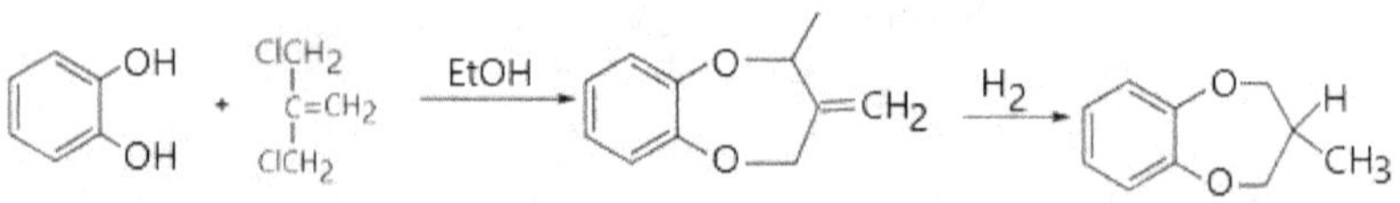

REACTIONS

4-Ethoxy-4,5-dihydro-H-2,3-dioepin-1-ol gives homophthalic acid on being oxidised with hydrogen peroxide. Reduction with zinc in ethanol - water - acetic acid resulted in homo-phthaldehyde (a yellow oil); however, **reduction** with lithium aluminium hydride yielded homophthalyl alcohol (colourless oil). Reduction with 5% sodium hydroxide is an exothermic reaction and gives o-carboxyphenylacetaldehyde which is unstable and finally turns into acid.

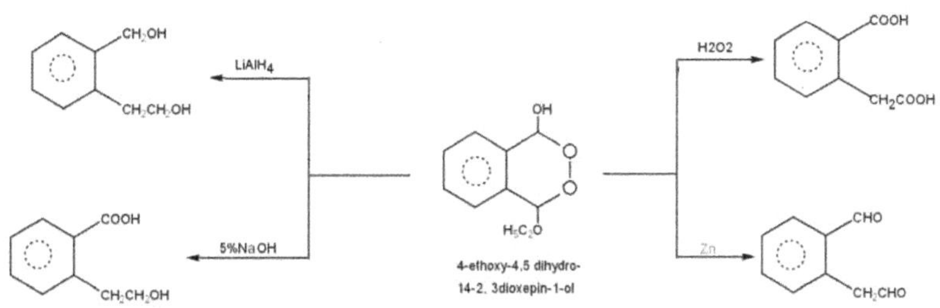

3.3.3.5.3 TRIOXEPINE RING SYSTEMS

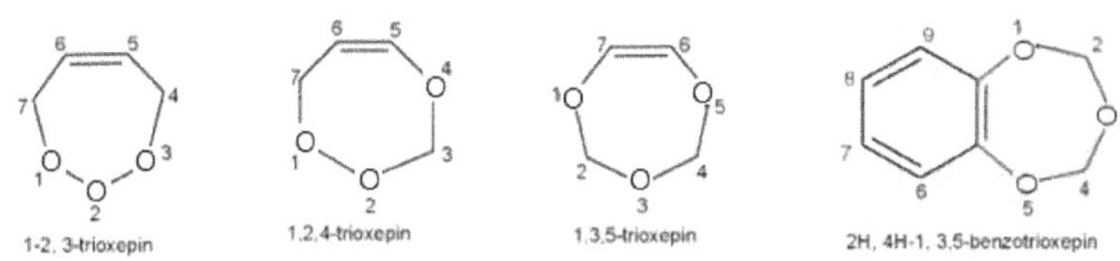

3.4. THIEPIN

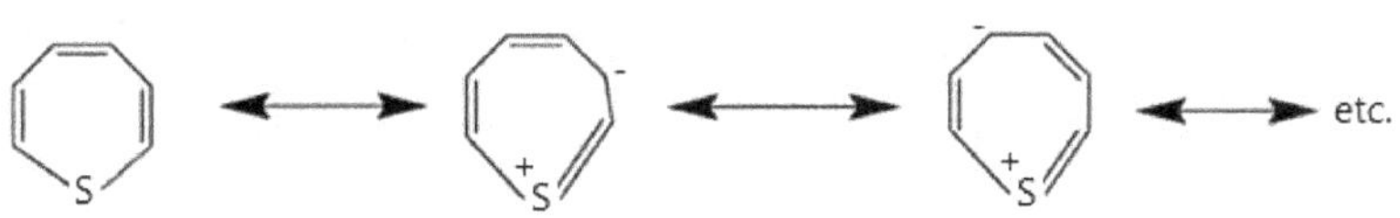

3.4.1. SYNTHESIS OF THIEPIN

Thiepin 1,1-dioxide (sulfone) is synthesized by the following equation as achieved by Mock.

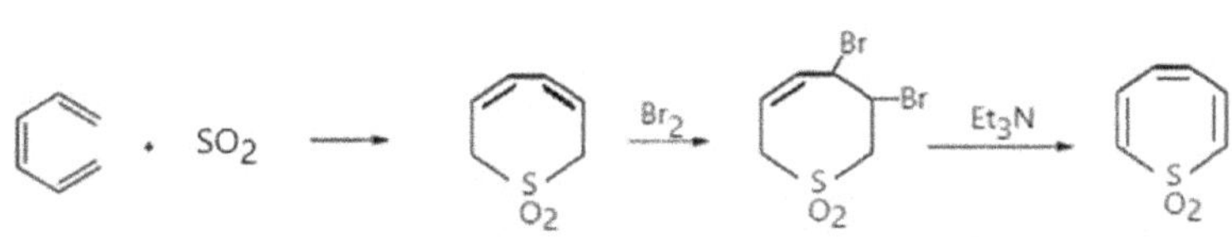

3.4.2. PROPERTIES OF THIEPIN

Sulfone absorbs three moles of hydrogen in the presence of palladium-on-charcoal in ethyl acetate to obtain thiepane 1,1-dioxide. It **decomposes** with heat (above melting point) and forms benzene via an intermediate releasing sulphur dioxide.

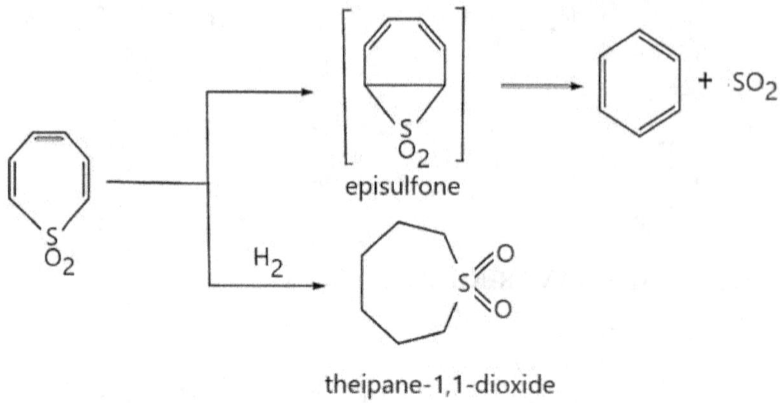

3.4.3 TYPES OF THIEPIN

3.4.3.1 DIHYDRO THIEPIN

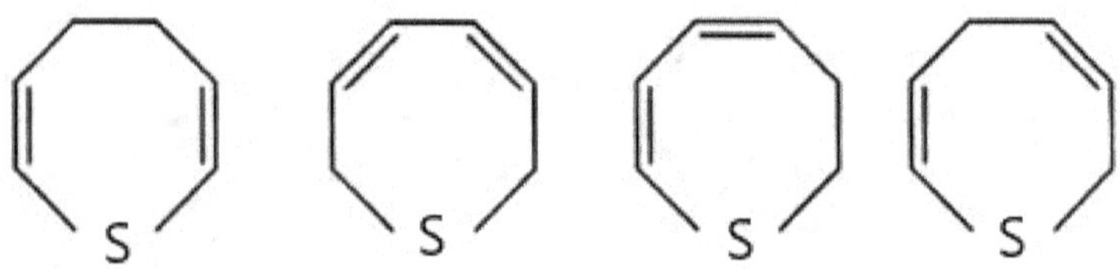

SYNTHESIS

Cis-1,2-divinylthiirane rearranges thermally to result in 4,5-dihydrothiepin.

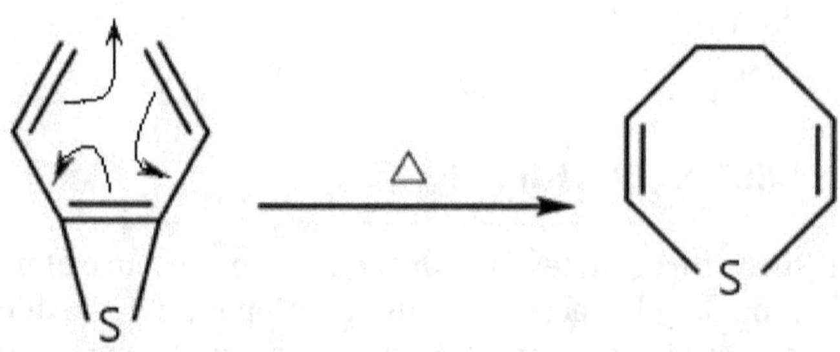

The cyclization of hexatriene leads to the formation of 2,7-dihydrothiepin-1,1-dioxide in the presence of sulphur dioxide in ether at room temperature. This reaction is reversible, and reactants are regenerated at 150-160°C. **Reduction** forms saturated sulfone.

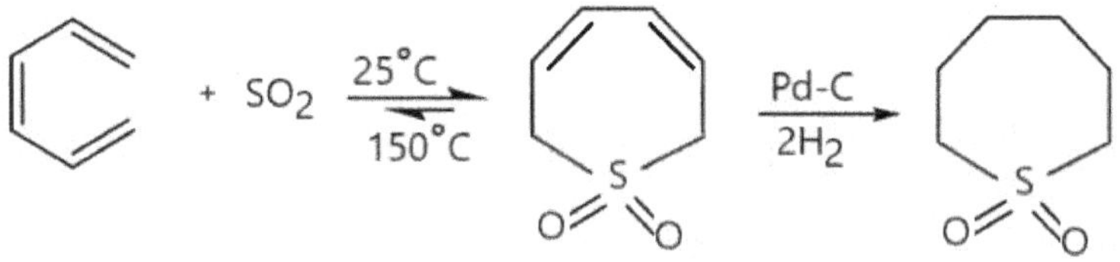

When 3,4-diphenyl-1,3,5-hexatriene reacts with sulphur monoxide, it forms 4,5-diphenyl-2,7-dihydrothiepin-1-oxide; this product is oxidised by m-chloroperbenzoic acid to get sulfone; on reduction with lithium aluminium hydride gives 4,5-diphenyl-2,7-dihydrothiepin.

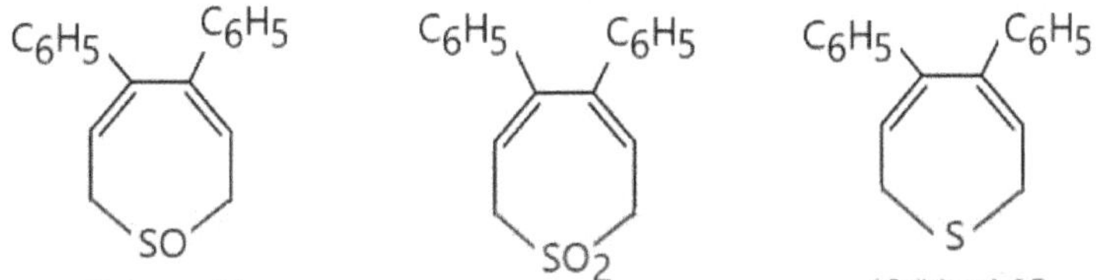

4,5-diphenyl-2,7-dihydrothiepin-1-oxide

Sulfone

4,5-diphenyl,-2,7 dihydrothiepin

3.4.3.2 TETRAHYDROTHIEPIN

2,3,6,7-tetrahydro thiepine-1,1-dioxide

2,3,4,7-tetrahydro thiepine-1,1-dioxide

3.4.3.3 THIEPANES

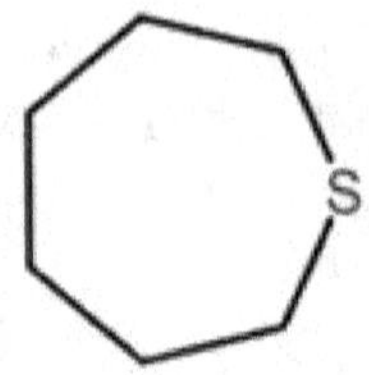

SYNTHESIS

Thiepane was originally synthesized from potassium disulphide and 1,6-diiodohexane. The same cyclization can also be brought about using sodium sulphide and 1,6-dibromohexane.

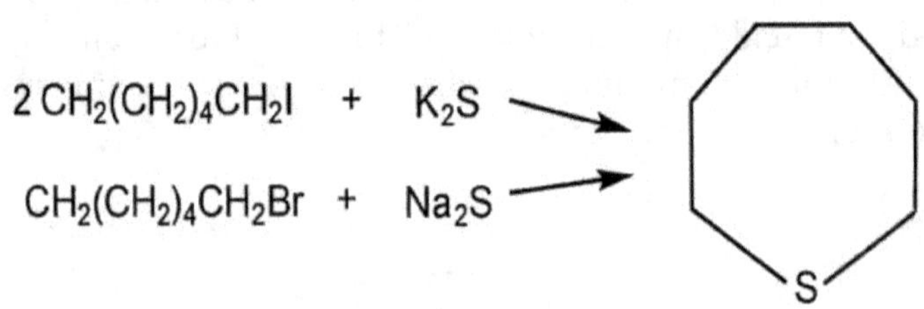

The reaction of 1,5-hexadiene with SCl_2 in hexane at 5-10°C gives 3,6-dichlorothiepane.

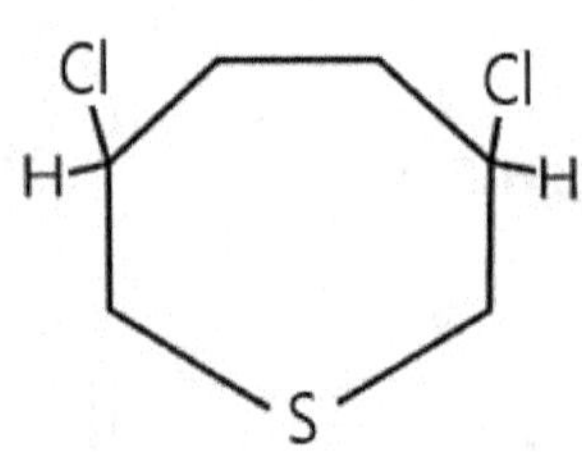

REACTIONS

Hydrolysis of 2-acetoxythiepane yielded 2-hydroxythiepane.

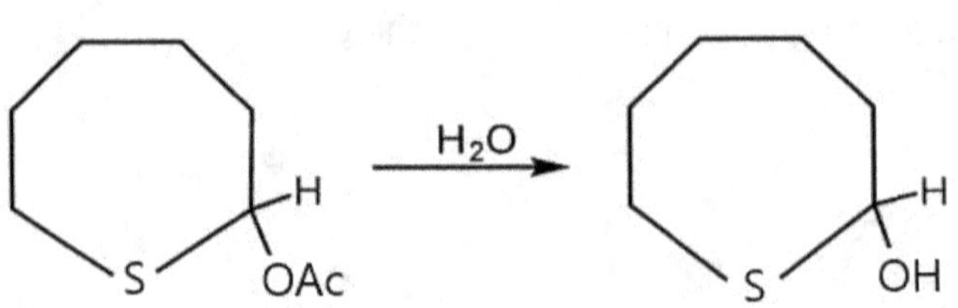

Thiepane reacts with N-chloro succinimide in carbon tetrachloride or benzine to form 2-chlorothiepane which oxidizes to 2-chloro thiepane-1,1–dioxide with peracetic acid and with methyl and phenyl Grignard's reagent gets converted to 2-methyl thiepane (66%) and 2-phenyl thiepane (56%) respectively.

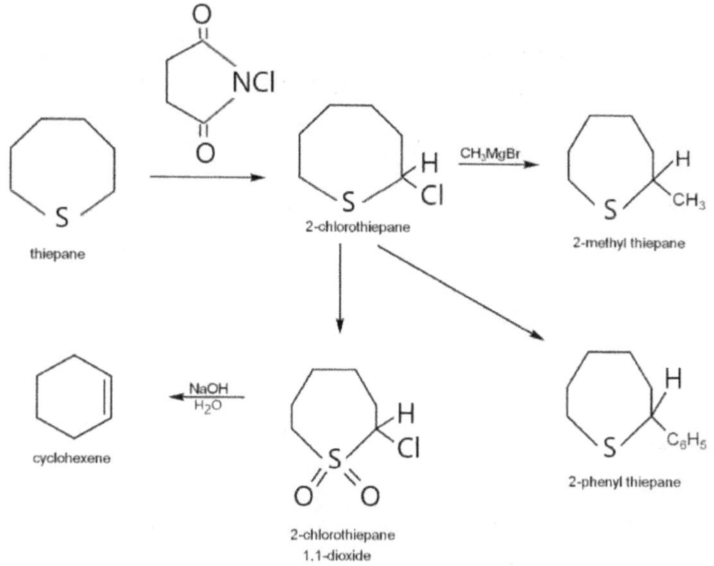

3.4.3.4. THIEPANONES

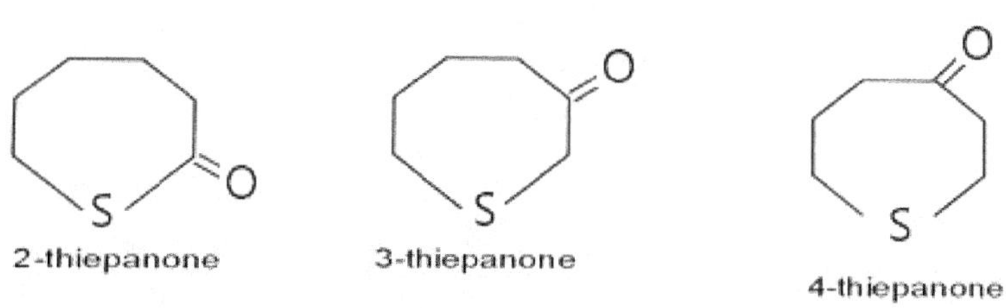

SYNTHESIS

Addition of diazoalkanes to a thiacyclohexane-3-one gives a mixture of substituted 3-thiepanone and 4-thiepanone.

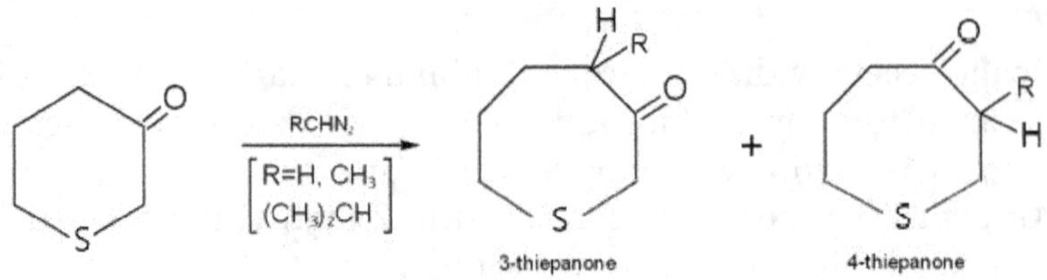

3-thiepanone 4-thiepanone

Tetrahydro-1,4-thiapyrone gives 4-thiepanone by ring enlargement. An epoxide was also formed; the two liquids separated by **distillation**.

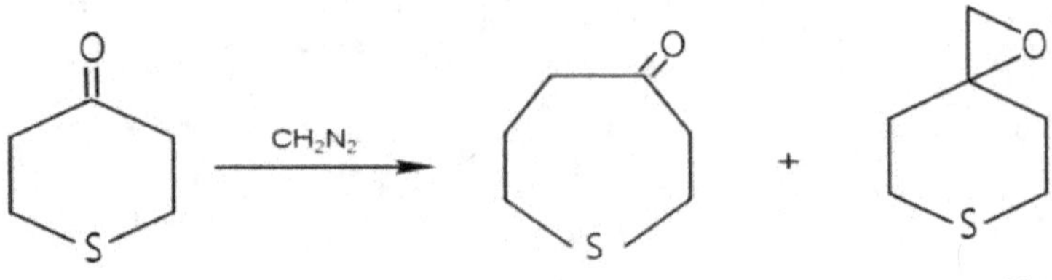

4-thiepanone epoxide

REACTIONS

Ring contraction of 3-thiepanone leads to the formation of 2-methyl-thiacyclohexane.

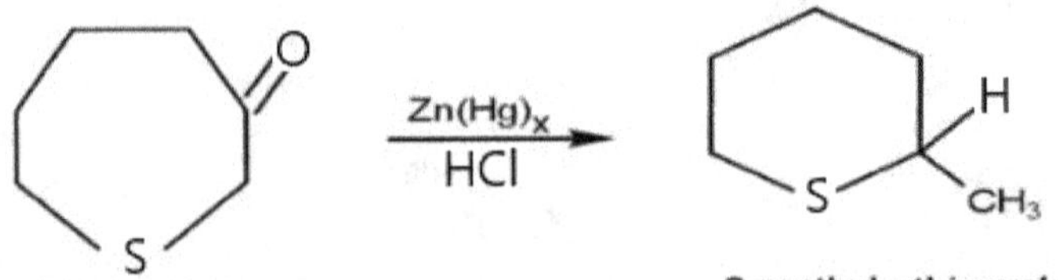

3-thiepanone 2-methyl - thiacyclo hexane

Oxidation of 4-thiepanone with hydrogen peroxide gives sulfone while oxidation with sodium metaperiodate gives sulfoxide which is hygroscopic in nature.

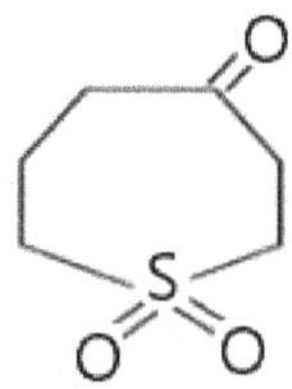

sulfone

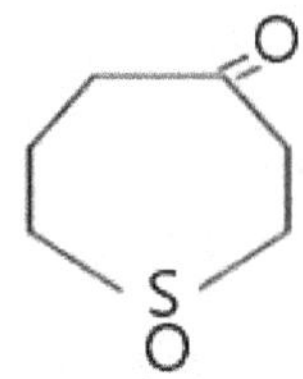

sulfoxide

Mono-hydrazone when oxidized with manganese dioxide gives an α-diazoketone which rearranges on spontaneous decomposition.

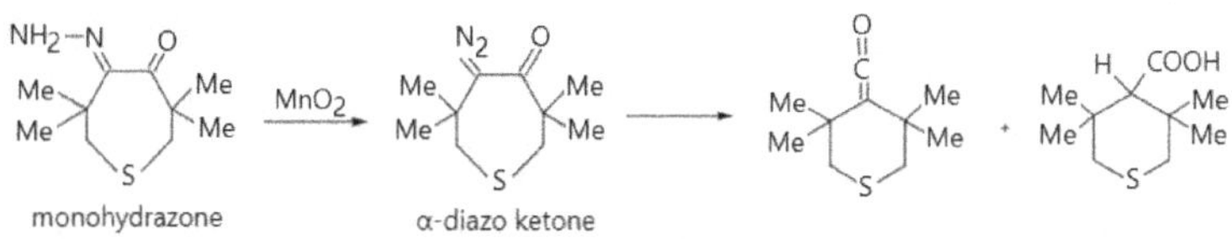

An abnormal **rearrangement** is observed on treatment with triphenyl phosphine dibromide in boiling dimethyl formamide (DMF).

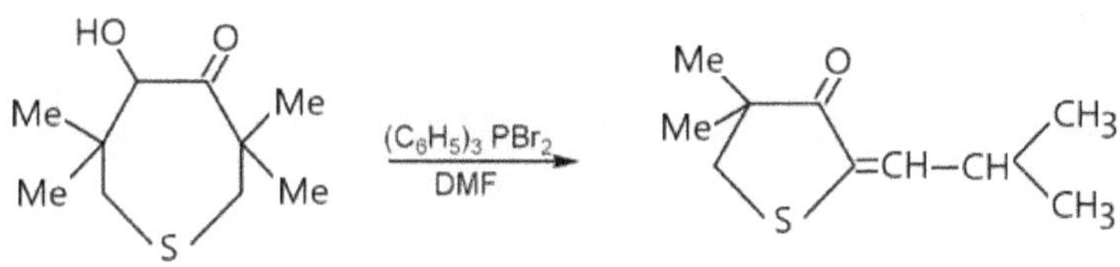

3.4.3.5. RINGS CONTAINING TWO SULPHUR ATOMS

Dithiepin are seven-membered heterocyclic compounds containing two sulphur atoms. These compounds can be arranged in three ways:

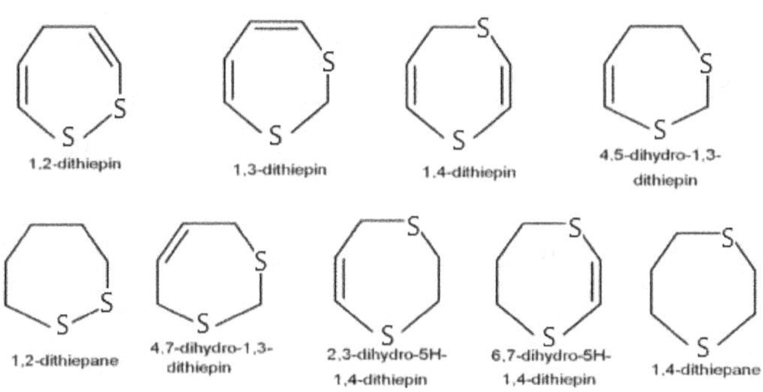

SYNTHESIS

1,2-dithiepone-3-propionic acid was obtained as a minor product after a multistep reaction.

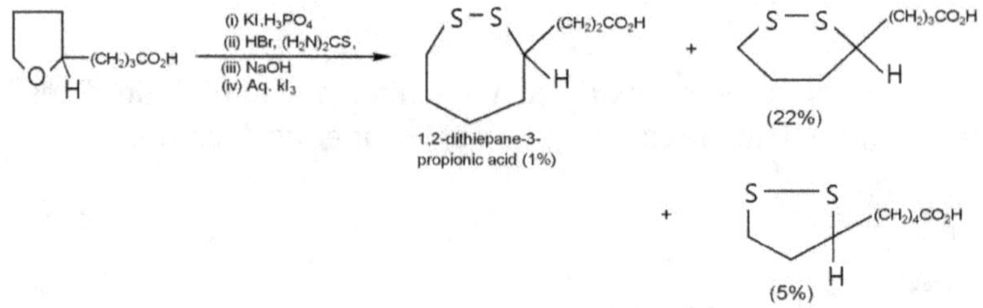

Among other methods, 1,5-addition of sulphur to perfluoro-1,4-pentadine is reported as a good method to yield two sulphur ring compounds.

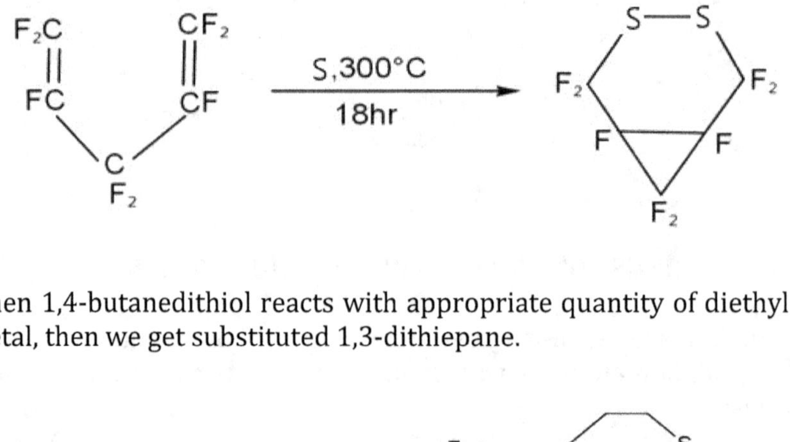

When 1,4-butanedithiol reacts with appropriate quantity of diethyl acetal, then we get substituted 1,3-dithiepane.

2-Methyl-6,7-dihydro-5H-1,4-dithiepin is formed on cyclization of allylic sulphide in the presence of potassium tert-butoxide.

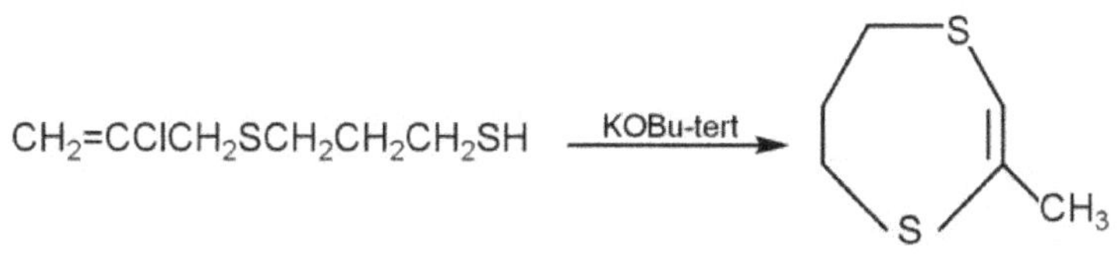

2-methyl-6,7-dihydro-5H-1,4-dithiepin

Dihydrothiepin derivatives can also be got by **double displacement** of chloride from chloromethyl-allyl chloride. In the same way, 6,7-dihydra-5H-1,4-dithiepin system is also got by reaction of propylene thioketal with P_2O_5 in pyridine.

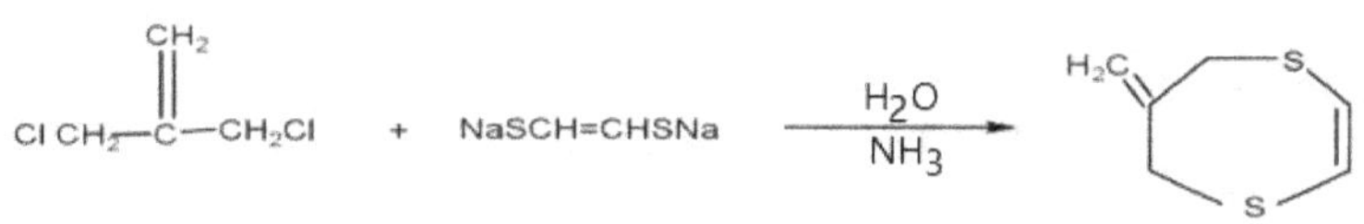

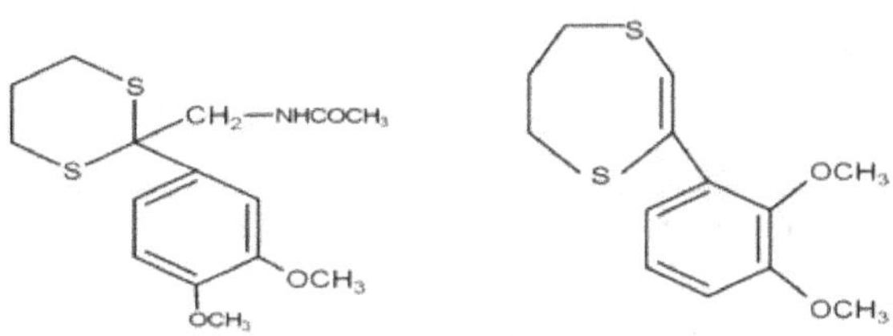

6,7-dihydro-5H-1,4-dithiepin system

1,4-dithiepane is isolated by the reaction of 1,2-ethanedithiol and 1,3-dibromopropane. On oxidation with potassium permanganate, the product forms disulfone.

REACTIONS

Irradiation of 1,2-dithiepane leads to the formation of dithiol while in the presence of alkali, mono-sulphide is formed.

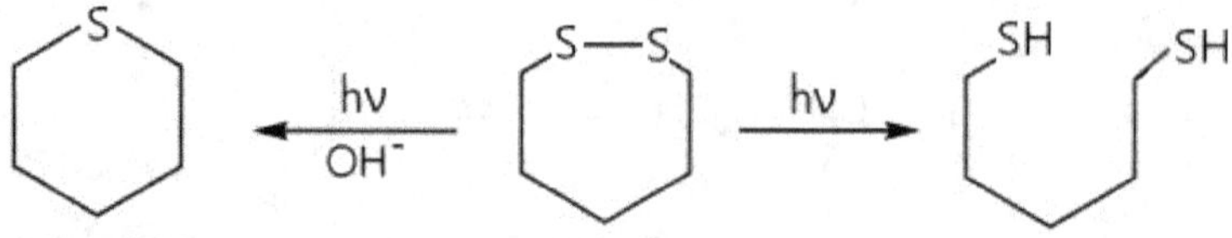

mono sulfide 1,2-dithiepane dithiol

Oxidation reaction of 1,4-dithiepan-6-one with 2 moles of hydrogen peroxide gives a 1,4-disulphoxide while with 4 moles of hydrogen peroxide, leads to the formation of 1,4-dithiepan-6-one –1,1,4,4-tetraoxide.

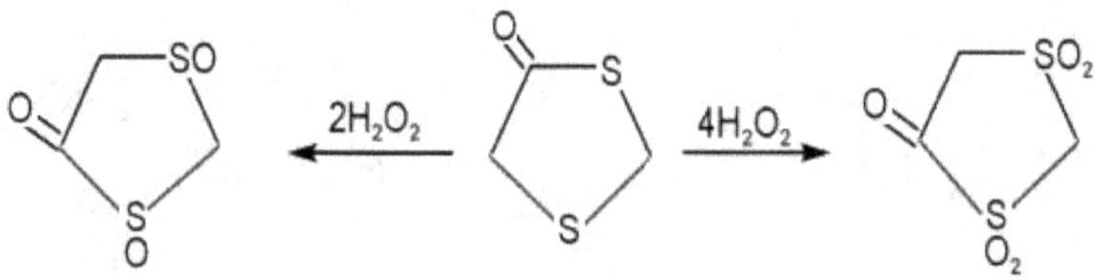

dioxide tetraoxide

3.4.3.6 RINGS CONTAINING THREE OR MORE SULPHUR ATOMS

The seven membered heterocyclic compounds containing three sulphur atoms are called trithiepin (with double bonds) or trith-iepanes (without double bonds).

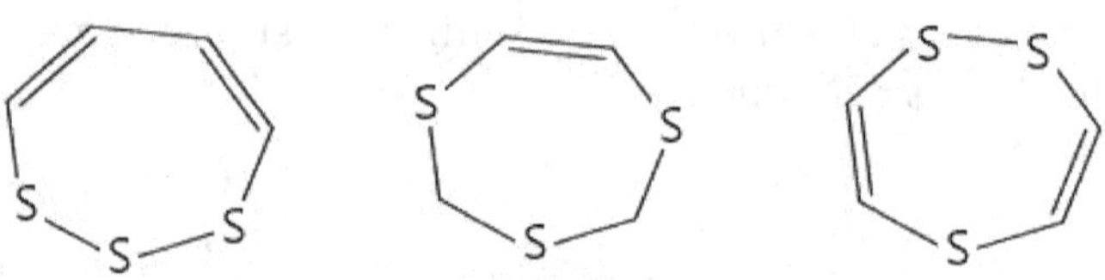

1,2,3-trithiepin 1,3,5-trithiepin 1,2,5-trithiepin

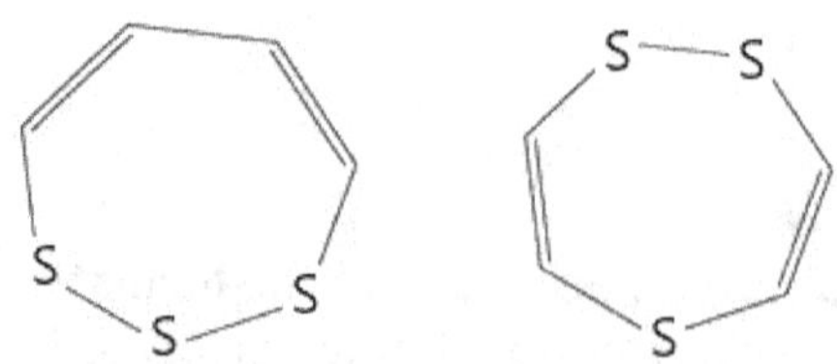

1,2,3-trithiepane 1,2,5-trithiepane

SYNTHESIS

1,2,3-Trithiepin compound is completely reported as saturated system(1,2,3-trithiepane) by the reaction of acetonylacetone with hydrogen sulphide at considerably high pressure.

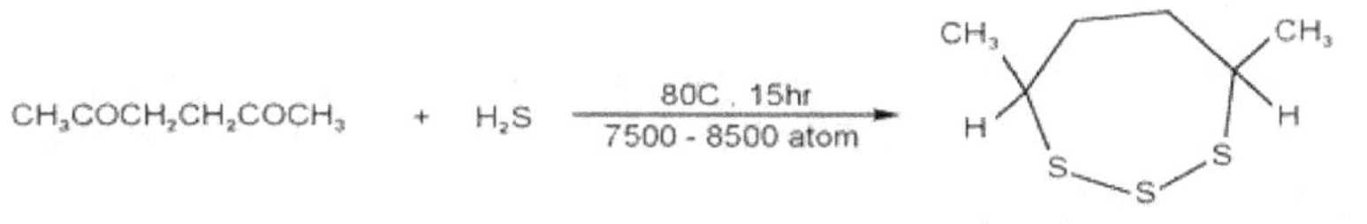

4,7-dimethyl-1,2,3-trithiepane

Tetrafluoro ethylene on **sulfonation** gives 1,2,5- trithiepane.

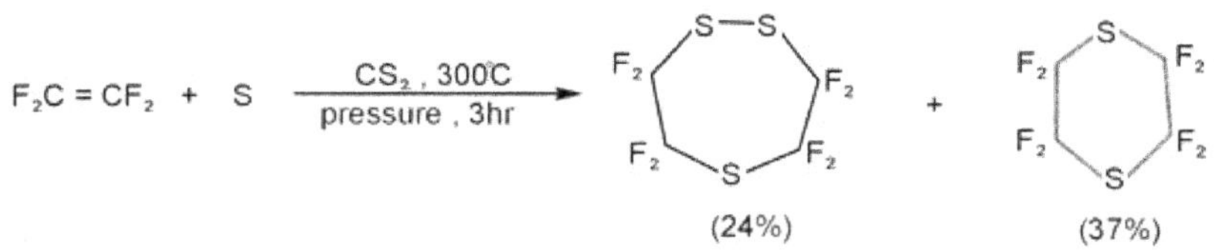

Reaction of disodium dimercaptomaleonitrile and bis(chloromethyl) sulphide forms 2,3- dicyano-1,4,6-trithia-2-cycloheptene.

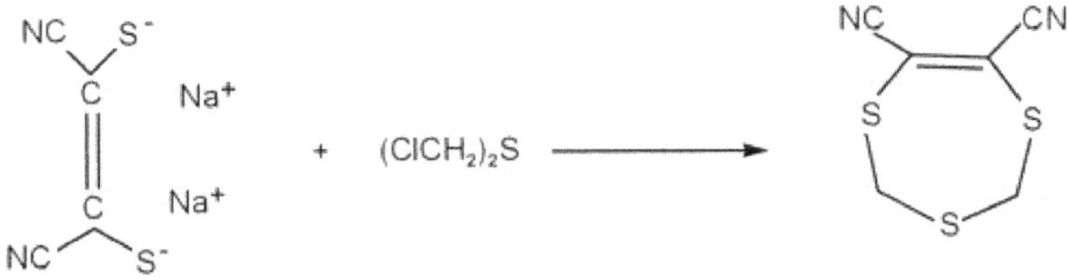

REACTIONS

Octafluro-1,2,5-trithiepane, on treatment with chlorine, gave sulfenyl chloride.

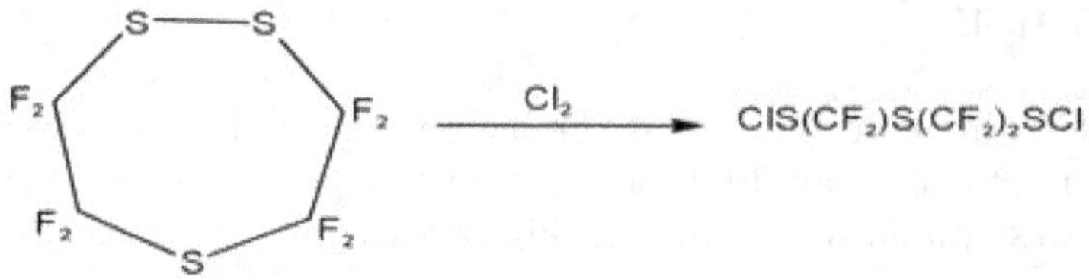

There are four possible arrangements of seven membered hetero-cyclic compounds containing four sulphur atoms; 1,2,3,4; 1,2,3,5; 1,2,4,5; 1,2,4,6 but of these only 1,2,4,6 ring system has been confirmed.

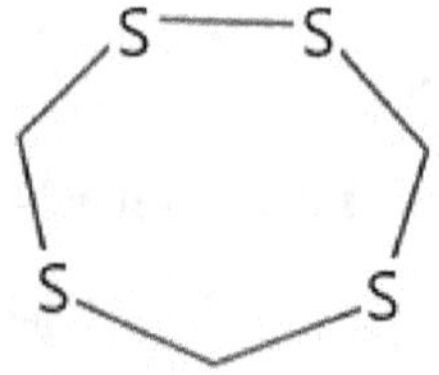

1,2,4,6- tetrathiepane

1,2,4,6-tetrathiepane has been prepared by the reaction of formaldehyde with hydrogen sulphide at low temperature.

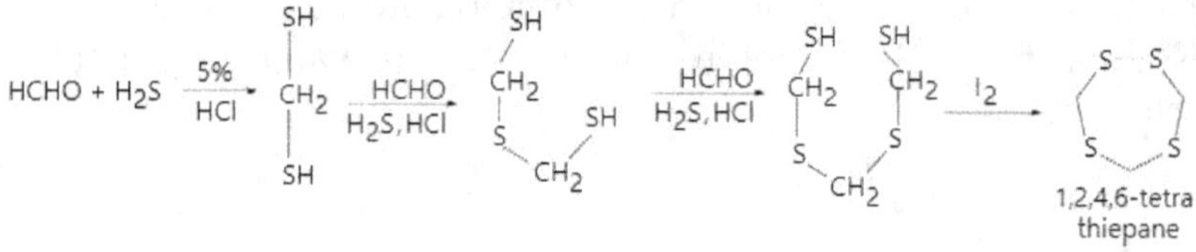

Of the three possible penta-thiepane ring systems (1,2,3,4,5; 1,2,3,4,6; 1,2,3,5,6) only 1,2,3,5,6 arrangement of sulphur atom is known.

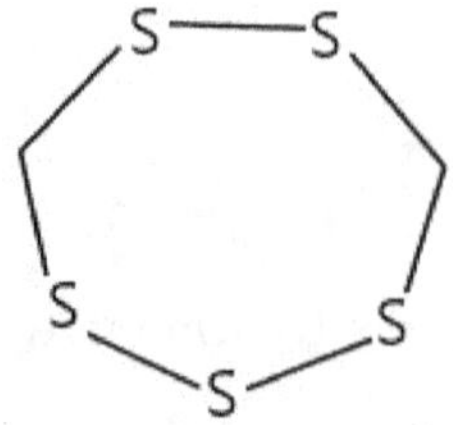

1,2,3,5,6- pentathiepane

A mixture of cycloheptanone and concentrated solution of ammonium disulphide on being allowed to stand for a week, resulted in a solid with formula $C_{14}H_{24}S_5$.

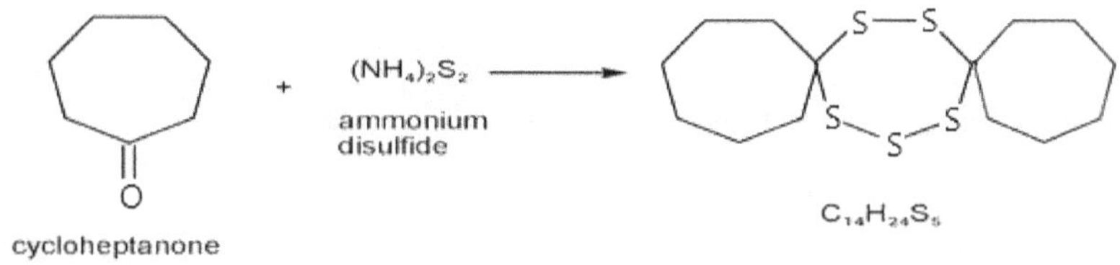

The reduction of bis(penta-methylene) penta-thiepane under two different experimental conditions has been investigated and reported with separate results; with zinc amalgam under acidic conditions, cyclohexane-1,1-dithiol was formed and with lithium aluminium hydride, cyclohexane thiol was formed.

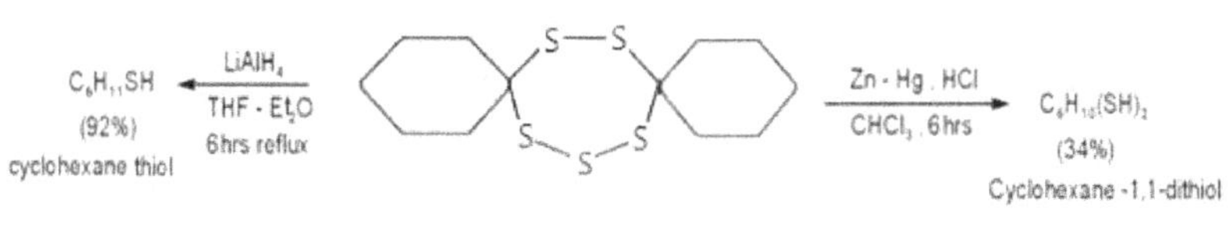

There is only one arrangement for seven membered heterocyclic compounds containing six sulphur atoms i.e. 1,2,3,4,5,6-hexathiepane.

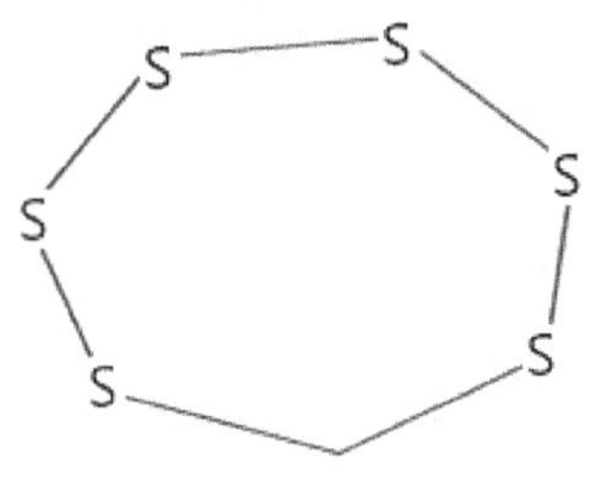

1,2,3,4,5,6 - hexathiepane

3.4.3.7 CONDENSED THIEPINS

Condensed thiepins and derivatives are compounds which have carboxylic and heterocyclic rings fused together.

3.4.3.7.1 2H-CYCLOPENTA[b]THIEPIN

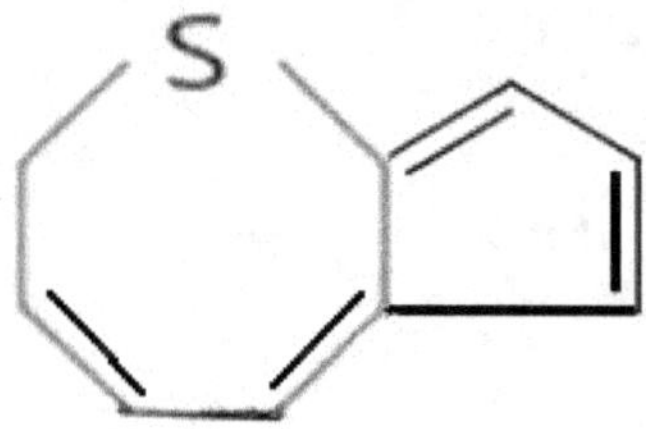

1,2,3,4,5,6-hexathiepane

2-(4-bromobutyl) cyclopentanone, when treated with potassium hydroxide(alcoholic), saturated with hydrogen sulphide, gets converted into 3,4,5,6,7,8-hexahydro-2H-cyclopenta [b] thiepin.

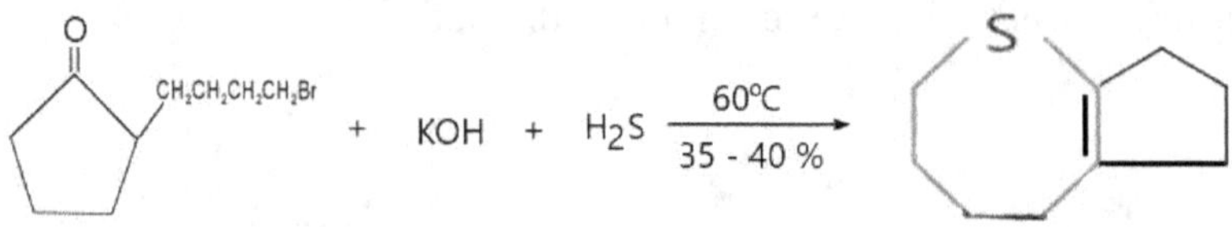

3,4,5,6,7,8- hexahydro-2H-cyclopenta- thiopin

3.4.3.7.2 1-BENZOTHIEPIN

1-benzothiepin in its original form has been reported as reaction intermediate and substituted 1-benzothiepins have been reported. When 2,4-dichloro=2,3-dihydro-1-benzothiepin reacted with potassium tertiary butoxide, 4-chloro-1-benzothiepin was formed, which further gave β-chloronapthalene on heating and 4-chloro-1-benzothiepin-1,1-dioxide by oxidation.

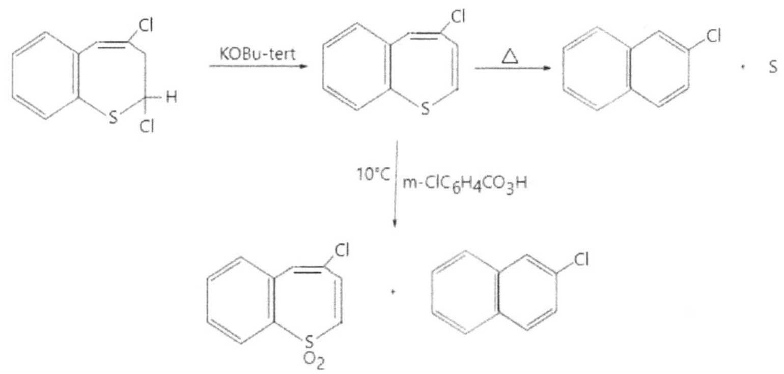

In the same way, formation of 1-chloronapthalene and 1,1'-napthyl disulphides proceed via **intermediary** of 1-benzothiepin.

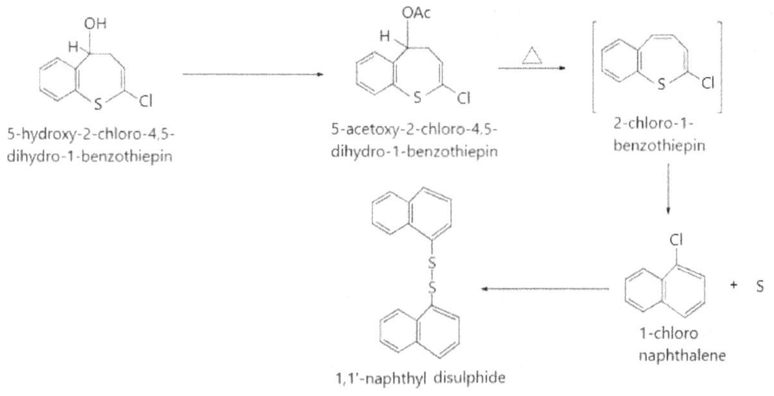

4H-1-benzo thiopyran produced a cyclopropyl adduct which on heating with quinoline at 210°C gives 2-chloronaphthalene.

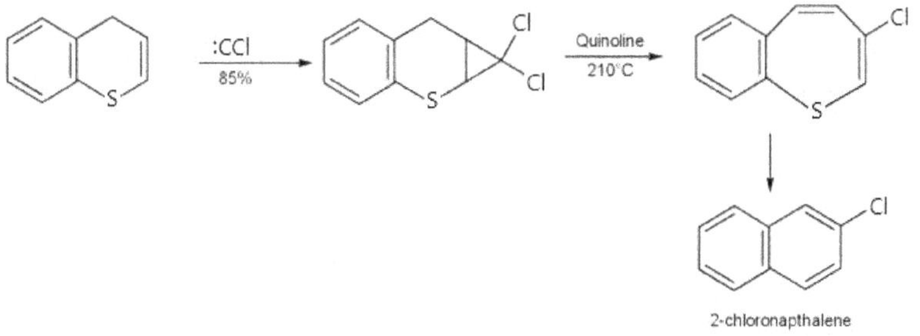

During **thermal conversion**, at temperature higher than 200°C, 1-benzothiepin intermediates the reaction for the formation of dimethyl napthalene-1,2-dicarboxylate.

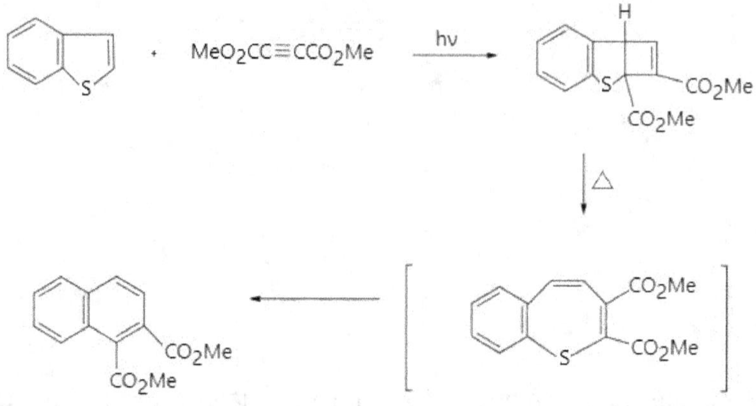

Apart from being intermediates, 1-benzothiepin have been isolated as substituted product in the form of 3,5-diacetoxy-4-phenyl-1-benzothiepin via **ring enlargement**.

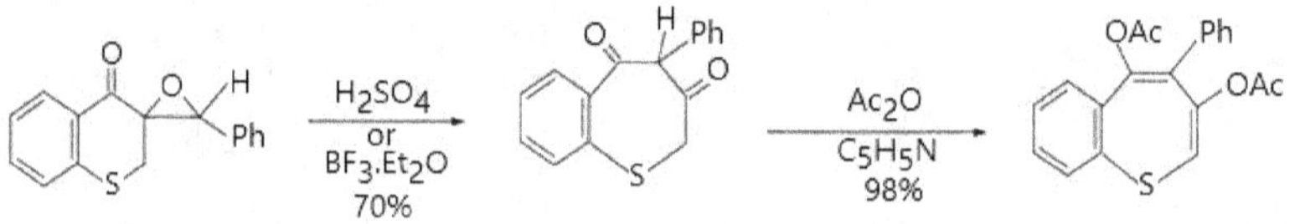

1-Benzothiepin-1-oxide in the form of chloro substituted compound (2-chloro-1-benzothuiepin-1-oxide) has been reported as an intermediate during the formation of 1-chloronaphthalene.

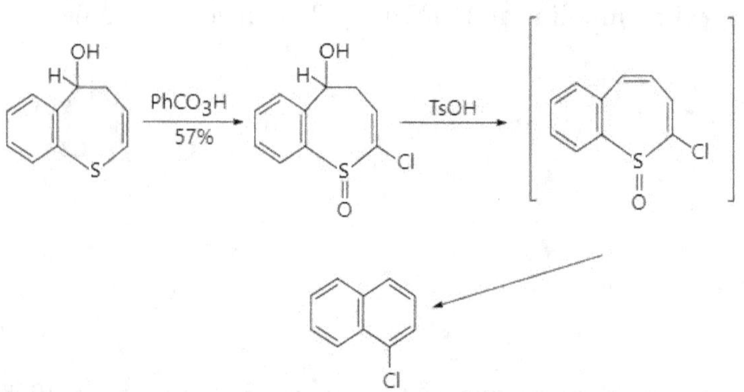

However, dioxide has been generated by **dehydration** of 5-hydroxy-2-chloro-4,5-dihydro-1-benzothiepin-1,1-dioxide.

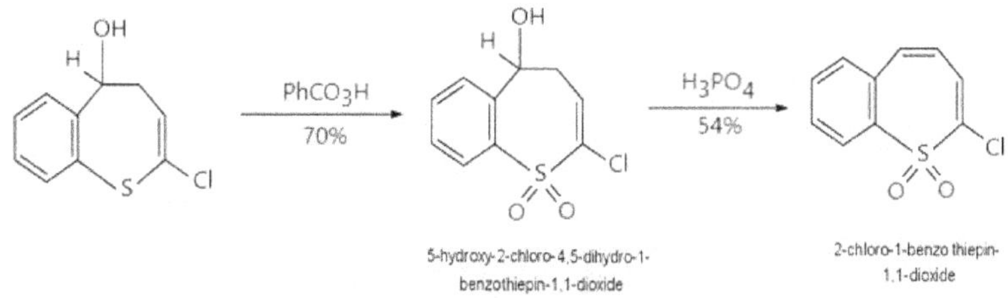

On analysing chemical properties of 1-benzothiepin-1,1-dioxide that **nitration** with fuming nitric acid gives two mono-nitro derivatives with nitro group substituted in benzene ring.

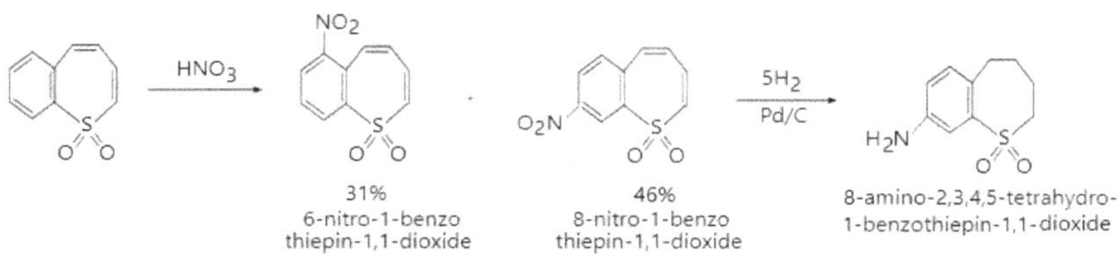

There are three possible arrangements of dihydro-1-benzothiepin.

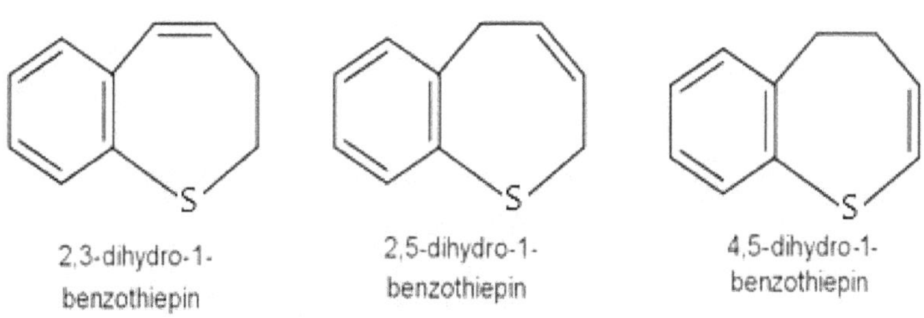

The chemical properties of 2,3-dihydro-1-benzothiepin have been investigated extensively. On **hydrogenation** and **bromination**, it gives tetra hydro derivation and 4,5-dibromo-2,3,4,5- tetrahydro-1-benzothiepin. It gives a low yield of 4-nitro-2,3-dihydro-1-benzothiepin with fuming nitric acid, which on partial reduction tends to form 2,5-dihydro-1-benzothiepin-4-(3H)-one-1,1- dioxide oxime and sulfone derivative on oxidation.

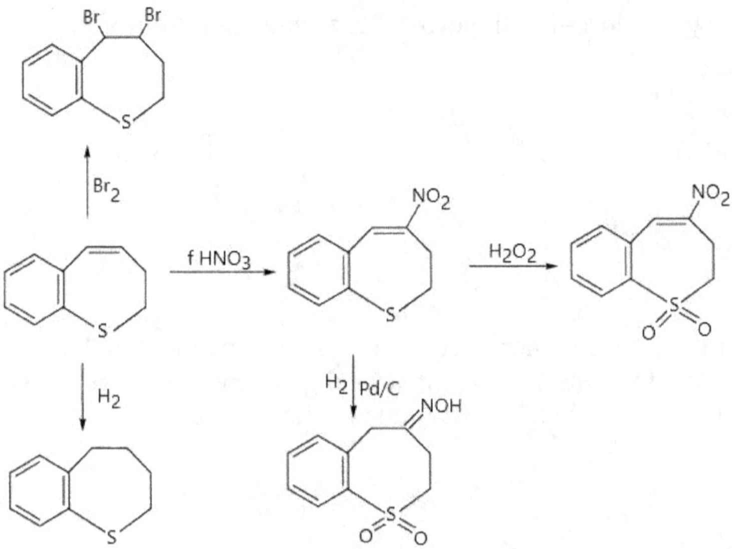

2,5-dihydro-1-benzothiepin-4(3H)-one-1,1-dioxide

Homo-thiochromanone (3,4-dihydro-1-benzothiepin-5(2H)-one) is the most studied ketone. Substituted varieties have been prepared of polyphosphoric acid as ring closure agent.

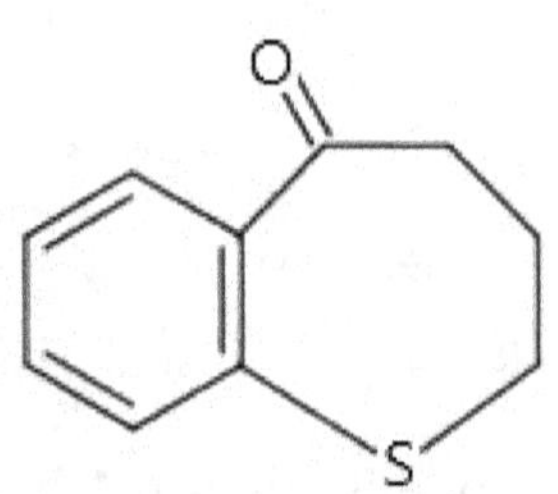

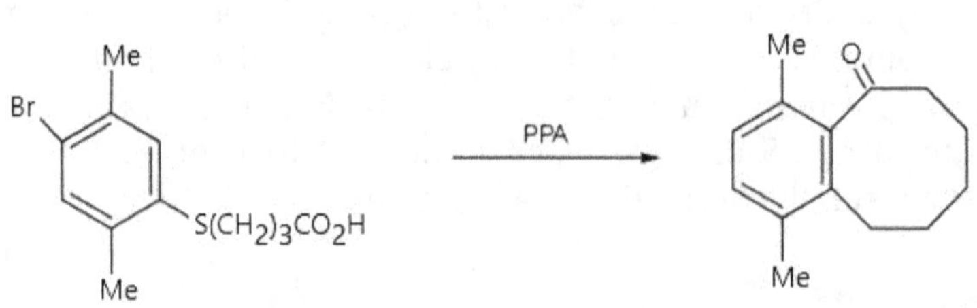

The **base catalysed condensation** reaction between ketone and benzaldehyde leads to the formation 4-(α-hydroxybenzyl)-3,4-dihydro-1-benzothiepin-5(2H)-one **(A)**, 4-(α-hydroxybenzyl)-3,4-dihydro-1-benzothiepin-5(2H,3H)-one **(B)** or 4,4'-benzylidene bis(3,4-dihydro-1-benzothiepien-5(2H) one) **(C)** depending on temperature and alcohol solvent used in the reaction. Reaction of parent ketone with ethyl formate and sodium methoxide gave hydroxy methylene derivative **(D)** which got concerted to enamine **(E)** on reaction with morpholine and further to 4-ethylidene-1-benzothiepin-5-(2H,3H)-one **(F)** methyl magnesium bromide and **(B)** with phenyl magnesium bromide. On mild hydrogen peroxide oxidation of **(B)**, we get sulfoxide **(G)** whereas vigorous oxidative conditions produce sulfone **(H)**. An alternate method of synthesis of **(H)** is by base catalysed condensation of 3,4-dihydro-1-benzothiepin-5-(2H)-one-1,1-dioxide **(I)** and benzaldehyde. Direct bromination of **(I)** leads to 4-bromo-3,4-dihydro-1-benzothiepin-5(2H)-one **(J)** and chloro-ketone **(K)** with sulfenyl chloride.

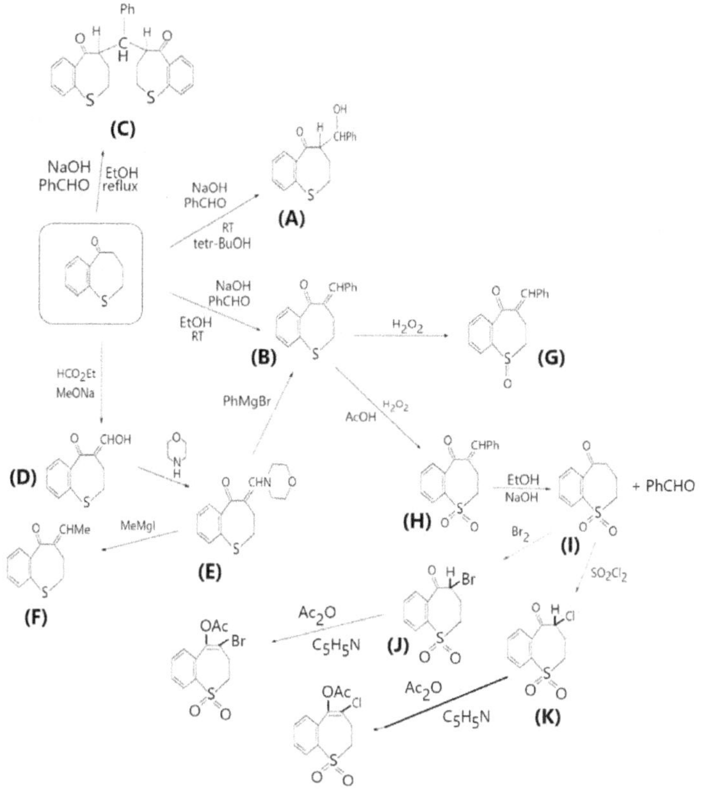

1-Benzothiepin-3,5-(2H,4H)-dione have been synthesized involving ring expansion of an α– β-epoxy ketone to produce 4-phenyl derivatives.

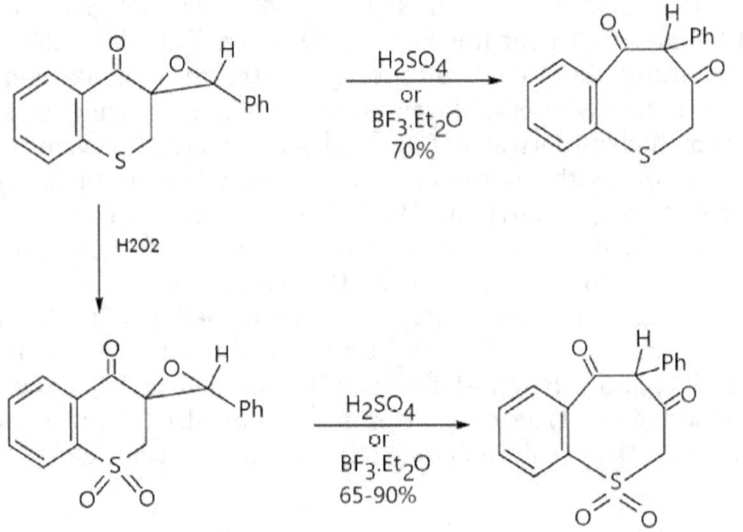

2,3,4,5-Tetrahydro-1-benzothiepin-1,1-dioxide is prepared by the reaction of 4-phenyl-1-butane sulfonyl chloride with aluminium chloride in nitrobenzene; however, nitration of the dioxide gives 8-nitro-2,3,4,5-tetrahydro-1-benzothiepin-1,1-dioxide which is **reduced catalytically** to amine and to 4-nitro-2-sulfobenzoic acid on oxidation.

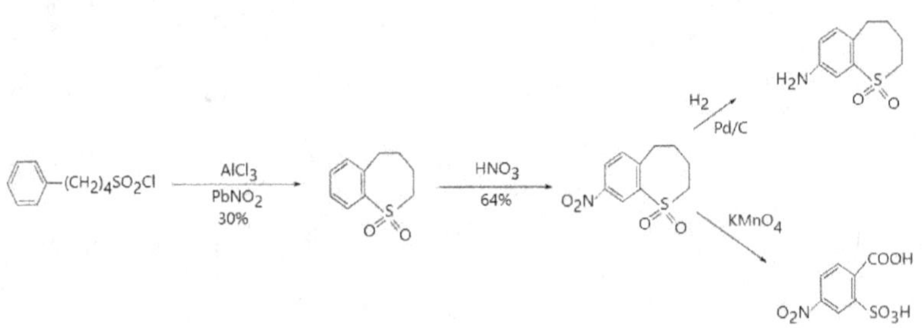

3.4.3.7.3 2-BENZOTHIEPIN

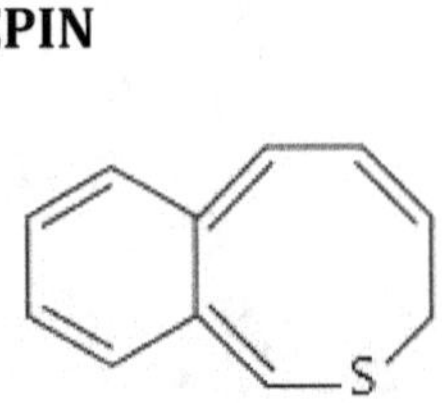

1,3,4,5-tetrahydro-2-benzothiepin has been prepared by reaction of sodium sulphide with substituted benzene and by reduction of ketone.

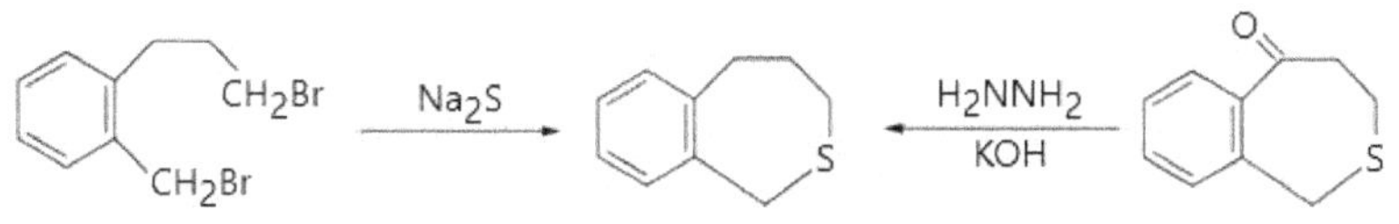

3.4.3.7.4 3-BENZOTHIEPIN

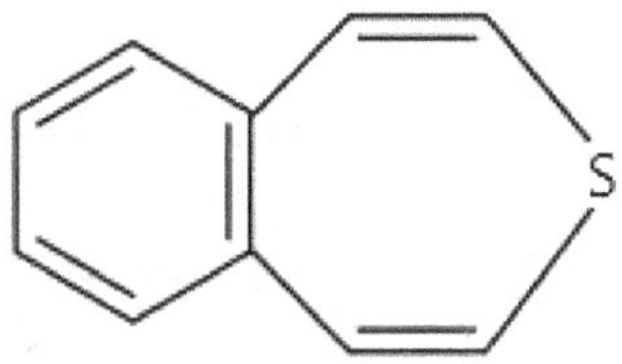

Base catalysed condensation reaction of o-phthalaldehyde and ethyl thiodiacetate lead to the formation of 3-benzothiepin-2,4-dicarboxylic acid which on standing lost sulphur and formed naphthalene-2,3-dicarboxylic acid. This dicarboxylic acid got converted to methyl ester via diazomethane and ethyl ester via esterification of the diacid chloride.

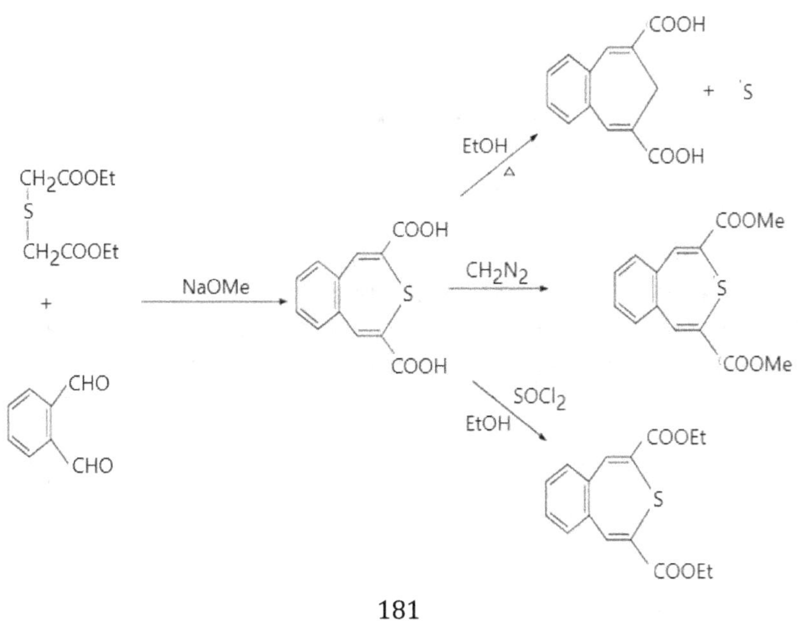

Intramolecular Friedel Crafts ring closure of 5-phenyl-3-thiapentanoyl chloride leads to the formation of 4,5-dihydro-3-benzothiepin-1(2H)-one.

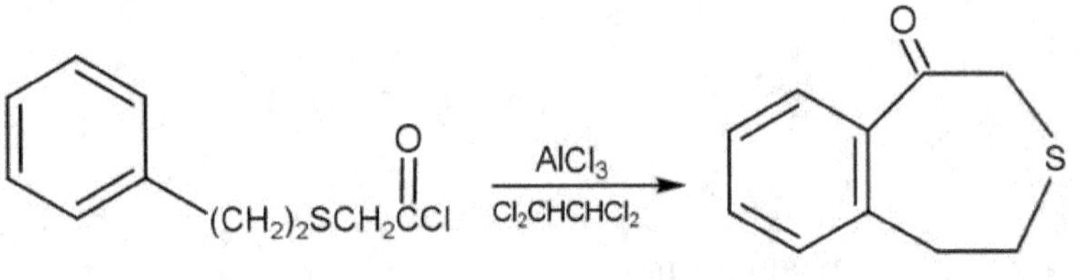

The preparation of 1,2,4,5-tetrahydro-3-benzothiepin involves the reaction of o-bis (beta- bromoethyl) benzene and sodium sulphide.

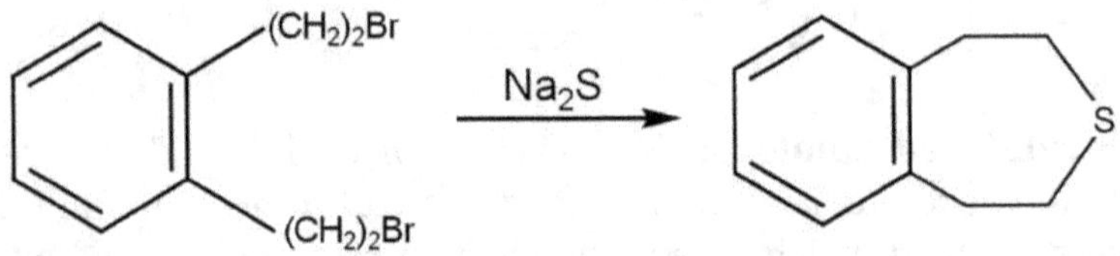

3.4.3.7.5 NAPHTHOTHIEPIN

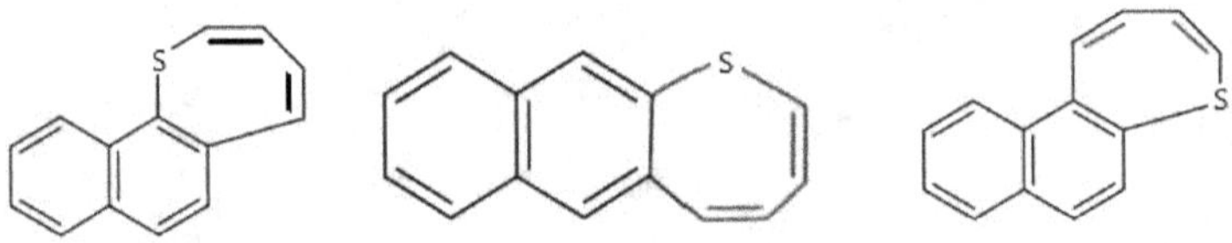

Naphtho[1,2-b]thiepin Naphtho[2,3-b]thiepin Naphtho[2,1-b]thiepin

Cyclohydration 4-(1-naphthylthio) butyric acid with polyphosphoric acid or via Friedel Crafts ring closure of the acid chloride leads to the formation of 3,4-dihydronaphtho [1,2-b]-thiepin-5(2H)-one.

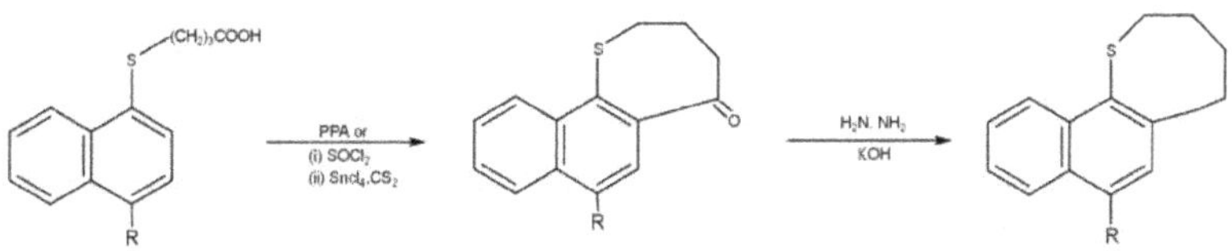

Ring closure reactions of 4-(2-naphthylthio) butyric acid gives the possibility of two sites, one at α-carbon and other at β-carbon of the naphthalene nucleus. Both the modes of ring closure are dependent on experimental conditions i.e., if the acid is treated via **Friedel Crafts ring closure**, then 2,3-dihydronaphtho[2,3-b] thiepin-5(2H)one is obtained and with polyphosphoric acid, it gives 3,4-dihydronaphtho[2,3-b]thiepin-5(2H)one. Further, wolf Kishner reduction of products yielded new products.

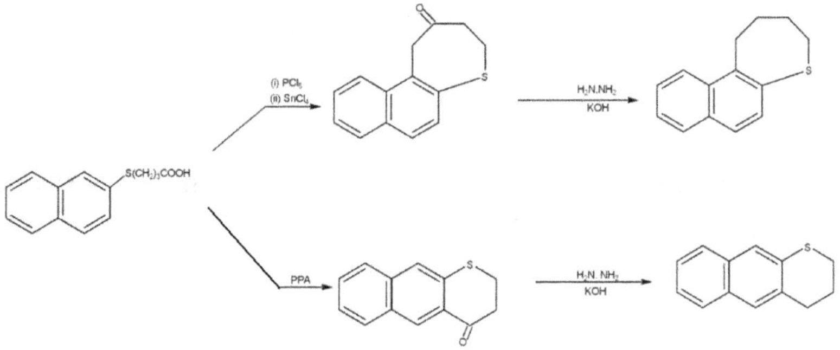

3.4.3.7.6. DIBENZO[b, d] THIEPIN

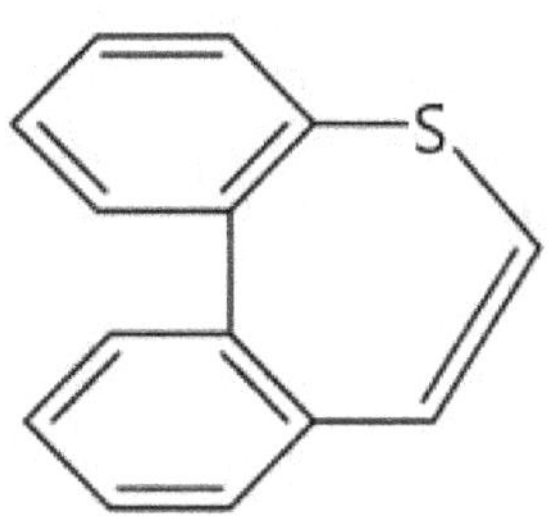

Ring closure of ethyl 2'-methylsulfonylbiphenyl-2-carboxylate lead to the synthesis of dibenzo[b, d] thiepin-4(6H)-one-5,5-dioxide.

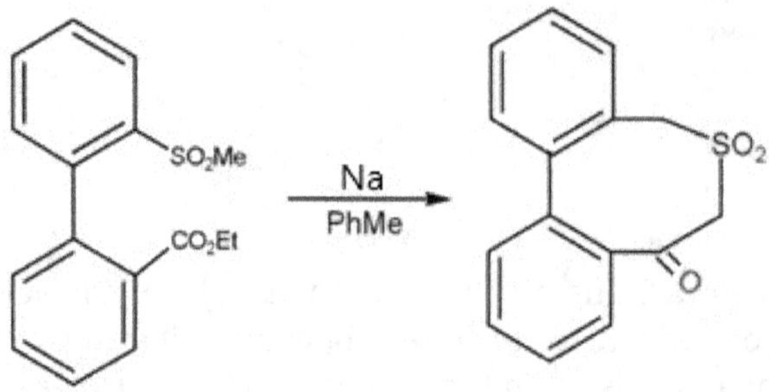

3.4.3.7.7. DIBENZO [b, f] THIEPIN

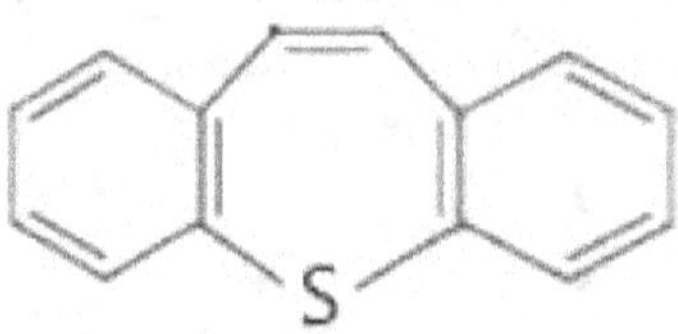

Ring expansion of 9-(α-hydroxybenzyl)-thioxanthene gives 10-phenyldibenzo [b, f] thiepin as product.

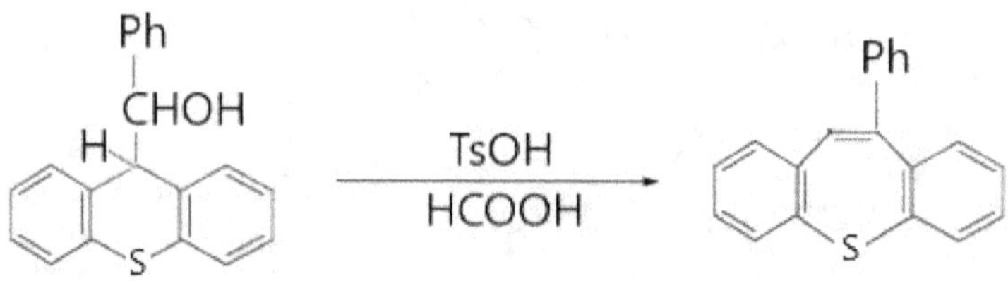

One of the dibenzo [b, f] thiepin derivative in ketonic form is prepared by by polyphosphoric acid ring closure to produce 2-nitro-8-methyldibenzo [b, f] thiepin-10(11H)-one.

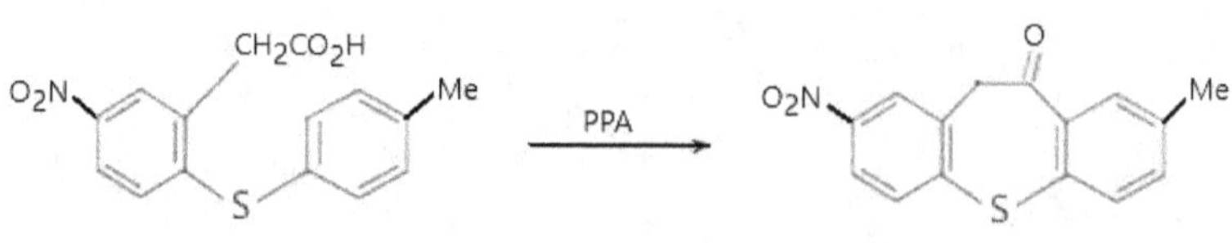

The chemical properties of the dibenzo [b, f] thiepin-10(11H)-one are typical for carbonyl compounds with an α-methylene group.

Reduction with sodium borohydride leads to alcohol and **Wolf-Kishner reduction** removes carbonyl functional group.

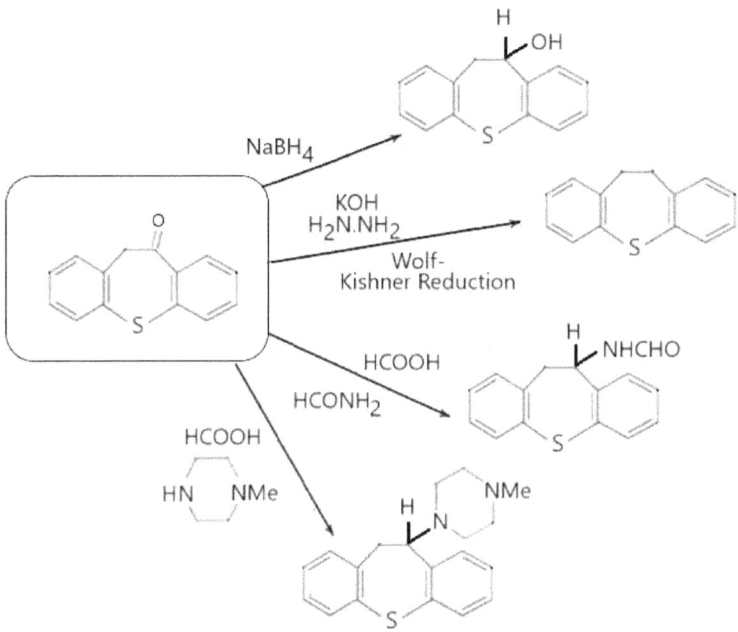

3.4.3.7.8. TRIBENZO[b,d,f] THIEPIN

Thermal decomposition of 2-chlorosulfonyl-2'-phenyl diphenyl sulphide at 250-260°C with cuprous chloride gave tribenzo (b, d, f) thiepin (a colourless crystalline solid).

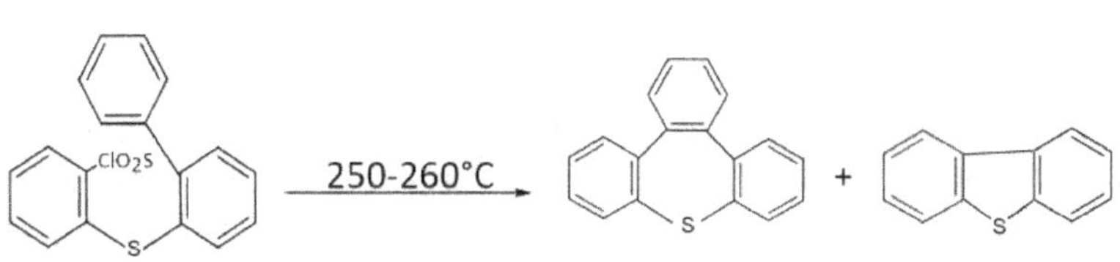

Amongst chemical properties, under the effects of copper bronze at 380°C for 5 hours, sulphur extrusion take place resulting in the formation of triphenyl; undergoes **sulphurization** with Raney Nickel to from o-triphenyl and oxidized with hydrogen peroxide to tribenzo [b, d, f] thiepin 9-9-dioxide.

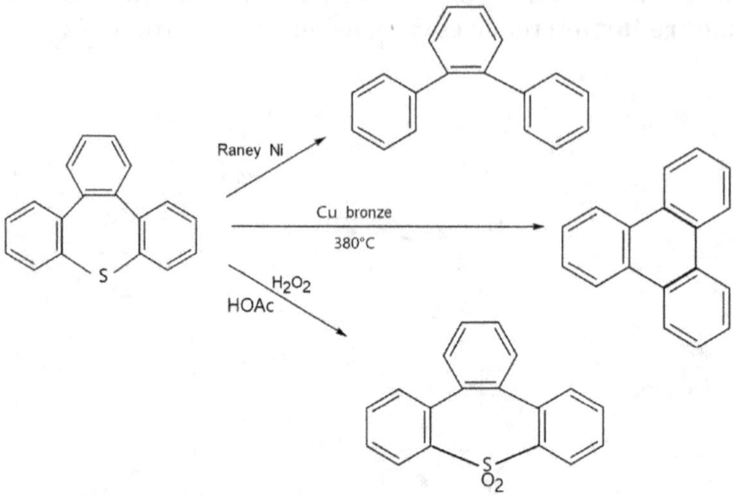

3.4.3.7.9. FUROTHIEPIN

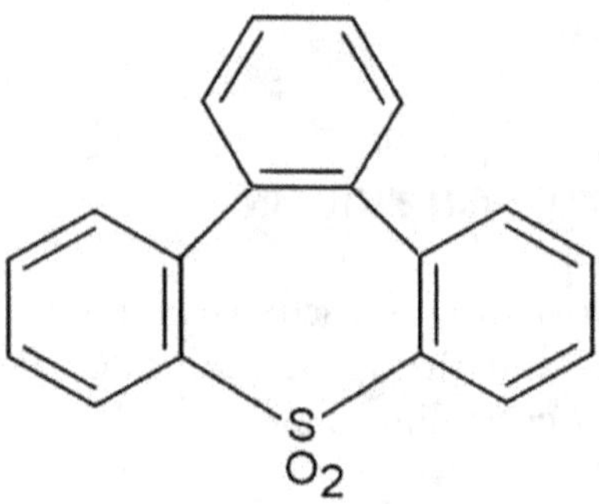

Condensation of dimethyl thioglycolate with furan dialdehyde leads to the resulting diacid which is converted to dimethyl-1,3-dimethyle-furo[3,4-d] thiepin-5-7-dicarboxylate and is finally converted to a highly reactive furothiepin diol.

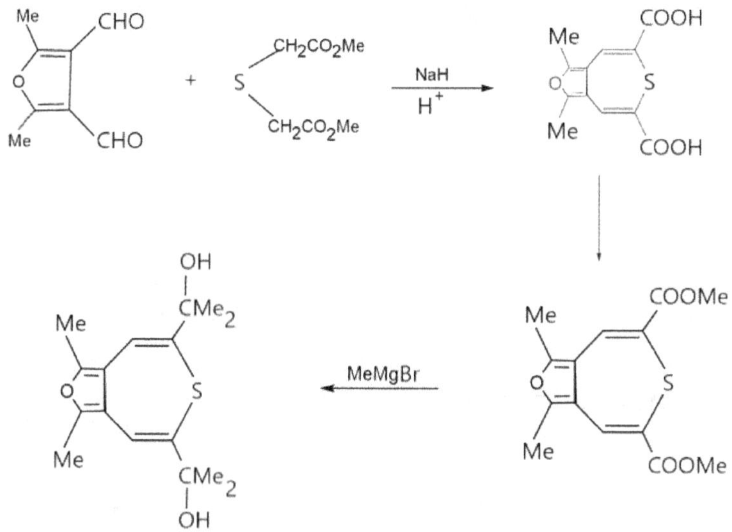

3.4.3.7.10 BENZODITHIEPIN

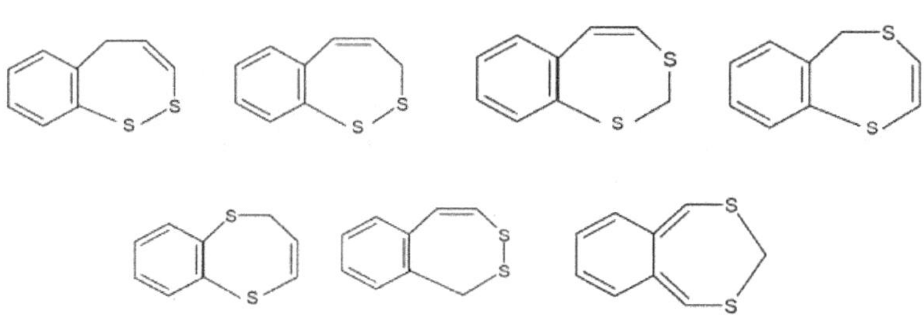

7-5-Dihydro-3H-2,4-benzodithiepin Is isolated by two methods: one by methylene iodide and other by the reaction with carbonyl compound in the presence of dry HCl.

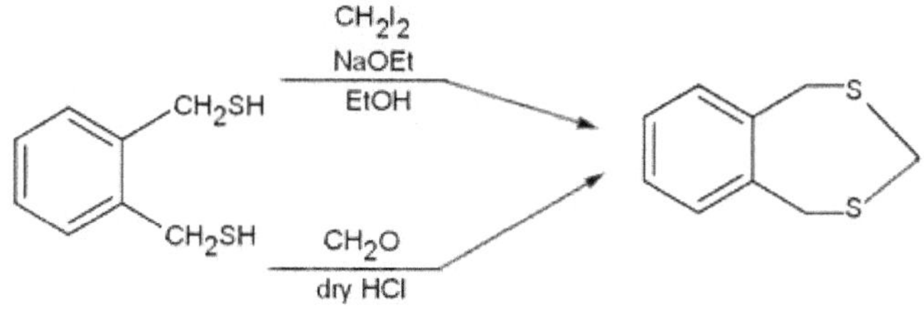

3.4.3.7.11 BENZOTRITHIEPIN

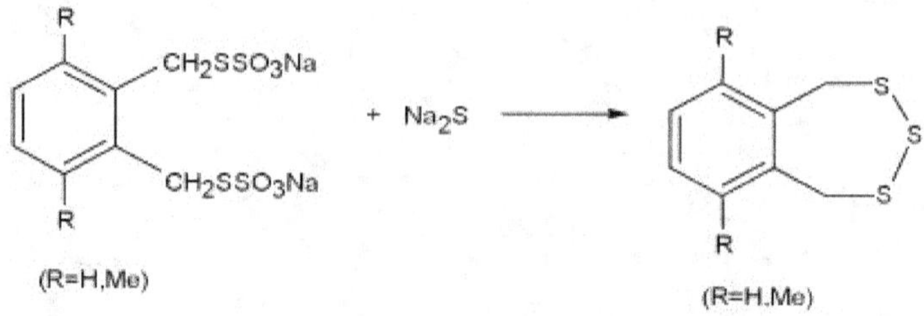

3.4.3.7.12 DINAPHTHOTRITHIEPIN

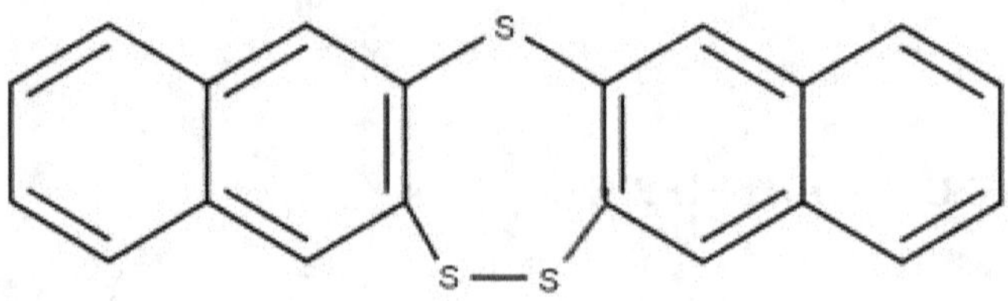

Dinaphthotrithiepin has been prepared by the reaction of dibromo naphthazarin and sodium sulphide.

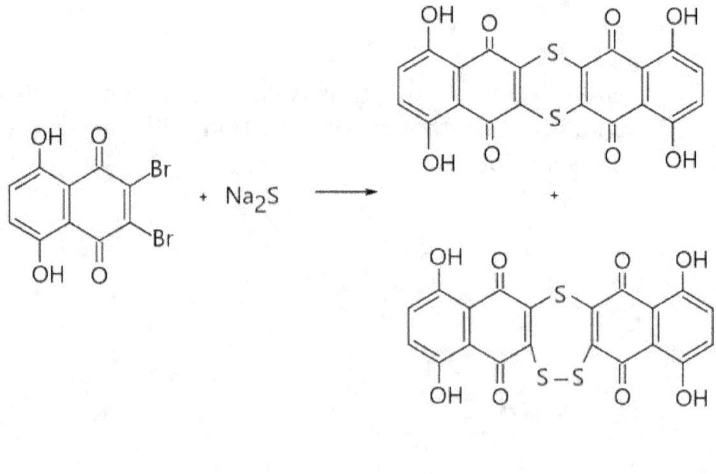

REFERENCES

1. Vishnu Ramji, Arun Sethi, Mahendra nath, Ramendra Pratap (2019); The chemistry of Hetrocycles, chap.3, 393-425.
2. R. Mallepalli, L. Yeramanchi, R. Bantu, L. Nagarpu (2011); Synlett, 2730-2732.
3. E. Vogel, R. Sehubart and W.A. Bocll (1964); Angew. Chem., **76**, 535.
4. E. Vogel and H. Guenther (1967); Angew. Chem., **79**, 429.
5. H. Guenther, R. Schubart. and E. Vogel (1967); Z. Naturforsch., **25**, 22b.
6. E. E. van Tamelon and D. Carty (1967); J. Amer. Chem. Soc., **89**, 3922.
7. H. Prinzbach, M. Arguelles and E. Druckrey (1966); Angew. Chem., **78**, 1057
8. E. Vogel, W. A. Boell and H. Guenther (1965); Tetrahedron Lett., **65**, 609.
9. L. A. Paquette and J. H. Barrett (1966); J. Amer. Chem. Soc., **88**, 1718.
10. J. Meinwald, D. W. Dicker and N. Danieli (1960); J. Amer. Chem. Soc., **82**, 4087.
11. G. Pohl (1961); Dissertation; Marburg University, cited by W. Luerker and E. Mueller (1966) in methoden der organischen chemie (Houben-Weyl), 4th ed., E. Mueller, Ed., Georg Thieme Verlag, Stuttgart, vol.6/4, p.466.
12. I. W. Matsechinlkaya, W. A. Barchasch and A. T. Pruschenko (1961); Zh. Obshch. Khim., **30**, 2362.
13. S. J. Rhoads and R. D. Cockroft (1969) J. Amer. Chem. Soc., **91**, 2815.
14. E. E. Schweizer and W. E. Parham. J. Amer. Chem. Soc., **82**, 4085.
15. R. A. Braun (1963); J. Org. Chem., **28**, 1383.
16. R. H. Snyder, H. J. Shine, K. A. Leibbrand and P. O. Tawney (1959); J. Amer. Chem. Soc., **81**, 4299.
17. J. Meinwald, D. W. Dicker and N. Danieli (1960); J. Amer. Chem. Soc., **82**, 4087.
18. J. Meinwald and H. Nozaki (1958); J. Amer. Chem. Soc., **80**, 3132.
19. H. J. Shine and R. H. Snyder (1914); J. Amer. Chem. Soc., **35**,1431.
20. A. Franke and F. Lieben (1914); Monatsh. Chem, **35**, 1431.

21. L. Michiels (1913); Bull. Soc. Chim. Belges., 27, 25, Chem. Abstr., **7**, 3602.
22. F. Nederl, J. Buddrus, W. Brodowski and P. Weyerstand (1986); Tetrahedron Lett., **5**, 385.
23. E. J. Corey, J. D. Bass, R. Le Mahiew and R. B. Mitra (1994); J. Amer. Chem. Soc., **86**, 5570.
24. H. Obara, Nippon kagaku Zasshi (1962); 82, 65, Chem. Abstr., **57**, 16427.
25. R. Sudo, A. Kaneda and N. Iyob (1967); J. Org. Chem, **32**, 1844.
26. K. Dimroth, G. Pohl and H. Mann (1996); Chem. Ber., **99**, 634.
27. R. N. Len and B. E. Roy (1930); J. Indian Chem. Soc., **7**, 401.
28. G. Fontaine (2008); Ann. Chim., **3**, 469.
29. G. Fontwine and P. Matte (1964); C. R. Acad. sci., Pairs, **258**, 4583.
30. O. Dann and W. D. Arndt (2004); Justus Liebigs Ann. Chem., **587**, 38.
31. A. Rieche and H. Gross (1962); Chem. Ber., **95**, 91.
32. K. Dimroth and H. Freyschlag (1957); Chem, Ber, **90**, 1623.
33. C. Normant (1967); C. R. Acad. Sci., Paris, **264**,1665.
34. K. Dimroth and H. Freyschlag (2007); Angew. Chem, **69**, 95.
35. R. Huisgen, E. Laschtavka, I. Ugi and A. Kammerneier, Justus Liebigs Ann. Chem., **630**, 128.
36. A. Skeglitz and H. Koch (1985); Chem. Ber., **588** ,78.
37. G. Bert (1951); Gazz. Chim. Ital., **81**, 570
38. M. Lamchem (1962); J. Chem. Soc., 4695.
39. F. A. L. Anet and P. M. G. Bavin (1987); Can. J. Chem., **35**, 1084.
40. E. D. Bergmann, I. Shahak and Z. Eizenstat (1969); Tetrahedron Lett., 2007.
41. K. Kawak, Nippon Yakurigaku Zasshi (1990); **56**, 971.
42. A. H. Rees (1957); Chem. Ind. (London), **76**.
43. T. Kametani and K. Fukuneoto (1963); Chem. Pharm. Bull. (Tokyo), **11**, 1322.
44. L. Oyster and H. Adkins (2001); J. Amer. Chem. Soc., **43**, 208.
45. J. D. Brooks (1955); J. Appl. Chem. (London), **5**, 250.
46. F. Bell and F. B. Briggs (1938); Chem. Soc., 1561.
47. P. Cagniant and C. Charaux (1966); Bull. Soc. Chem. Fr., 3349.
48. J. N. X. Chatterjea (1957); J. Indian Chem. Soc., **34**, 299.

49. M. S. Newman and H. S. Whitehouse (1949); J. Amer. Chem. Soc., **71**, 3664.
50. T. K. Ametani, S. Shibuya, S. Seino and K. Fukumoto (1964); Tetrahedron Lett., **25**, 964.
51. O. Dann and W. D. Arndt (1954); Justus Liebigs Ann. Chem., **587**, 88.
52. T. Kametani, K. Fukumoto and K. Mesuko (1963); Yakugaku Zasshi, **83**, 1052.
53. T. Kubola, N. Ichikawa, W. Brodowski and P. Weyerstahl (1976); Tetrahedron Lett, 5385.
54. F. Mathys, V. Prelog and R. B. Woodward (1956); Helv. Chim. Acta, **39**,1095.
55. E. Hope and J. S. Anderson (1936); J. Chem. Soc., **14**, 74.
56. A. Zinke and L. Ammerer (1993); Monatsh. Chem., **84**, 423.
57. R. De Ridder and R. H. Martin (1960); Bull. Soc. Chim. Belges, **69**, 534.
58. K. C. Brannock and G. R. Lappin (1956); J. Org. Chem., **21**, 1366.
59. A. Seib (1988); German Patent 855, 864; Chem. Abstr; **52**, 9197.
60. K. Ziegler and A. Liittringhaus (1939); German Patent 671, 840; Chem. Abstr., **33**, 6532.
61. R. Hartmnt, S. Klaus and M. Manfred (1998); Z. Chem., **8**, 220.
62. W. L. Mock (1967); Abstracts of the 153rd Meeting of the American Chemical Society, Spring, Paper 157.
63. W. L. Mock (1967); J. Amer. Chem. Soc., **89**, 1281.
64. E. L. Stogryn and S. J. Brois (1969); J. Amer. Chem. Soc., **89**, 605.
65. R. M. Dodson and J. P. Nelson (1969); J. Chem. Soc., **D**, 1159.
66. J. V. Brawn (1910); Chem. Ber., **43**, 3220.
67. E. Gricherich-Trokhimorskii (1916); J. Russ. Phys. Chem. Soc., **48**, 944, Chem. Abstracts, 11, 786(1917).
68. D. L. Tuleen and R. H. Bennett (1969); J. Heterocycl. Chem., **6**, 115.
69. R. Borsdorf, H. Kesper, and H.-D. Repp (1967); Angew. Chem., 2nd. Ed. Engl., **6**, 872.
70. C. G. Overberger and A. Katchman (1956); J. Amer. Chem. Soc., **78**, 1965.
71. N. I. Leonard and J. Figueras, Jr. (1952); J. Amer. Chem. Soc., **74**, 917.

72. N. I. Leonard and W. L. Rippie (1963); J. Org. Chem., **28**, 1957.
73. A. De Groot, J. A. Boerma, J. De Walk and H. Wynberg (1968); J. Org. Chem., **33**, 4025.
74. A. De Groot, J. A. Boerma and H. Wynberg (1969); Rec. Trav. Chim. Pays-Bas, **88**, 994.
75. M. W. Bullock, J. A. Brochman, Jr., E. L. Patterson, J. V. Pieree and E. L. R. Stokstad (1952); J. Amer. Chem. Soc., **74**, 1868.
76. M. W. Bullock, J. A. Brockman, Jr., E. L. Patterson, J. V. Pierce and E. L. R. Stokstad (1952); J. Amer. Chem. Soc., **74**, 3455.
77. C. J. Benning (1961); U.S. Patent **2**, 968, 659; Chem. Abstr., **55**, 13460.
78. S. Oal, W Tagaki and A. Ohno (1964); Tetrahedron., **20**, 427, 437.
79. A. T. Bottini and E. F. Bitter (1966); J. Org. Chem., **311**, 506.
80. H. Richter K. Sehulze and M. Miihlstadt (); Z. Chem., **8**, 220.
81. J. L. Massingill, Jr., M. G. Reinecke and J. E.Hodgekins (1970); J. Org. Chem., **35**, 823.
82. J. R. Meadow and E. E. Reid (1934); J. Amer. Chem. Soc., **56**, 2177.
83. N. B. Tucker and E. E. Reid (1933); J. Amer. Chem. Soc., **55**, 775.
84. E. G. Howard and R. V. Lindsey, Jr. (1960); J. Amer. Chem. Soc., **82**, 158.
85. T. L. Carins G. L. Evans. A. W. Larchar and B. C. Mekusick (1952); J. Amer. Chem. Soc., **74**, 3982
86. C. G. Krespan and W. R. Brasen (1962); J. Org. Chem., **27**, 3995.
87. A. V. Schoor, E. Jacobi, S. Lust and H. Flemming (1961); German Patent 1, 060, 655; Chem. Abstr., **55**, 7748.
88. C. G. Krespan (1963); U. S. Patent 3, 099, 688; Chem. Abstr., **59**, 10096.
89. E. Baumann (1890); Chem. Ber., **23**, 1869.
90. A. Fredga (1958); Acta Chem. Scand., **12**, 891.
91. F. O. Bobbio and P. A. Bobbio (1965); Chem. Ber., **98**, 998.
92. R. Mayer and I. Leibster (1958); Angew. Chem., **70**, 105.
93. V. J. Traynelis and J. R. Livingston, Jr. (1964); J. Org. Chem., **29**, 1092.
94. W. E. Parham and R. Koncos (1961); J. Amer. Chem. Soc., **83**, 4034.

95. D. Neckers, J. Dopper and H. Wynberg (1969); Tetrahedron Lett., **34**, 2913.
96. H. Hoffmann and H. Westernacher (1966); Angew. Chem. Int. Ed. Engl., **5**, 958.
97. H. Hofmann and H. Westernacher (1969); Chem. Ber., **102**, 205.
98. V. J. Traynelis and R. F. Lore (1964); J. Org. Chem., **29**, 366.
99. P. Cogniant and D. Cogniant (1966); Bull. Soc. Chim. Fr., 3674.
100. R. Ireland and P. Shiess (1963); J. Org. Chem., **28**, 6.
101. R. Otto (1887); J. Prakt. Chem., [2] **36**, 450.
102. W. E. Truce, D. D. Emrick and R. E. Miller (1953); J. Amer. Chem. Soc., **75**, 3359.
103. J. ron Braun (1925); Chem. Ber., **58** B, 2165.
104. P. Cogniant and D. Cogniant (1998); Bull. Soc. Chim. Fr., 1959.
105. G. P. Scott (1993); J. Amer. Chem. Soc., **75**, 6332.
106. K. Dimroth and G. Lenke (1996); Angew. chem., **68**, 519.
107. K. Dimroth and G. Lenke (2006); Chem. Ber., **89**, 2608.
108. J. Ron Braun and K. Weissbach (1999); Chem. Ber., **62** B ,2416.
109. W. E. Truce and F. J. Lots Peich (1993); J. Amer. Chem. Soc., **78** ,848.
110. P. Cogniant and D. Cogniant (2006); Bull. Soc. Chim. Fr., **6**, 203.
111. D. D. Enrick and W. E. Truce (2000); J. Org. Chem., **25** ,1103.
112. E.D. Bergmann and M. Rabinovite , Israel J. chem. ., **1**, 125 (1963).
113. J. D. Loudon, A. B. D. Sloan and L. A. Summers (1997); J. Chem. Soc., **38**, 149.
114. J.O. Jilek, V. Seidlova, E. Svatek, M. Protiva, J. Pomykacek and Z. Sediry , Monatsh. Chem., **96**,182 (1965).
115. A. J. Neale, T. J. Rawlings and E. B. McCall (1989); Tetrahedron, **21**, 1299.
116. J. Hoffman, Jr. and R. Schlessinger (1970); J. Amer. Chem. Soc., **92**, 5263.
117. W. Autenreith and R. Hennings (2002); Chem. Ber., **35**, 1388.
118. B. Milligan and J. M. Swan (1965); J. Chem. Soc., 2901.
119. K. Brass, R. Pfluger and K. Honsberg (1936); Chem. Ber., **69**, 80.

CHAPTER 4

HETEROCYCLIC COMPOUNDS CONTAINING ELEMENTS OTHER THAN N, S and O

4.1 INTRODUCTION

Compound having nitrogen, sulphur, and oxygen as part of the ring system to comprise of heterocyclic compounds have been in practice since times immemorial, however the chemistry of heterocycle rings containing elements, apart from the above mentioned, is gaining ground now.

4.2 HETEROCYCLIC COMPOUNDS CONTAINING PHOSPHORUS

4.2.1 THREE-MEMBERED RINGS

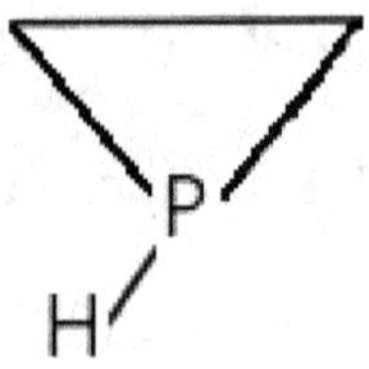

These compounds (phosphirane) can easily be prepared as under:

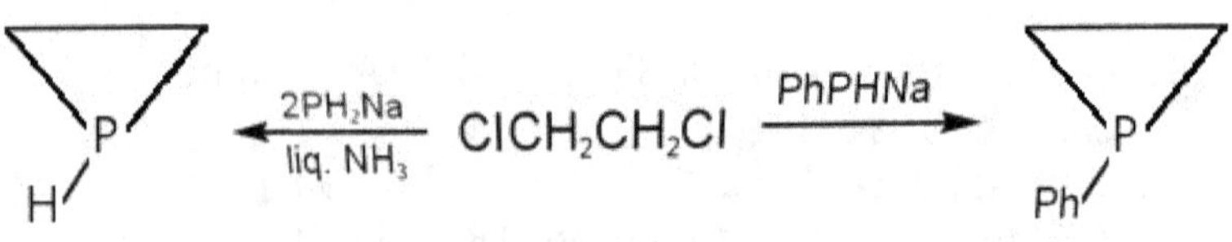

Of the 1-substituted phosphiranes, 1-phenyl phosphirane is the most stable. In a different method phosphirane derivative is formed by diazomethane addition.

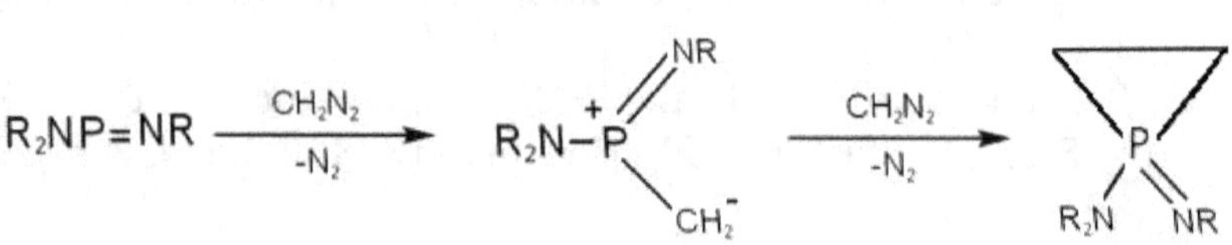

4.2.2 FOUR-MEMBERED RINGS

The method of synthesis of phosphetane involves addition of PCl_2^+ at 0°-10°C.

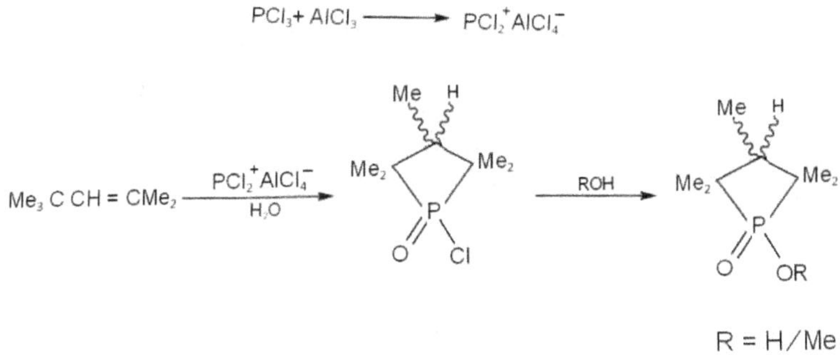

4.2.3 FIVE-MEMBERED RINGS

The five membered rings system containing phosphorus has some structural aspects of interest.

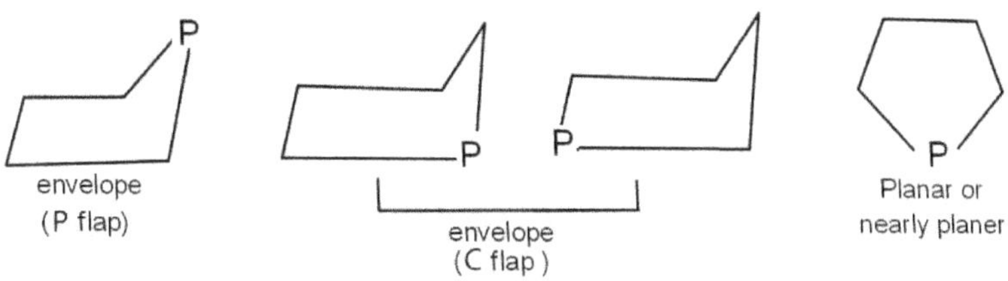

The **rings strain** is also greater in five membered varieties, being 95°-97° As C-P-C bond angle of phospholane. The synthesis involves reaction of bis-Grignard reagent of 1,4-dibromobutane with $PhPCl_2$. Many other methods have also been mentioned. 1-substituted phospholanes have properties like tertiary phosphines. Phospholane can also be prepared by various other methods.

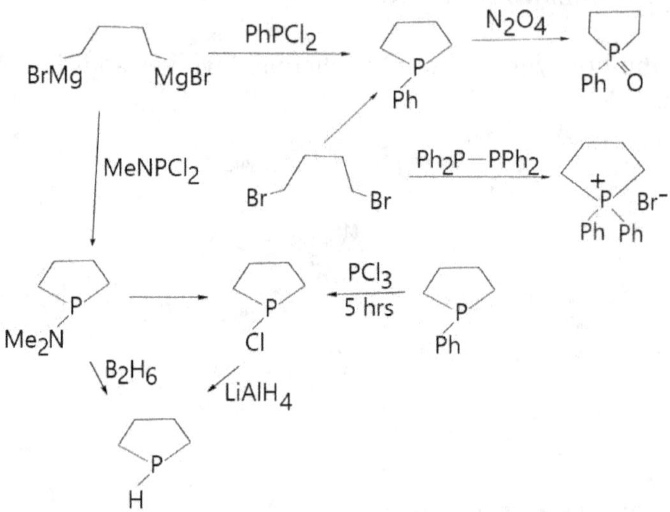

The unsaturated five-membered heterocyclic compound, phospholes, have many interesting ways of synthesis.

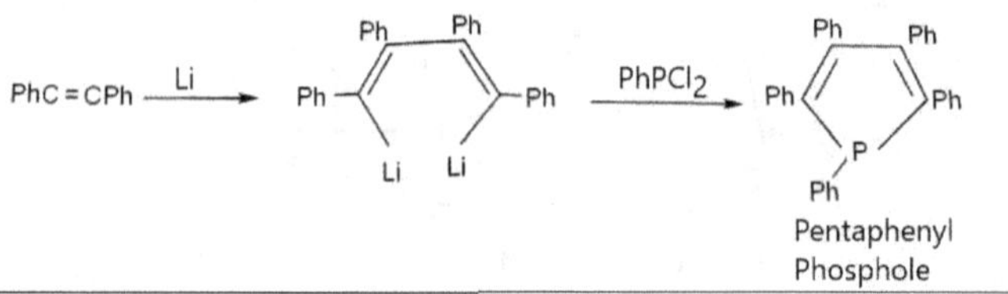

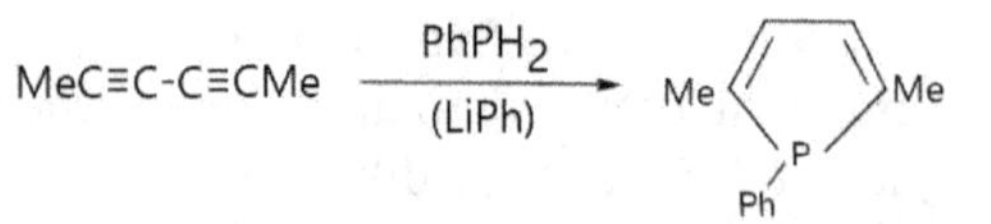

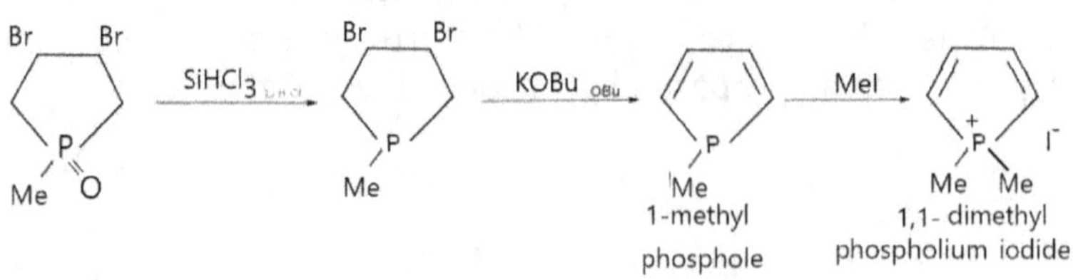

4.2.4. SIX-MEMBERED RINGS

These compounds are mostly air-sensitive and non-crystalline. The synthetic methods involve reaction of 1,5–dibromo pentane with dichlorophenylphosphine resulting in the formation of 1-phenylphosphorinane. The product is air sensitive, colorless oil and has a characteristic smell of phosphine. Another method involves 1,5-dibromo-3,1-butylpentane.

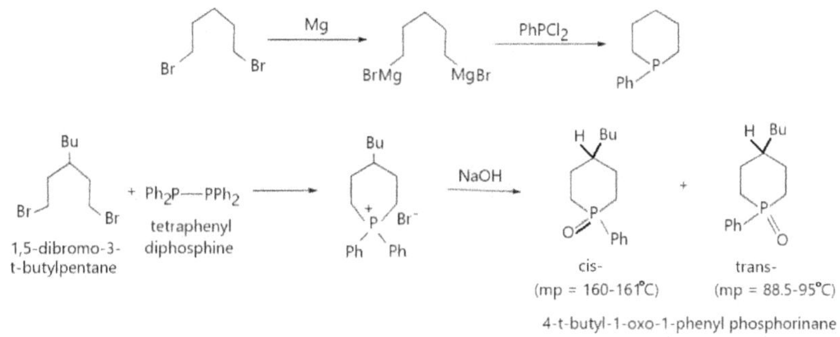

In another route, phenyl phosphorinane, on heating with PCl_3 at 80°C gives 1-chlorophosphorinane which further, after a series of reactions, yields the parent phosphorinane.

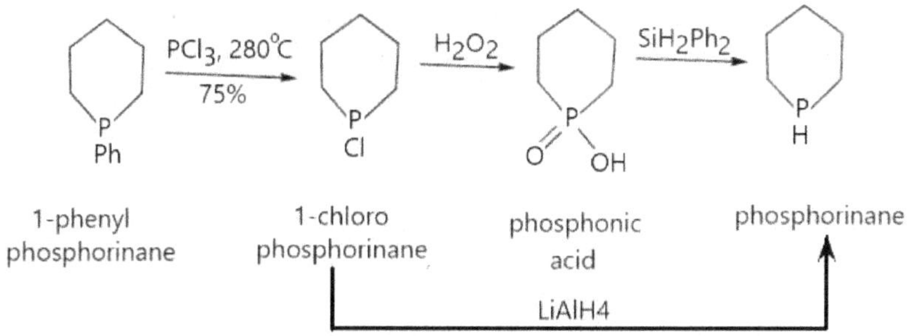

Synthesis of benzo-phosphorinanes can be achieved by using phosphorus substituted benzenes.

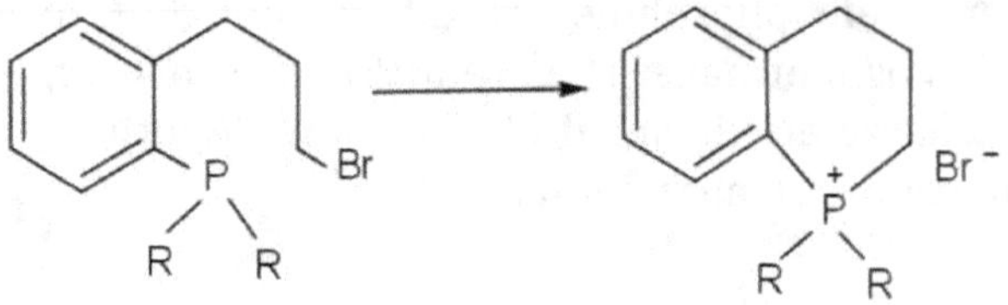

R = Ph or Et.

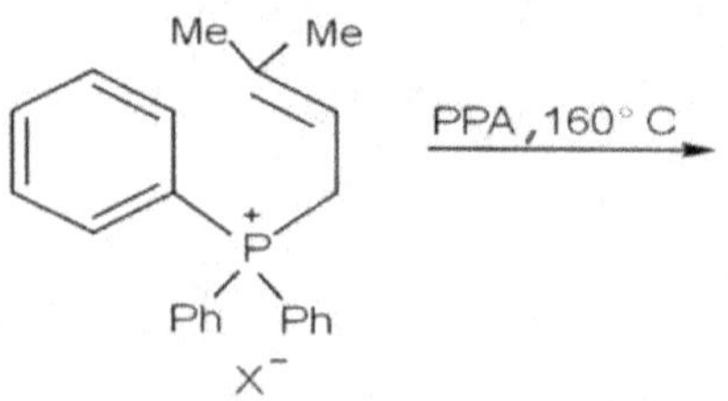

4.2.5 SEVEN AND HIGHER-MEMBERED RINGS

These higher members show no unusual properties. Synthesis of these compounds takes place by oxidation with O_3 at -78°C in CH_2Cl_2 and then reduction of ozonide yields diketone derivative.

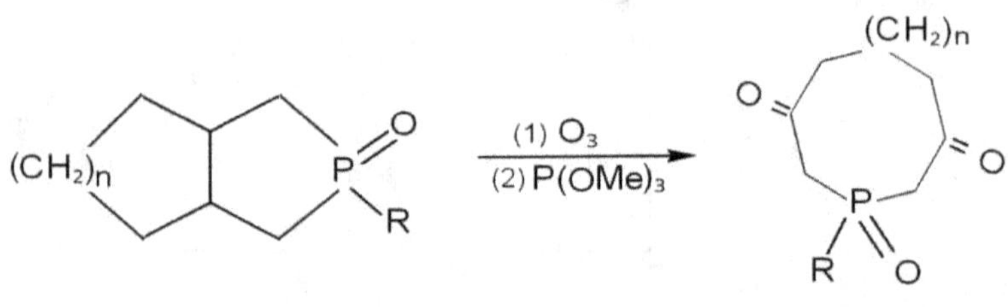

n= 0,1,2

4.3 HETEROCYCLIC COMPOUNDS CONTAINING ARSENIC, ANTIMONY or BISMUTH

4.3.1 FOUR-MEMBERED RINGS

Four membered ring compounds containing arsenic have been reported and are prepared by **reductive cyclization**. 1-methyl acetate (bp = 58-64° C/ 20mm Hg) and 1-phenyl arsetane (bp = 80-82°C/ 1mm Hg) are air-sensitive pale-yellow oils.

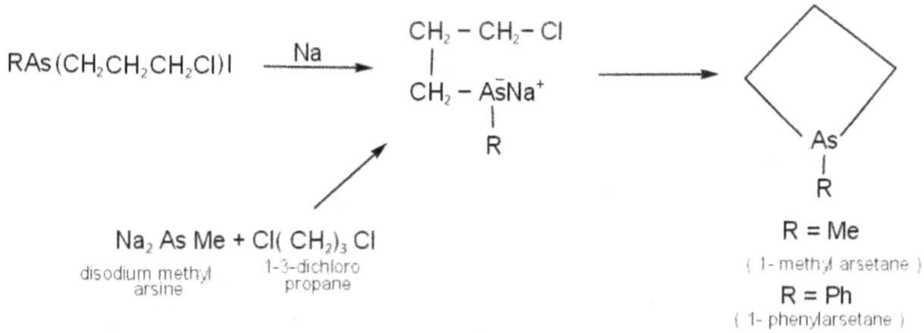

4.3.2 FIVE-MEMBERED RINGS

The compounds reported in this category are very selective for arsenic and antimony (arsolane and stibolane).

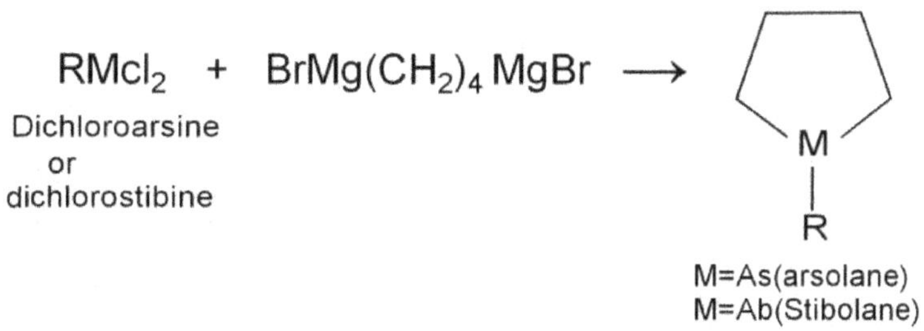

The compounds containing arsenic from dihalide form adduct with mercuric chloride which on heating gives 1- chloroarsolane.

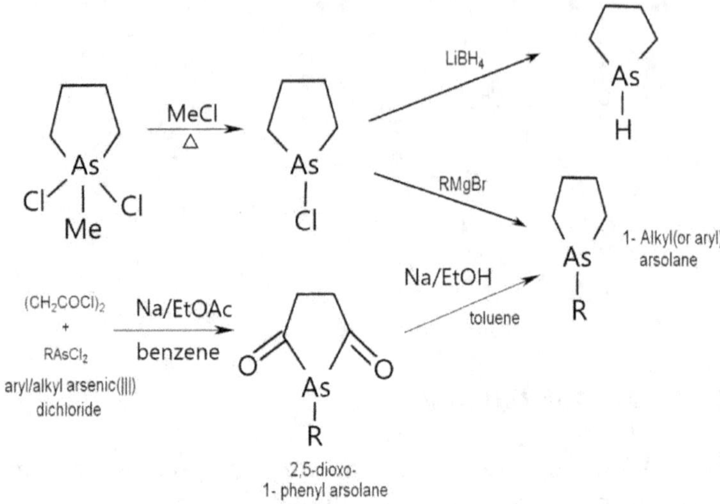

S. No.	Alkyl/Aryl Group	Arsolane Derivatives 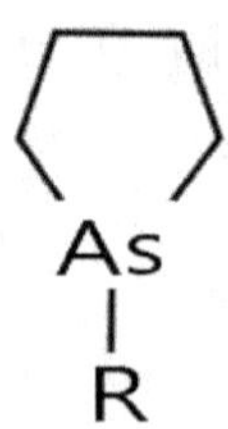 BP (°C/mm Hg)	Stibolane Derivatives 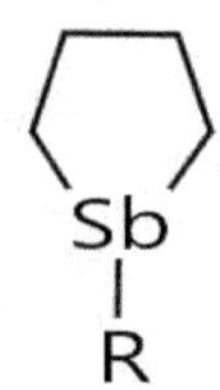 BP (°C/mm Hg)
1.	Me	65 – 66 / 15	77 – 79 / 17
2.	Ph	128.5 / 15	120 – 122 / 0.4

The five membered ring compound containing antimony shows bond cleavage of methyl-antimony giving place to addition of sodium. This 1-sodium derivative forms 1,1-bistibolane which further reacts with chlorine of sulphuryl chloride forming various chloro-derivatives.

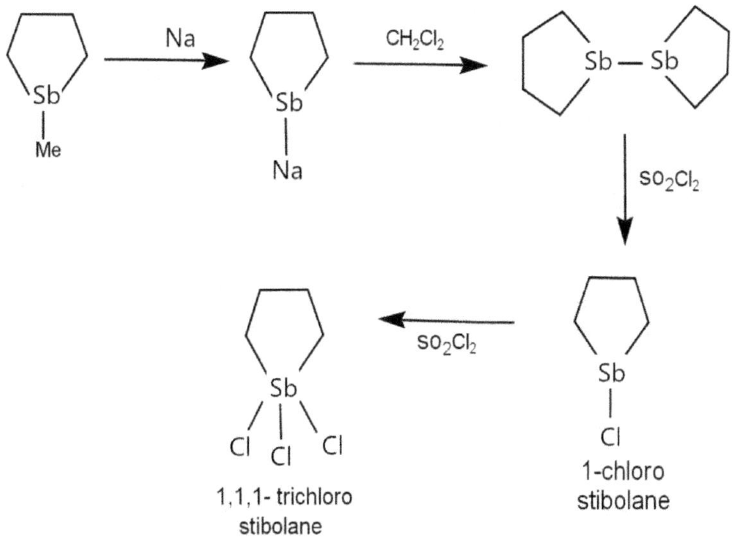

Various other five membered ring compounds containing one or two double bonds and condensed ring system are also known.

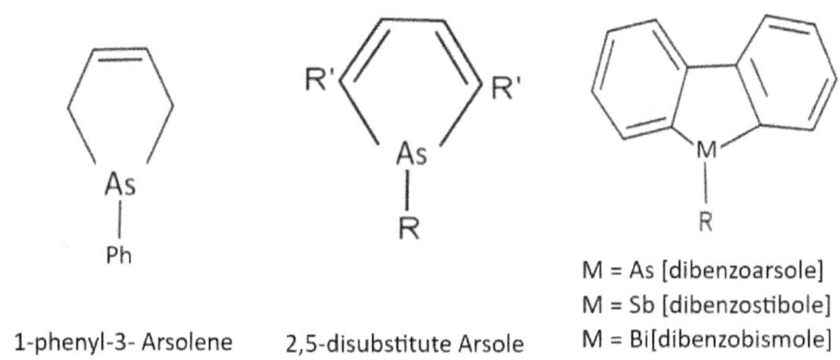

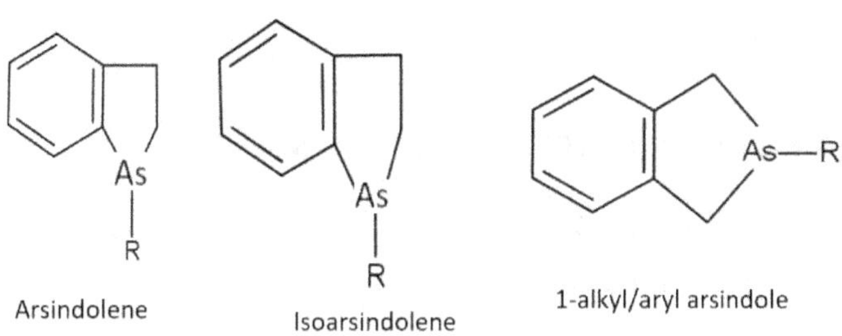

4.3.3 SIX-MEMBERED RINGS

Six membered ring systems exhibit occurrence in saturated, unsaturated, and condensed forms.

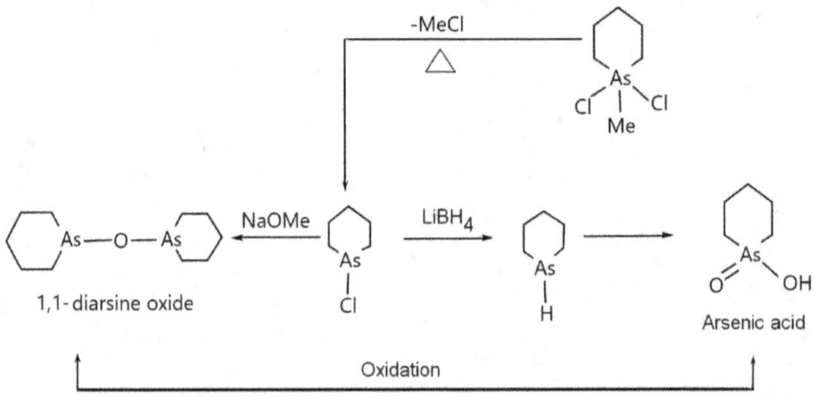

The above-mentioned scheme of reactions shows the formation of 1-chloroaresenane which through various routes changes into arsenic acid. This acid on oxidation of 1,1- diarsine oxide.

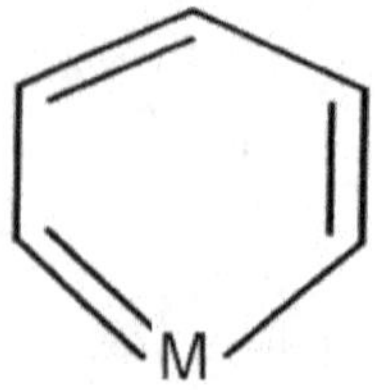

M = As [Arsenin]
M= Bi [Bismin]
M= Sb[Antimonin]

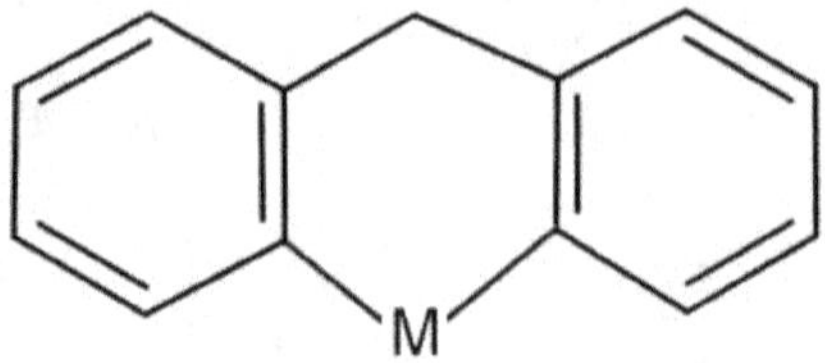

M=As } condensed
M=Sb } ring systems

4.3.4 SEVEN AND HIGHER-MEMBERED RINGS

Seven membered ring systems are called arsepins and stibepin and structurally illustrated as:

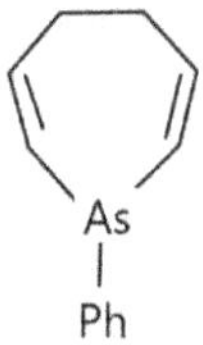

1-phenyl-4,5-dihydro arsepin

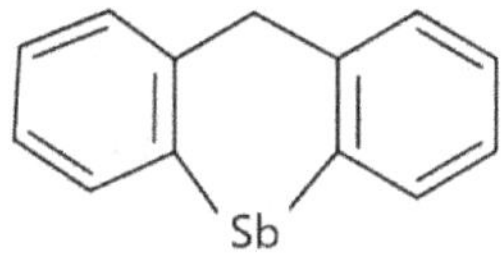

dibenzstibepin derivative

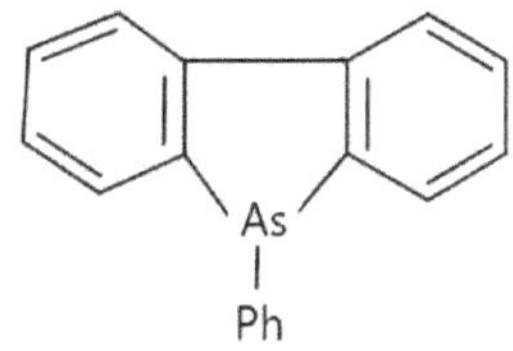

dibenzarsenin derivative

4.4 HETEROCYCLIC COMPOUNDS CONTAINING SILICON, GERMANIUM, TIN or LEAD

The widest occurrence of these elements is reported for heterocyclic ring containing silicon.

4.4.1 THREE MEMBERED RINGS

The three membered heterocyclic rings containing silicon (Sila cyclopropane or silirane) were earlier reported as intermediates in the formation of silanes. It has been individually prepared form bis (bromomethyl) silane.

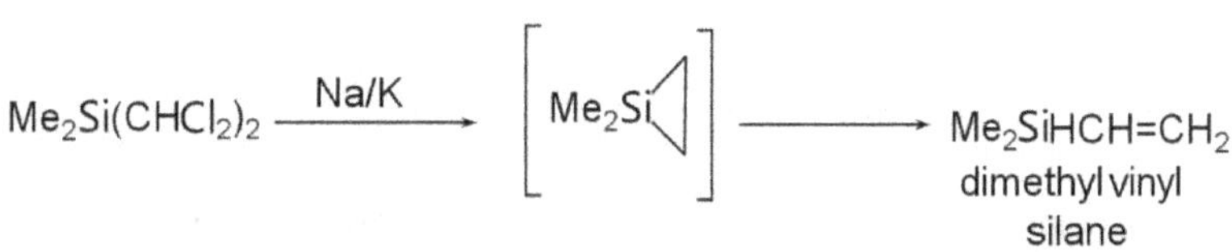

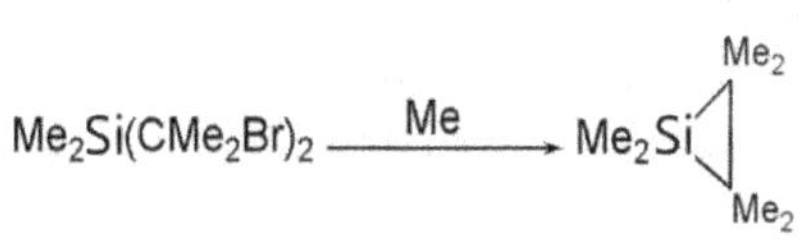

bis(bromomethyl)silane Hexamethylsilirane

There are certain ring systems having two members as silicon (disilacyclopropane) and ring containing double bond (silirenes).

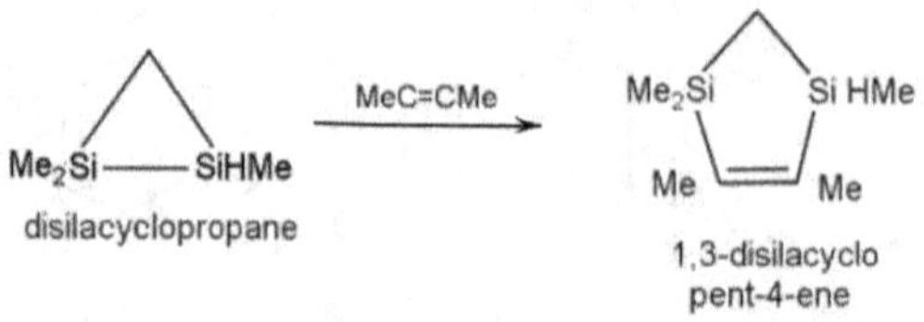

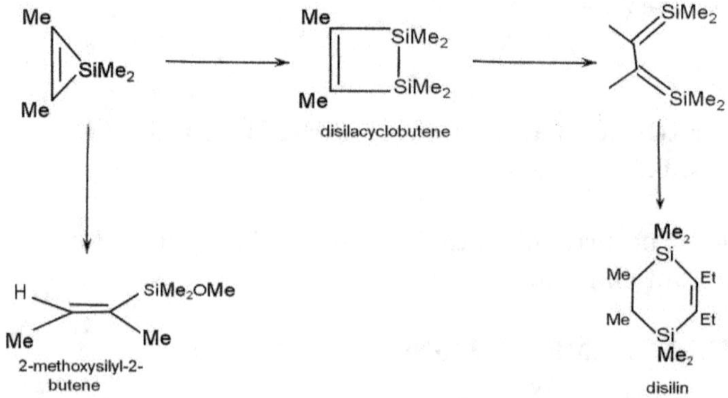

The double bond three membered silicon compound (silacyclopropene) which forms disilacyclobutene which leads to ring opening in the same way as the ring opens when treatment is done with methanol. The former ring open compound (1,4-disilabuta-1,3- diene) later results in disilin.

4.4.2 FOUR MEMBERED RINGS

Silacyclobutane, silacyclobutenes, germacyclobutanes and germacyclobutenes have been reviewed extensively. The preparation involves reaction of 3-bromo propyldimethylchlorosilane with magnesium resulting in ring closure to form Silacyclobutane.

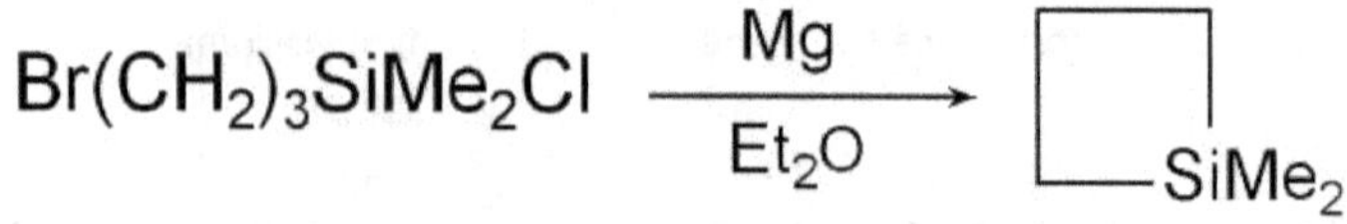

Germacyclobutane forms 1,2-digermacyclopentane by reinsertion on heating at high temperature.

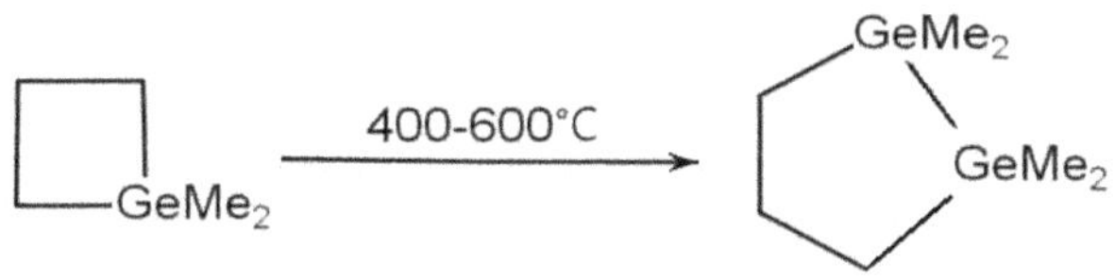

When halogen comes into play at silicon, it further reduces the reactivity of ring. Further aluminium chloride leads to **ring expansion**.

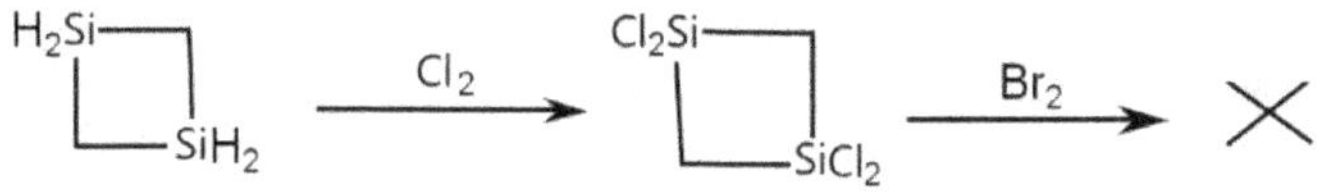

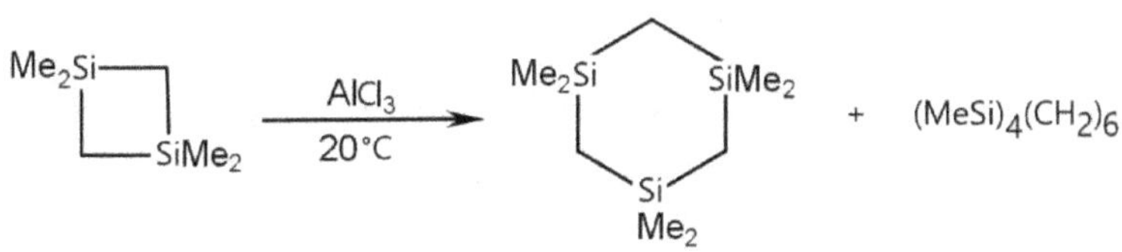

Flash vacuum pyrolysis (FVP) yields double bond four membered silicon compounds by two methods.

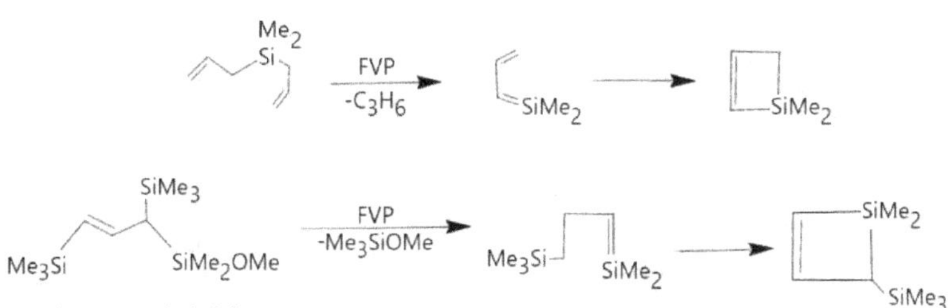

1,3-bis(trimethylsilyl)-3-methoxysilylpropene

1,2-Disilacyclobutene gives various reactions under different conditions these by forming a series of compounds.

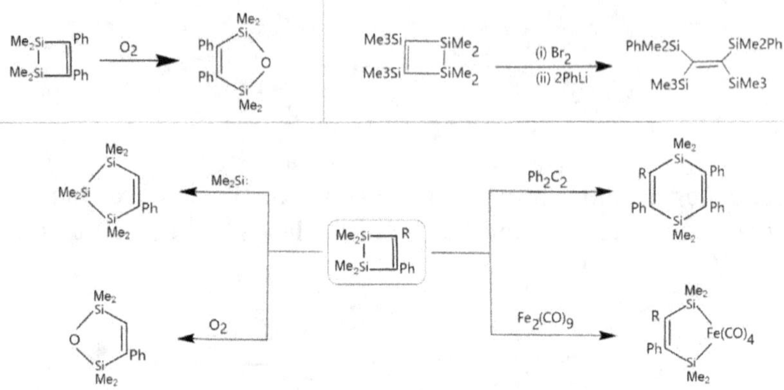

4.4.3 FIVE MEMBERED RINGS

When five membered silicon compounds are considered; they are prepared in the presence of a metal (for e.g., lithium, magnesium etc.).

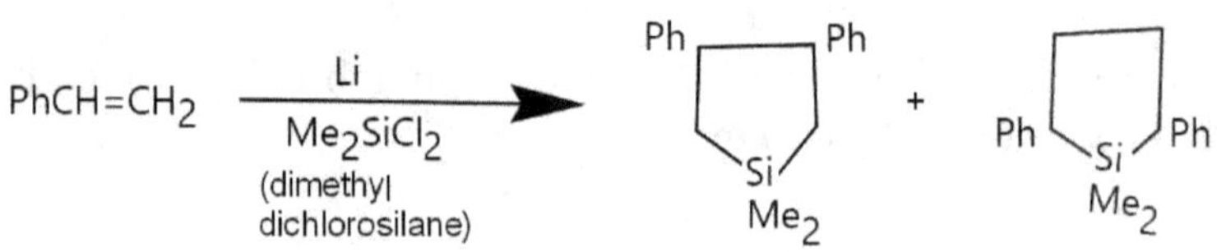

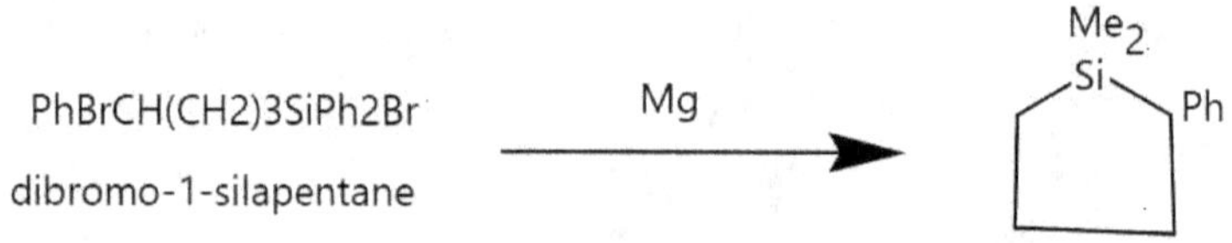

If chloroplatinic acid (H_2PtCl_6) is used as catalyst, it brings about **cyclisation** in enyl silanes. Cyclopentanes of silicon, germanium, tin, and lead can also be prepared from di-Grignard's derivatives with dihalide.

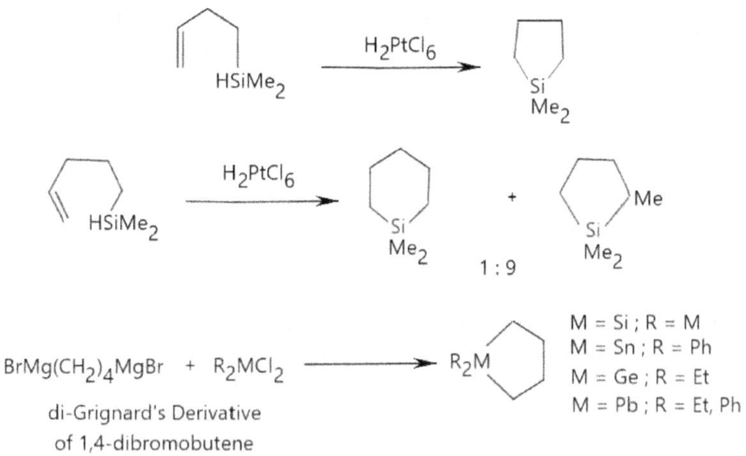

In the presence of aluminium chloride, **polymerization** is induced in Sila cyclopentanes. Stannocyclopentanes polymerize by sulfodiimide.

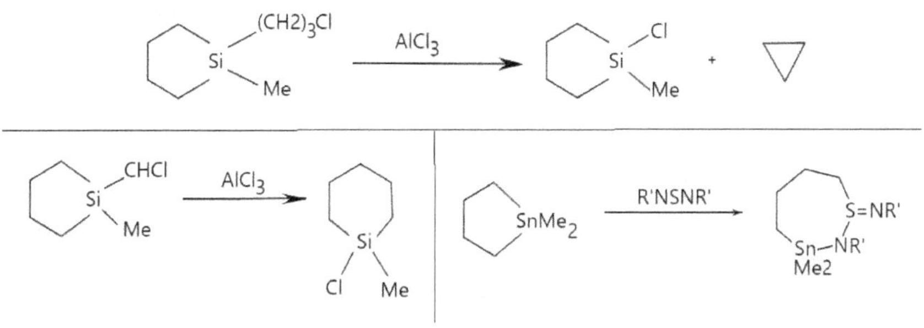

Double bond compounds of the same category also finds easy synthesis. Silacyclopent-3-enes and silacyclopent-2-enes can be synthesized by **pyrolysis** of ethyl trichlorosilane or methyl trichlorosilane and 4-chlorobut-2-enyltrichlorosilane, respectively.

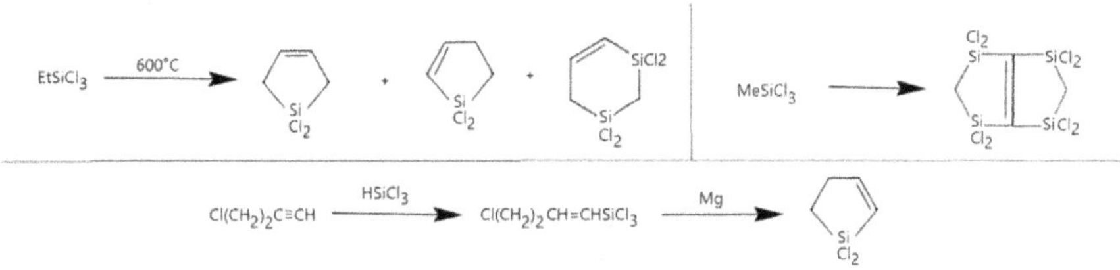

Dimethyl removal to facilitate formation of same kind of compounds with germanium is due to greater electronegativity of this elements.

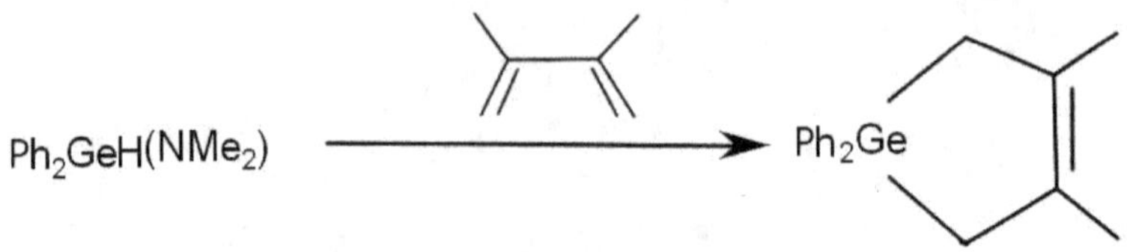

There are also a series of these compounds containing two double bonds and these compounds are named siloles (silicon), germoles (germanium), stannoles (tin) and plumbols (lead) respectively. Siloles can be prepared by reacting diphenyldichlorosilane with ferrole (a product formed by treating, diphenylacetylene and iron pentacarbonyl). It can more conveniently be prepared by several other routes also. Reaction of phenyl germanium trichloride with 1,4-dilithiotetraphenylbuta-1,3-diene gives germole and digermole.

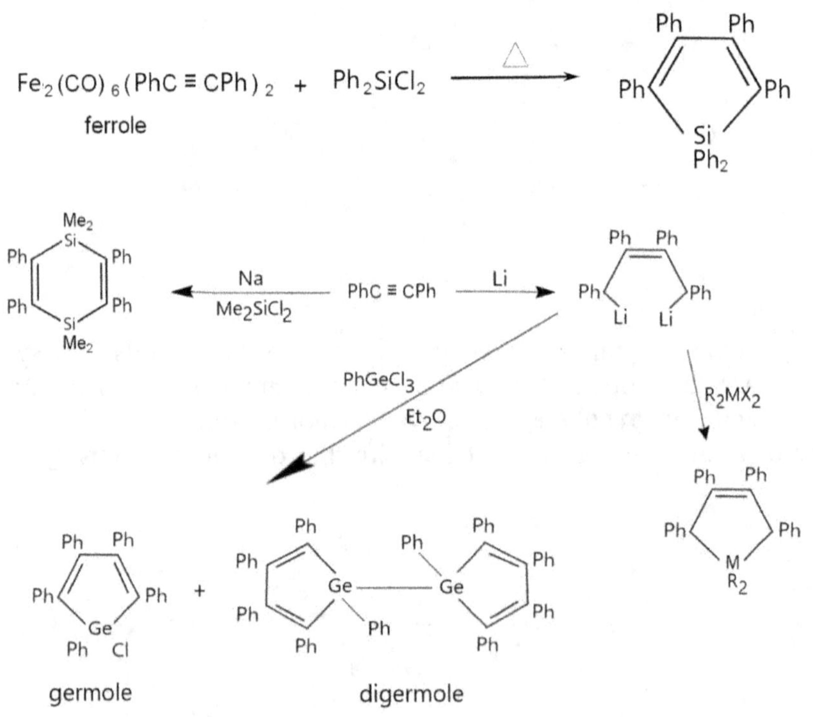

4.4.4 SIX AND HIGHER MEMBERED RINGS

If silicon is substituted with chlorine, then **halogenation** brings about ring substitution but if not then it leads to ring opening.

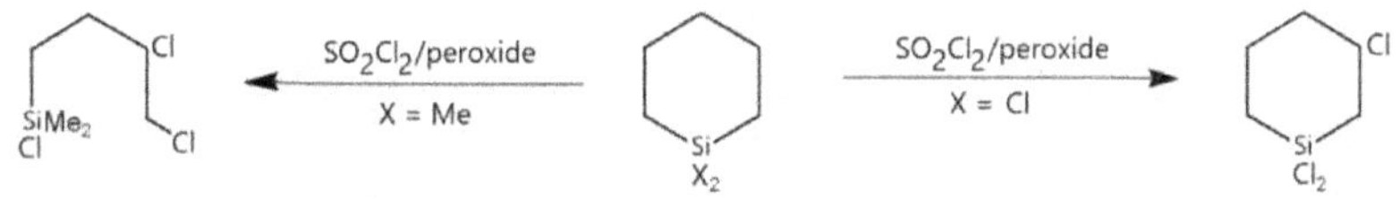

1,2-Disilacyclohexanes can be prepared from 1,2-dichlorosilane and di-Grignard reagent, or by 1,4-disilabutane which cyclizes using alkali metals. 1,3,5-Trisilapentanes, which may be substituted, by using magnesium or Grignard's reagent.

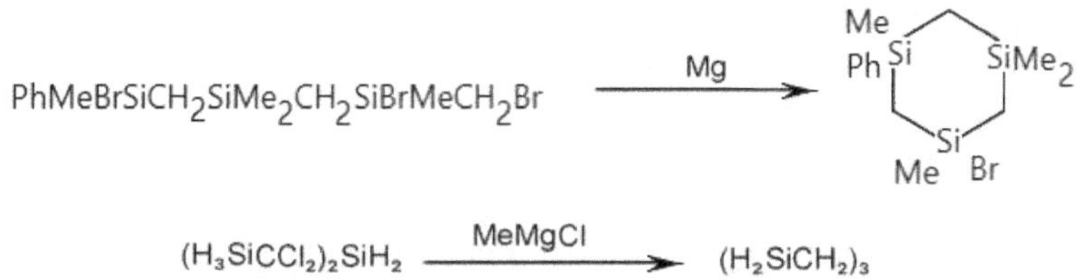

Perchloro –1,3,5–trisilacyclohexane **rearranges** in presence of Grignard's reagent.

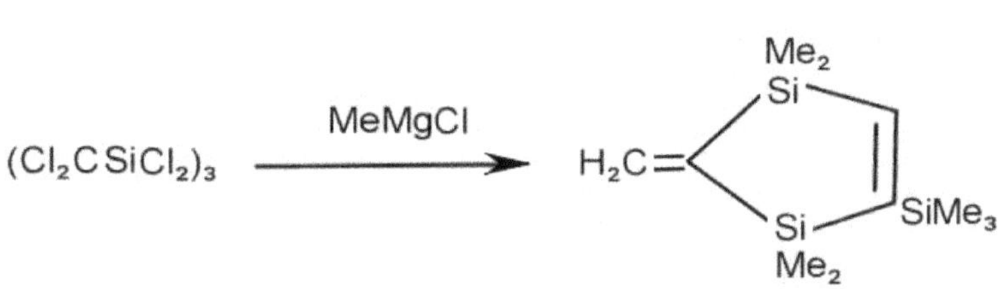

Silacyclohexadienes have been prepared by various routes.

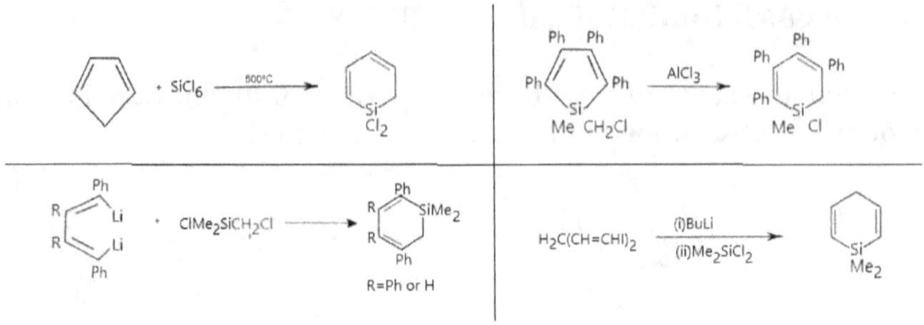

Higher membered rings are formed by cyclization of 2,2'-dilithiodibenzyl with dihalide.

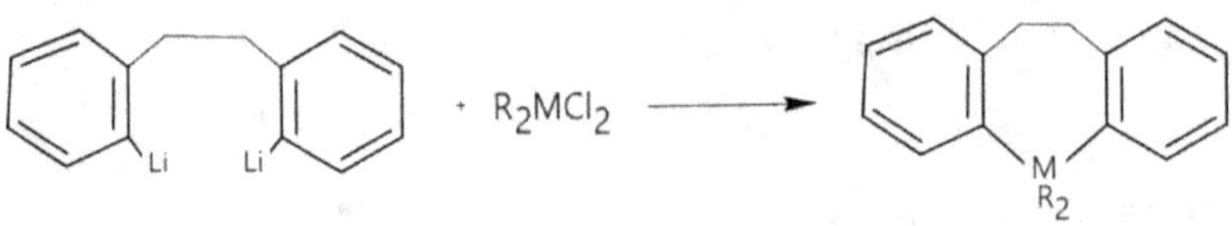

M=Si, Ge, Sn, Pb : R=Me, Ph

4.5 HETEROCYCLIC COMPOUNDS CONTAINING BORON

When diphenylacetylene and dibromophenylborane undergo addition reaction in presence of lithium, triphenyl borine (a three membered compound of boron) is formed as an intermediate.

4.5.1 FIVE MEMBERED RINGS

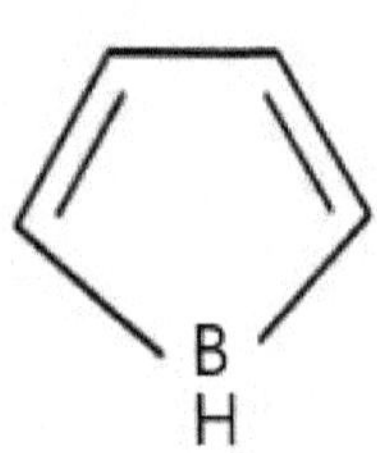

Borole

The reaction of five membered substituted tin compounds with dichlorophenylborane yields pentaphenyl borole.

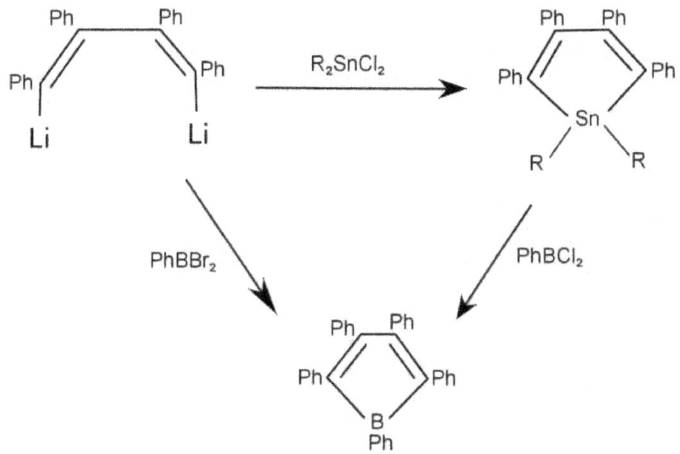

There have been various other ways used to synthesize these compounds in which a boron compound (which can be trialkyl borane, boroxine, tris(dimethyl amino) borane, alkylbis (alkylthio) borane or sodium borohydride) is used to supply boron atom to the product ring and other molecule provides only other carbon atoms of the ring.

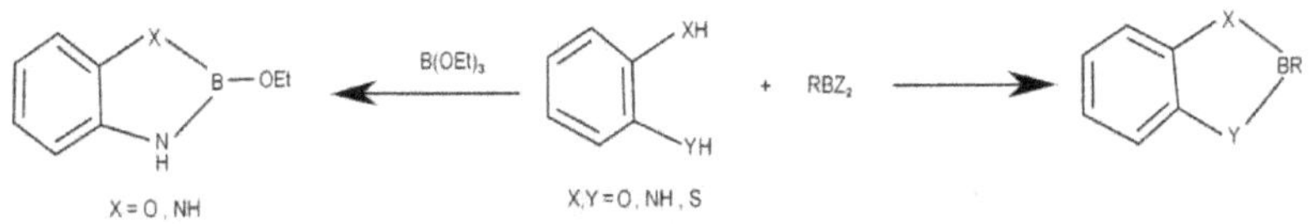

2-Dimethyl amino-1,2-dihydro-3H-1,3,2-benzo diazaborale is prepared from o-phenylene diamine and tris(dimethyl amino) borane.

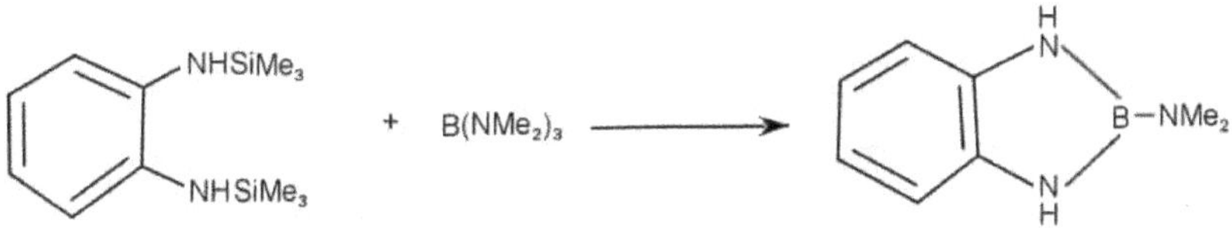

Almost all these preparations depict a general pattern of a five membered ring system getting attached to an already existing ring; however individual rings have also been prepared like, 2,4,5–

triphenyl-1,3,2-dioxaborole and 2,3,5-triphenyl-1,2- dihydro-1,3,4,2-oxadiazaborole can be prepared from reaction of benzene and N-phenylbenzhydride with dichlorophenylborane. Substituted thiaboroles can also be prepared from dimercaptoethylene.

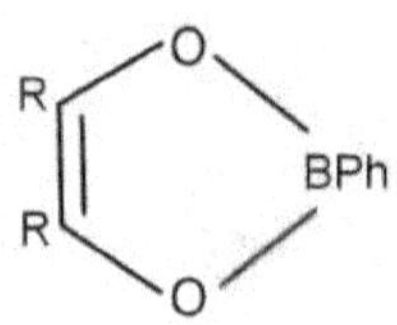

2,4,5 – triphenyl – 1,3,2 – Dioxaborole

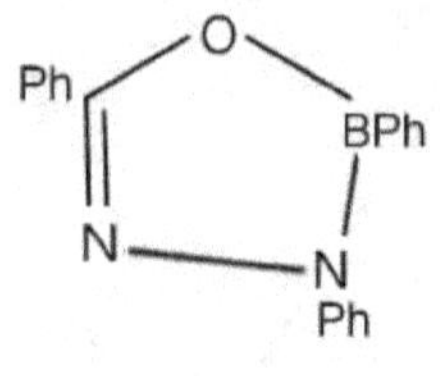

2,3,5 - triphenyl - 1,2,- dihydro - 1,3,4,2 - oxadiaza-borole

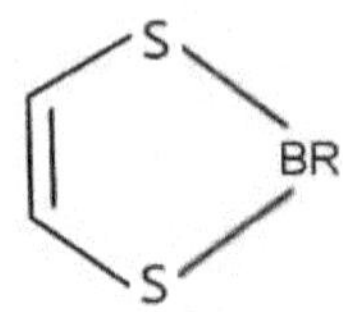

1,3,2 - dithia borole

1,2-Dihydro-3H-1,3,2-diazaborole is prepared by diimines. First boron dihalide is added in petroleum/ether and then **reduction** is brought about with sodium amalgam.

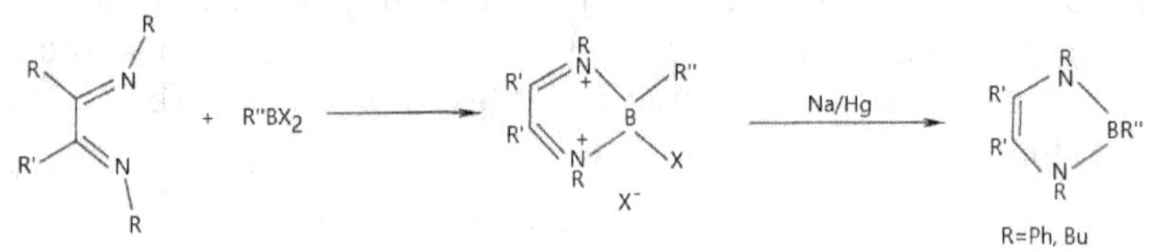

R=Ph, Bu
R'=H, Me
R''=Me, Ph, Cl, Br

An unsaturated ring compound (1,2-dihydro-3H-1,3,2-diazoboroles) is prepared by **dehydrogenation** in presence of palladium from a saturated reactant.

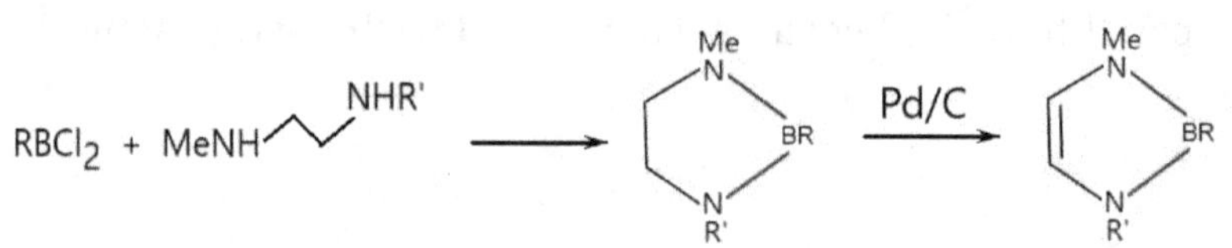

R=Ph
R'=H, Me

There are various observations for benzo-fused ring systems; for example, benzo-1,3,2-diazoborole was reported stable in alcohol by one group and unstable by another group. In the same way,

1,3,2-benzodioxaboroles can be easily **hydrolyzed** at room temperature. The same compound (R=H) is oxidized by dry air at room temperature.

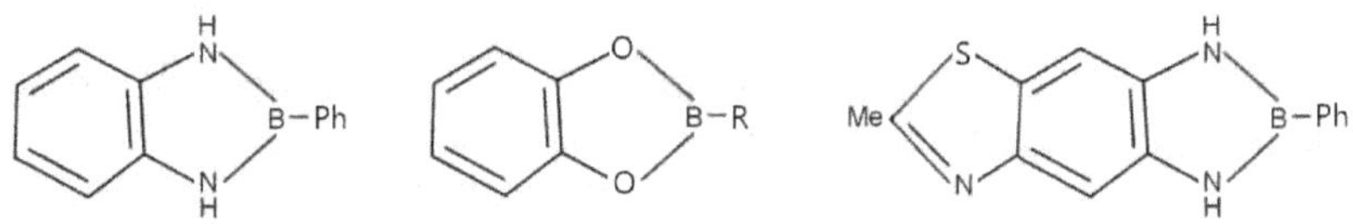

Benzo – 1,3,2 – diazoborole

1,3,2 - Benzo dioxaborole

4.5.2 SIX MEMBERED RINGS

The six membered heterocyclic rings containing boron are called borin. The dibenzoborin (also named 9-Bora anthracene) is the only known neutral borin.

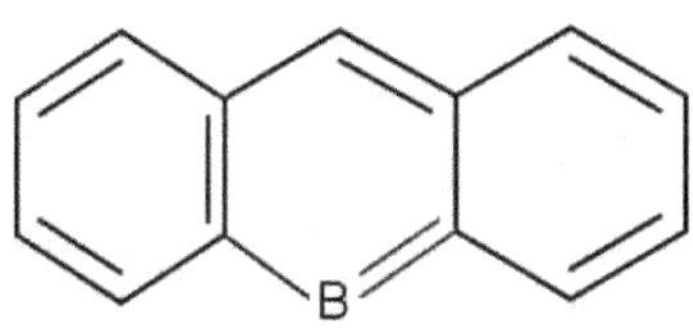

Borin(9-boraanthracene)

The best-known group of boron six membered heterocyclic compounds are those that contain boron and nitrogen together. 1,2-Dihydro-2,3-diphenyl-6-(2-methoxycarbonylethyl)-1,2-azaborine has been prepared from a thieno fused compound. The same procedure using thiophene as building block has also been used.

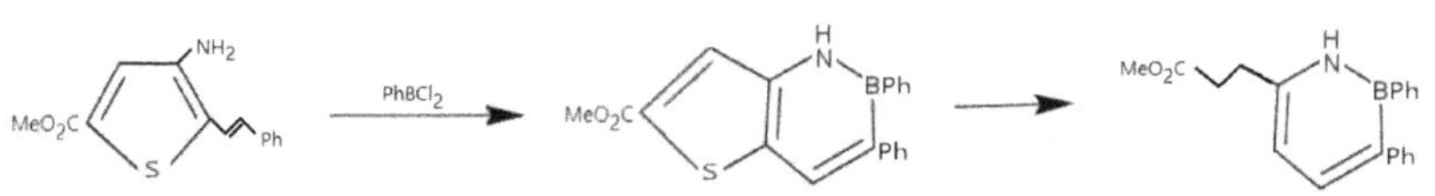

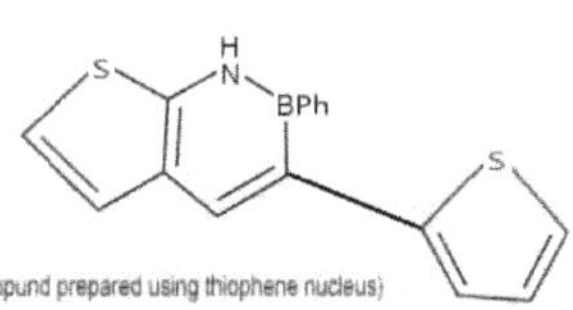

(compund prepared using thiophene nucleus)

1,2–Dihydro–1,2-azaborine has been obtained from 3-butylamine by hydroboration carried about by two agents viz, trimethylamine–phenyl borane and diborane. The product 1,2–dihydro–1,2–azaborine (compound 1) gives various products under different conditions.

I) If (R = OC_4H_5BN) in compound 1; alkaline hydrolysis gives anhydride of compound 1(R = OH)

2) if (R=OH) in compound 1, reduction with Lithium Aluminium Hydride gives parent compound (R=H)

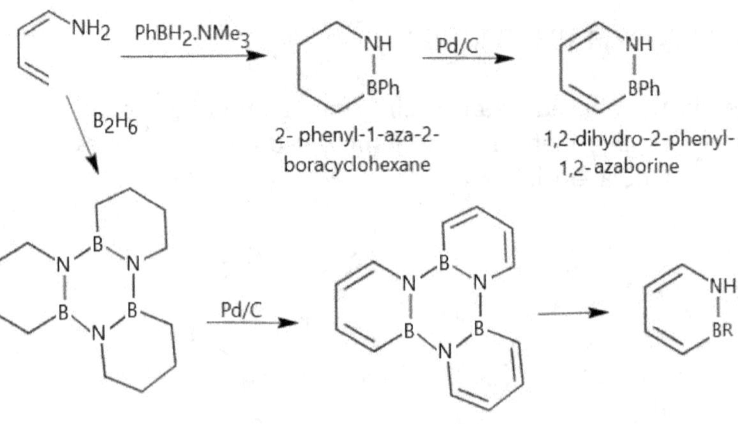

tribenzoborazine

For preparation of fused ring compound of boron and nitrogen, boron trichloride or dichlorophenyl borane is treated with substituted aniline.

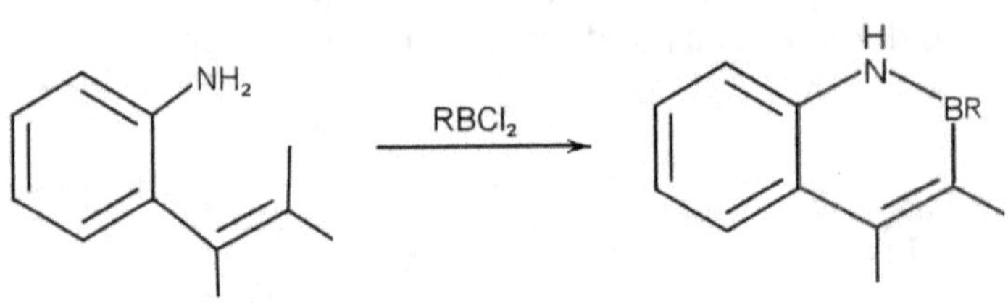

R = Ph , Cl

Photochemical ring closures have also been reported as a method of preparation using iodinated and non-iodinated anilines.

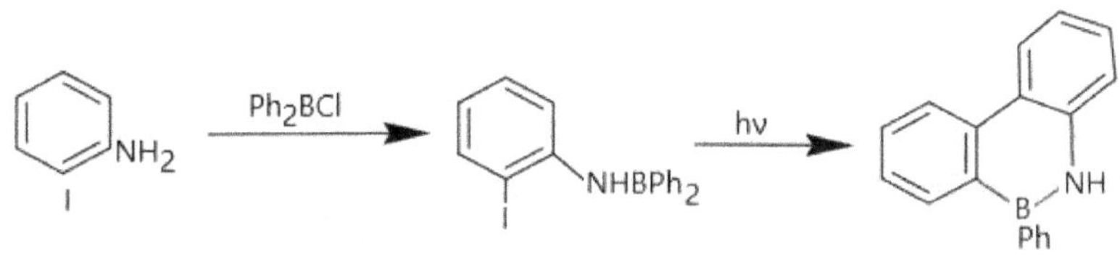

Many compounds containing nitrogen and boron in monocyclic structure as well as fused state has also been prepared and exhibit considerable reactivity. Some examples are as under:

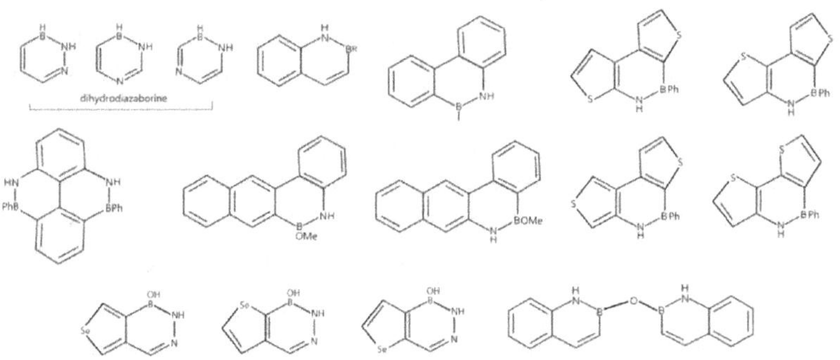

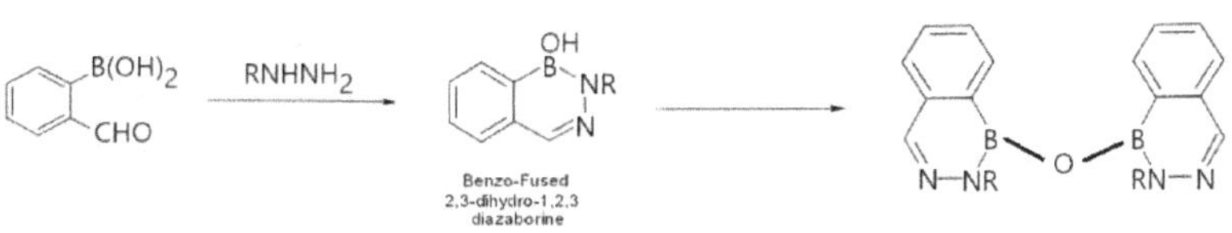

2-Aminobenzonitrile and boron trihalides are used to obtain derivatives of 1,2-dihydro-1,3,2-diazaborine. A somewhat different compound is obtained if 2-aminobenzo nitrile is replaced by 2-aminobenzamide.

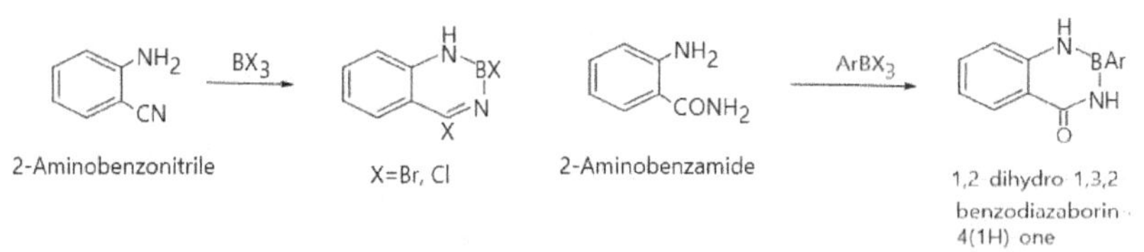

There are various compounds in these categories that have been prepared.

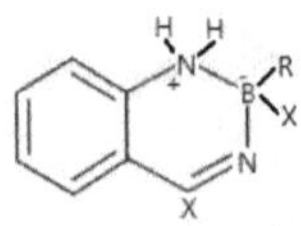

a) R=Ph; X=Cl
b) R=Me; X=Br

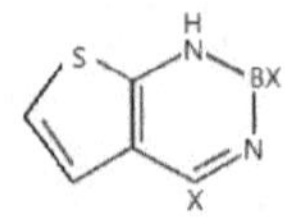

X=Br, Cl

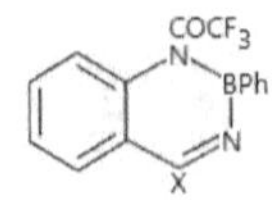

X=Br, Cl

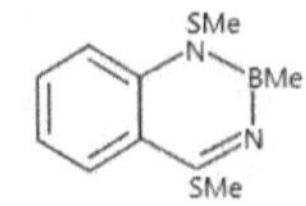

Treatment with tris(acetonitrile)tricarbonyl chromium gave hexa-hapto complexes. A similar synthesis from 1,3-dichloro-4,5-dimethyl-2,1,3-thiadiborole and 1,2-dimethyl-1,2-bis (tri methyl silyl) hydrazine gave the derivative (**A**). Benzo and thieno fused 1,2,3,4-tetrahydro-2,3-diaza-1,4-diborine (**B**) and (**C**) were obtained from fused thiadiborole. Other benzo-fused tetrahydrodiaza-diborines are also known. The addition product between o-toluidine and boron trichloride gave on heating with triethylamine, compound (**D**) (X=Cl) as the main product along with borazine.

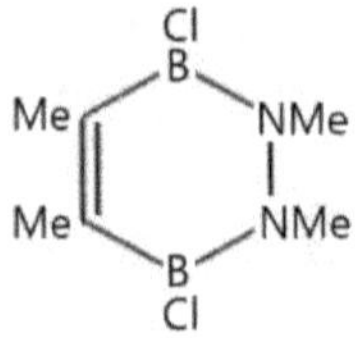

(A)

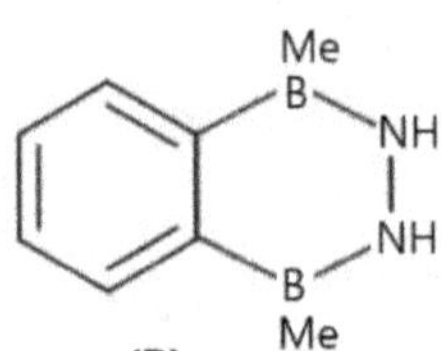

(B)

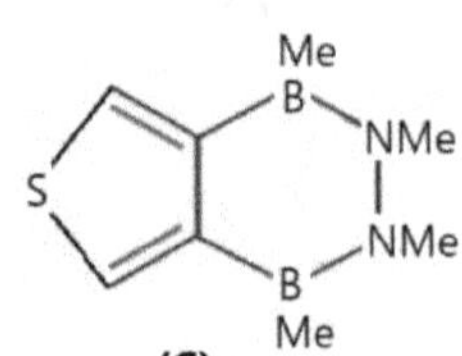

(C)

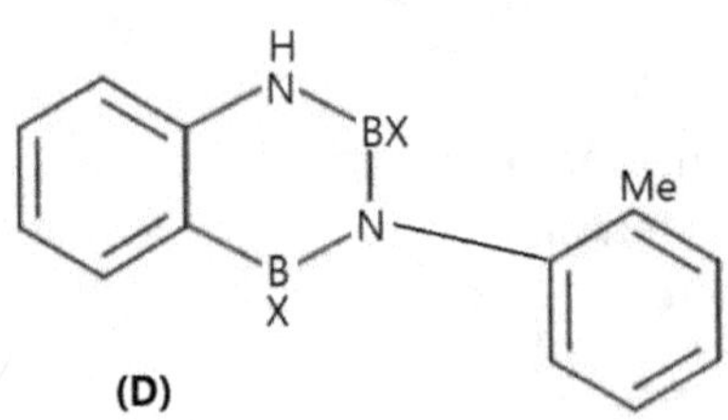

(D)

X=Cl, I, Me, Ph

Six membered compounds containing oxygen or sulphur along with boron are not as stable as nitrogen. These compounds can be prepared from boron trichloride or by **photo-chemical cyclization**.

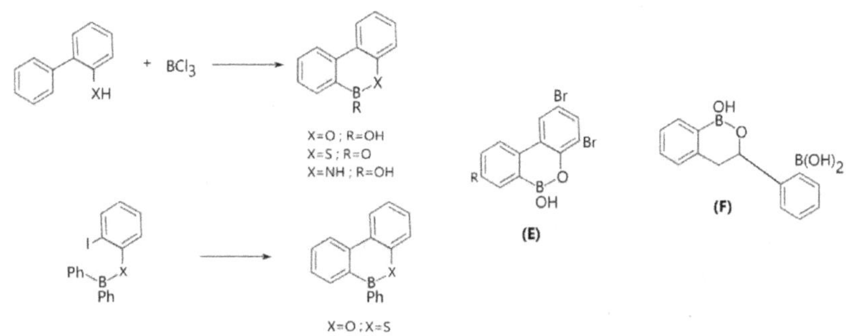

Brominated products (**E**) are also obtained. Di-boronic acid cyclized in presence of alkali to give product (**F**).

4.5.3 SEVEN MEMBERED RINGS

Seven membered unsaturated heterocyclic compounds containing boron are called borepin. The earlier compounds were prepared using 2,2-dilithiobibenzyl.

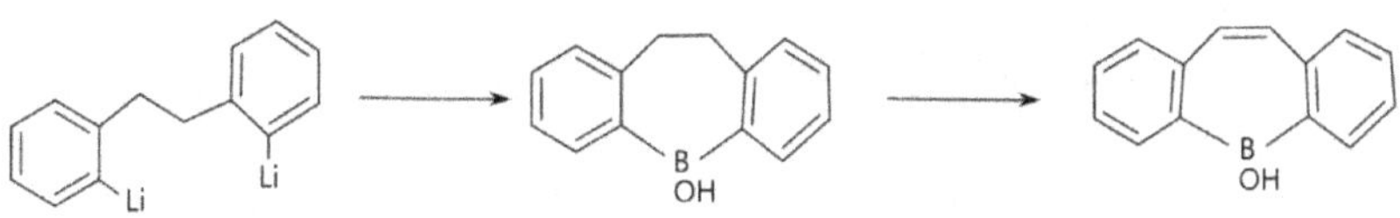

REFERENCES

1. R.I. Wagner, L. D. Freeman, H. Goldwhite and D.G Roswell (1967); J. Am Chem. Soc, **89**, 1102.
2. E. Niecke and W. Flick (1975); Angew. Chem., Int. Ed. Engl.; **14**, 363.
3. E. Jungermann, J. J. McBridge Jr., R. Clutter and A. Mais (1962); J. Org. Chem., **27**, 606.
4. G. Griithner and E. Krause (1916); Ber., **49**, 437.
5. L. D. Quin; Hetrocyclic chemistry of Phosphorus (1981); Wiley, New York.
6. A. B. Burg and P. J. Slota Jr. (1960); J. Am. Chem. Soc., **82**, 2148.
7. B. Fell and H. Bahrmann (1974); Synthesis, 119.

8. K. Sommer (1970); Z. Anorg. Allg. Chem., **56**, 379.
9. F. C. Leavitt, T. A. Manuel, F. Johnson, L. U. Matternas and D. S. Lehman (1960); J. Am. Chem. Soc., **82**, 5099.
10. G. Merkl, F. Lieb and A. Merz (1967); Angew. Chem. Int. Ed. Engl., **6**, 87.
11. L. D. Quin and S. G. Borleske and J. F. Engel. (1967); J. Org. Chem. Soc., **89**, 5984.
12. L. D. Quin and J. G. Bryson (1967); J. Am. Chem. Soc., **89**, 5984.
13. G. Markl (1963); Angew. Chem, Int. Ed. Engl., **2**, 620.
14. K. L. Marsi, J. L. Jasperse, F. M. Llort and D. B. Kanne (1977); J. Org. Chem., **42**, 1306.
15. K. Sommer; Z. Anorg. Allg. Chem. (1972); **379**, 56.
16. M. H. Beeby and F. G. Mann; J. Chem. Soc. (1951); 411.
17. G. Markl (1963); Angew. Chem. Int. Ed. Engl., **2**, 153.
18. W. R. Purdum, G. A. Dilbeck and K. D. Berlin (1975); J. Org. Chem., 3763.
19. L. D. Quin and E. D. Middlemas (1977); J. Am. Chem. Soc., **99**, 8370.
20. L. D. Quin, E. D. Middlemas and N. S. Rao (1982); J. Org. Chem., **42**, 905.
21. M. Mickiewicz and S. B. Wild (1977); J. Chem. Soc., Dalton Trans., 704.
22. H. N. Das–Gupta (1938); J. Indian Chem. Soc., **15**, 495.
23. W. Steenkopf, I. Schbart and J. Roch (1932); Ber., **65**, 409.
24. H. A. Meinema, H. F. Martens and J. G. Naltes (1976); J. Organomet. Chem., **110**, 183.
25. G. Griither and M. Wiernik (1915); Ber., **48**, 1473.
26. J. J. Monagle (1975); J. Org. Chem., **27**, 3851.
27. K. Issleib and A. Balszuweit (1975); Z. Anorg. Allg. Chem., **418**, 158.
28. H. A. Meinema and J. G. Nottes (1978); J. Organomet. Chem., **160**, 435.
29. E. Wibierg and K. Modritzer (1957); Z. Naturforsch., Teil B, **12**, 135.
30. I. Gorski, W. Schpanski and L. Muljan (1934); Ber., **67**, 730.
31. G. Markl and G. Dannhardt (1973); Tetrahedron Lett., **14**, 455.
32. M. H. Beeby, F. G. Mann and E. E. Turner (1950); J. Chem. Soc., **19**, 923.

33. P. S. Skell and E. J. Goldstein (1964); J. Am. Chem. Soc., **86**, 1442.
34. D. Seyferth and D. C. Annarelli; J. Am. Chem. Soc., **97**, 2273.
35. T. J. Barton an J. A. Kilgour (1976); J. Am. Chem. Soc., **98**, 7746.
36. R. T. Conlin and P. P. Gaspor (1976); J. Am. Chem. Soc., **98**, 3715.
37. R. Damrauer (1972); Organomet. Chem. Rev.(A), **8**, 67.
38. L. H. Sommer and G. A. Baum (1954); J. Am. Chem. Soc., **76**, 5002.
39. N. S. Nametkin, L. E. Gusel'nikov, R. L. Ushakova, V. Yu. Orlov, O. V. Kur'min and V. M. Vdovin (1970); Dokl. Akd. Nauk SSSR, **194**, 1096 (Chem. Abstr., 1971, 74, **42**, 440)
40. G. Fritz and E. Matern (1976); Z. Anorg. Allg. Chem. Soc., **426**, 28.
41. E. D. Babich, V. N. Karel'skii, V. M. Vdovin and N. S. Nametkin (1975); Izv. Akad. Nauk SSSR, Ser. Khim., 953 (Chem. Abstr., 1975, **83**, 97, 444).
42. E. Block and L. K. Revelle (1978); J. Am. Chem. Soc., **100**, 1630.
43. G. T. Burns and J. J. Barton (1981); J. Organomet. Chem., **216**, C5.
44. H. Sakurai, H. Tobita, M. kira and Y. Nakadaira (1980); Angew. Chem. Int. Edn. Engel., 19, 620.
45. H. Sakurai, T. Kobayashi and Y. Nakadaira (1978); J. Organomet. Chem., **162**, C43.
46. C. W. Cheng and C. S. Liu (1974); J. Chem. Soc. Chem. Commun., 1013.
47. Y. Chi and C. S. Liu (1981); Inorg. Chem., **20**, 3456.
48. D. R. Weyenberg, L. H. Toporcer and A. E. Bey (1965); J. Org. Chem., **30**, 4096.
49. H. Gilman and W. H. Atwell (1964); J. Am. Chem. Soc., **86**, 2687.
50. H. Sakurai, T. Hirose and A. Horomi (1975); J. Organoment. Chem., **86**, 197.
51. R. West (1954); J. Am. Chem. Soc., **76**, 6012.
52. P. Mazerolles, J. Dubac and M. Lesbre (1966); J. Organomet. Chem., **5**, 35.
53. H. Zimmer, C. W. Blewett and A. Brakas (1968); Tetrahedron Letters, 1615.

54. E. C. Juenge and S. Gray (1967); J. Organomet. Chem., **10**, 465.
55. G. Fritz and R. Huber (1976); Z. Anorg. Allg. Chem., **421**, 163.
56. G. Fritz, E. Matern, H. Bock and G. Brahler (1978); Z. Anorg. Allg. Chem., **439**, 173.
57. R. A. Benkeser, Y. Nagai, J. L. Noe, R. F. Cunico and P. H. Gund (1964); J. Am. Chem. Soc., **86**, 2446.
58. P. Riviere, G. Dousse and J. Satge (1974); Synth. React. Inorg. Metal- org. chem, 1974,4,281.
59. P. Riviere, M.Riviera- Baudet, C. Couret and J. Satge (1974); Synth. React. Inorg. Metal-Org. Chem., **4**, 295.
60. E. H. Braye and W. Hubel (1959); chem. Ind. (London), 1250.
61. M. V. George and R. Balasubramanian (1976); J. Organomet. Chem. Libr., **2**, 103.
62. R. A. Benkeser, S. D. Smith and J. L. Noe (1968); J. Org. Chem., **33**, 597.
63. G. Frictzf and W. Kemmerling (1963); Z. Anorg. Allg. Chem.,**322**,34.
64. G. Fritz and K-H. Schmid (1978); Z. Anorg. Allg. Chem., **441**, 125.
65. G. Fritz and N. Szczepanski (1967); Angew. Chem, Int. Ed. Engl., **6**, 1077.
66. E. A. Chernyshev, N. G. Komalenkova and S. A. Baskirova (1972); Dokl. Akad. Nauk SSSR, **205**, 868(Chem. Abstr., 1972, **77**, 152 265)
67. V. Hagen and K. Ruhlmann (1969); Z. Chem., **9**, 309.
68. G. Markl and P. L. Merz (1971); Tetrahedron. Lett., 1303.
69. G. Markl, H. Baier, P. Hofmeister, F. Kees and C. Soper (1979); J. Organomet. Chem., **173**, 125.
70. J. Y. Corey, M. Dueber and M. Malaidza (1972); J. Organomet. Chem., **36**, 49.
71. J. J. Eishch and L. J. Gonsior (1967); J. Ogranomet. Chem., **8**, 53.
72. J. J. Eishch, N. K. Hola and S. Kozima (1969); J. Am. Chem. Soc., **91**, 4575.
73. R. J. Brotherton and H. Steinberg (1961); J. Org. Chem., **83**, 4406.
74. S. G. Friedman (1962); J. Gen. Chem. USSR (Engl. Transl.), **33**, 201.
75. H. Beyer, K. Niedenzu and J. W. Dawson (1962); J. Org. Chem., **27**, 4701.

76. R. Goetze and H. Noth (1980); Z. Naturforsch, Teil B, **35**, 1212.
77. J. Goubeau and H. Schneider (1964); Liebigs Ann. Chem., **675**, 1.
78. K. A. Jensen and T. Pedersen; Acta, Chem. Scand.,**15**, 1780.
79. G. Smolinsky (1961); J. Org. Chem., **26**, 4915.
80. M. Pailer and H. Huemer (1964); Monatsh. Chem., **95**, 373.
81. G. Schmid and J. Schulze (1977); Chem. Ber., **110**, 2744.
82. K. Niedenzu and J. S. Merriam (1974); Z. Anorg. Allg. Chem., **406**, 251.
83. W. Weber and K. Niedenzu (1981); Synth. React. Inorg. Metal-Org. Chem., **11**, 211.
84. E. Nyilas and A. H. Soloway (1959); J. Am. Chem. Soc., **81**, 2681.
85. R. Hemming and D. G. Johnston (1966); J. Chem. Soc. (B), 314.
86. H. C. Brown and S. K. Gupta (1975); J. Am. Chem. Soc., **97**, 5249.
87. P. J. Grisdale, M. E. Glogowiski and J. L. R. Williams (1971); J. Org. Chem., **36**, 3821.
88. P. J. Grisdale and J. L. R. Williams (1969); J. Org. Chem., **34**, 1675.
89. S. Gronowitz and I. Ander and P. Zanitaro (1980); Chem. Ser., **15**, 23.
90. S. Gronowitz, I. Ander and P. Zanitaro (1981); Chem. Ser., **19**, 32.
91. D. Semard, C. Poulmier, J. Morel and P. Pastour (1972); Int. J. sulphurchem., Part A., **2**, 257.
92. A. Meller, W. Maringgele and G. Maresh (1974); Monatsh. Chem., **105**, 637.
93. H. L. Yale (1971); J. Heterocycl. Chem., **8**, 193.
94. C. Glennow (1976); University of Lend, Lend, Sweden.
95. A. Meller, W. Maringgele and H. Fetzer (1980); Chem. Ber., **113**, 1950.
96. W. Siebert, R. Full, H. Schinidt, J. von Seferl, M. Halstenberg and G. Huttner (1980); J. Organomet. Chem., **191**, 15.
97. W. Haubold and A. Gemmelr (1980); Chem. Ber., **113**, 3352.
98. B. Asgarouladi, R. Full, K-J. Schaper and W. Siebert (1974); Chem. Ber., **107**, 34.
99. F. A. Daries and M. J. S. Dewar (1968); J. Am. Chem. Soc., **90**, 3511.

100. P. J. Grisdale and J. L. R. Williams (1969); J. Org. Chem., **34**, 1675.
101. E. E. van Tamelen, G. Brieger and K. G. Untch (1960); Tetrahedron lett., **8**, 14.

www.ingramcontent.com/pod-product-compliance
Lightning Source LLC
LaVergne TN
LVHW052312200926
848481LV00002B/533
* 9 7 8 1 9 2 2 6 1 7 5 2 1 *